VOLUME FOUR HUNDRED AND FORTY-SIX

METHODS IN ENZYMOLOGY

Programmed Cell Death, The Biology and Therapeutic Implications of Cell Death, Part B

METHODS IN ENZYMOLOGY

Editors-in-Chief

JOHN N. ABELSON AND MELVIN I. SIMON

Division of Biology
California Institute of Technology
Pasadena, California

Founding Editors

SIDNEY P. COLOWICK AND NATHAN O. KAPLAN

VOLUME FOUR HUNDRED AND FORTY-SIX

Methods in
ENZYMOLOGY

Programmed Cell Death, The Biology and Therapeutic Implications of Cell Death, Part B

EDITED BY

ROYA KHOSRAVI-FAR
Harvard Medical School, BIDMC
Department of Pathology
Boston, MA, USA

ZAHRA ZAKERI
Professor of Biology
Queens College of City
University of New York
Flushing, NY, USA

RICHARD A. LOCKSHIN
Department of Biological Sciences
St. John's University
Jamaica, NY, USA

MAURO PIACENTINI
Department of Biology
University of Rome
Rome, Italy

AMSTERDAM • BOSTON • HEIDELBERG • LONDON
NEW YORK • OXFORD • PARIS • SAN DIEGO
SAN FRANCISCO • SINGAPORE • SYDNEY • TOKYO

ELSEVIER
Academic Press is an imprint of Elsevier

Academic Press is an imprint of Elsevier
525 B Street, Suite 1900, San Diego, California 92101-4495, USA
84 Theobald's Road, London WC1X 8RR, UK

This book is printed on acid-free paper. ∞

For information on all Elsevier Academic Press publications
visit our Web site at www.books.elsevier.com

ISBN-13: 978-0-12-374464-7

PRINTED IN THE UNITED STATES OF AMERICA
08 09 10 11   9 8 7 6 5 4 3 2 1

Contents

Contributors

Floribeth Aguilar
Department of Physiology, Faculty of Medicine University of Manitoba, Winnipeg, Manitoba R2H2A6

Alexandru Almasan
Department of Radiation Oncology, Cleveland Clinic, Cleveland, Ohio, and Department of Cancer Biology, The Lerner Research Institute, Cleveland Clinic, Cleveland, Ohio

Alakananda Basu
Department of Molecular Biology and Immunology, University of North Texas Health Science Center, Fort Worth, Texas

Joseph Bednarczyk
Department of Physiology, Faculty of Medicine University of Manitoba, Winnipeg, Manitoba R2H2A6

Federico Bernal
Department of Pediatric Oncology and the Program in Cancer Chemical Biology, Dana-Farber Cancer Institute, and the Division of Hematology/Oncology, Children's Hospital Boston, Harvard Medical School, Boston, Massachusetts

Gregory H. Bird
Department of Pediatric Oncology and the Program in Cancer Chemical Biology, Dana-Farber Cancer Institute, and the Division of Hematology/Oncology, Children's Hospital Boston, Harvard Medical School, Boston, Massachusetts

Jemima Burden
Medical Research Council Cell Biology Unit, MRC Laboratory for Molecular Cell Biology, and Anatomy and Developmental Biology Department, University College London, London, UK

M. Cecilia Caino
Department of Pharmacology and Institute for Translational Medicine and Therapeutics (ITMAT), University of Pennsylvania School of Medicine, Philadelphia, Pennsylvania

Virve Cavallucci
Laboratory of Molecular Neuroembryology, IRCCS Fondazione Santa Lucia, Rome, Italy and Dulbecco Telethon Institute at the Department of Biology, University of Rome "Tor Vergata", 00133 Rome, Italy

Francesco Cecconi
Laboratory of Molecular Neuroembryology, IRCCS Fondazione Santa Lucia, Rome, Italy and Dulbecco Telethon Institute at the Department of Biology, University of Rome "Tor Vergata", 00133 Rome, Italy

Irene L. Ch'en
Division of Biological Sciences and the Department of Cellular and Molecular Medicine, University of California, San Diego, La Jolla, California

Chi-Wu Chiang
Institute of Molecular Medicine and Center for Gene Regulation and Signal Transduction Research, National Cheng Kung University, Tainan, Taiwan

Marcello D'Amelio
Laboratory of Molecular Neuroembryology, IRCCS Fondazione Santa Lucia, Rome, Italy and Dulbecco Telethon Institute at the Department of Biology, University of Rome "Tor Vergata", 00133 Rome, Italy

Adamo Diamantini
Laboratory of Neuroimmunology, IRCCS Fondazione Santa Lucia, Rome

Kurt Degenhardt
Department of Pharmaceutical Sciences, St. John's University College of Pharmacy, Queens, New York, and Center for Advanced Biotechnology and Medicine, Rutgers University, Piscataway, New Jersey

Donna Denton
Hanson Institute, Institute of Medical and Veterinary Science, Adelaide, Australia

Wafik S. El-Deiry
University of Pennsylvania School of Medicine, Philadelphia, Pennsylvania

Bernal Federico
Department of Pediatric Oncology and the Program in Cancer Chemical Biology, Dana-Farber Cancer Institute, and the Division of Hematology/Oncology, Children's Hospital Boston, Harvard Medical School, Boston, Massachusetts

Niklas Finnberg
University of Pennsylvania School of Medicine, Philadelphia, Pennsylvania

Nathalie C. Franc
Medical Research Council Cell Biology Unit, MRC Laboratory for Molecular Cell Biology, and Anatomy and Developmental Biology Department, University College London, London, UK

Daming Gao
Department of Pathology, Beth Israel Deaconess Medical Center, Harvard Medical School, Boston, Massachusetts

Corinne Giusti
Centre d'Immunologie INSERM-CNRS-Univ.Medit. de Marseille-Luminy, Marseille, France

Pierre Golstein
Centre d'Immunologie INSERM-CNRS-Univ.Medit. de Marseille-Luminy, Marseille, France

Liti Haramaty
Institute of Marine and Coastal Sciences, Rutgers University, New Brunswick, New Jersey

Nicholas Harper
MRC Toxicology Unit, University of Leicester, Leicester, E1 9HN, UK

Stephen M. Hedrick
Division of Biological Sciences and the Department of Cellular and Molecular Medicine, University of California, San Diego, La Jolla, California

Alexander Hoffmann
Signaling Systems Laboratory, Department of Chemistry and Biochemistry, University of California, San Diego, La Jolla, California

Shimin Hu
Abramson Family Cancer Research Institute and Department of Cancer Biology, University of Pennsylvania School of Medicine, Philadelphia, Pennsylvania

Hiroyuki Inuzuka
Department of Pathology, Beth Israel Deaconess Medical Center, Harvard Medical School, Boston, Massachusetts

Vassiliki Karantza-Wadsworth
The Cancer Institute of New Jersey, New Brunswick, New Jersey, and Division of Medical Oncology, Department of Medicine, University of Medicine and Dentistry of New Jersey, Robert Wood Johnson Medical School; Piscataway, New Jersey

Cristina M. Karp
Department of Molecular Biology and Biochemistry, Rutgers University, Piscataway, New Jersey, and Center for Advanced Biotechnology and Medicine, Rutgers University, Piscataway, New Jersey

Marcelo G. Kazanietz
Department of Pharmacology and Institute for Translational Medicine and Therapeutics (ITMAT), University of Pennsylvania School of Medicine, Philadelphia, Pennsylvania

Roya Khosravi-Far
Department of Pathology, Beth Israel Deaconess Medical Center and Harvard Medical School, Boston, Massachusetts

Lorrie A. Kirshenbaum

Institute of Cardiovascular Sciences, St. Boniface Gen Hosp Res Centre, Faculty of Medicine, University of Manitoba, Winnipeg, Manitoba R2H2A6, and Department of Pharmacology and Therapeutics, Faculty of Medicine University of Manitoba, Winnipeg, Manitoba R2H2A6, and Department of Physiology, Faculty of Medicine University of Manitoba, Winnipeg, Manitoba R2H2A6

Artemis Kosta

Centre d'Immunologie INSERM-CNRS-Univ.Medit. de Marseille-Luminy, Marseille, France

Sharad Kumar

Hanson Institute, Institute of Medical and Veterinary Science, Adelaide, Australia

Kageaki Kuribayashi

University of Pennsylvania School of Medicine, Philadelphia, Pennsylvania

James LaBelle

Department of Pediatric Oncology and the Program in Cancer Chemical Biology, Dana-Farber Cancer Institute and the Division of Hematology/Oncology, Children's Hospital Boston, Harvard Medical School, Boston, Massachusetts

David Lam

Centre d'Immunologie INSERM-CNRS-Univ.Medit. de Marseille-Luminy, Marseille, France

Tae H. Lee

Department of Biology and Protein Network Research Center, Yonsei University, Seoul, South Korea

Wenhua Li

Department of Urology, Massachusetts General Hospital, Harvard Medical School, Boston, Massachusetts

Dan Liu

Cell-based Assay Screening Service Core, Baylor College of Medicine, Houston, Texas

Marie-Françoise Luciani

Centre d'Immunologie INSERM-CNRS-Univ.Medit. de Marseille-Luminy, Marseille, France

Marion MacFarlane

MRC Toxicology Unit, University of Leicester, Leicester, LE1 9HN, UK

Raj Mariappan

Department of Pathology, Beth Israel Deaconess Medical Center and Harvard Medical School, Boston, Massachusetts

Pilar Martinez-Chinchilla
Department of Biochemistry and Molecular Biology and Kimmel Cancer Center, Thomas Jefferson University

Robin Mathew
Center for Advanced Biotechnology and Medicine, Rutgers University, Piscataway, New Jersey, and University of Medicine and Dentistry of New Jersey, Robert Wood Johnson Medical School, Piscataway, New Jersey

Suparna Mazumder
Department of Cancer Biology, The Lerner Research Institute, Cleveland Clinic, Cleveland, Ohio

Kathryn Mills
Hanson Institute, Institute of Medical and Veterinary Science, Adelaide, Australia

Jose L. Oliva
Unidad de Biología Celular, Centro Nacional de Microbiología, Instituto de Salud Carlos III, Carretera Majadahonda-Pozuelo, Madrid, Spain

Aria F. Olumi
Department of Urology, Massachusetts General Hospital, Harvard Medical School, Boston, Massachusetts

Sareh Parangi
Department of Surgery, Massachusetts General Hospital and Harvard Medical School, Boston, Massachusetts, and Department of Surgery, Beth Israel Deaconess Medical Center and Harvard Medical School, Boston, Massachusetts

Sun-Mi Park
Current address: Ben May Department for Cancer Research, The University of Chicago, Chicago, IL 60637, USA, and Department of Biology and Protein Network Research Center, Yonsei University, Seoul, South Korea

Shalini D. Persaud
Department of Molecular Biology and Immunology, University of North Texas Health Science Center, Fort Worth, Texas

Kenneth Pitter
Department of Pediatric Oncology and the Program in Cancer Chemical Biology, Dana-Farber Cancer Institute and the Division of Hematology/Oncology, Children's Hospital Boston, Harvard Medical School, Boston, Massachusetts

Dragos Plesca
School of Biomedical Sciences, Kent State University, Kent, Ohio, and Department of Cancer Biology, The Lerner Research Institute, Cleveland Clinic, Cleveland, Ohio

Cristina Pop
Program in Apoptosis and Cell Death Research, The Burnham Institute for Medical Research, La Jolla, California

Kelly M. Regula
Department of Physiology, Faculty of Medicine University of Manitoba, Winnipeg, Manitoba R2H2A6

Natalia A. Riobo
Department of Biochemistry and Molecular Biology and Kimmel Cancer Center, Thomas Jefferson University

Guy S. Salvesen
Program in Apoptosis and Cell Death Research, The Burnham Institute for Medical Research, La Jolla, California

Fiona L. Scott
Program in Apoptosis and Cell Death Research, The Burnham Institute for Medical Research, La Jolla, California

James Shaw
Department of Pharmacology and Therapeutics, Faculty of Medicine University of Manitoba, Winnipeg, Manitoba R2H2A6

Elizabeth A. Silva
Medical Research Council Cell Biology Unit, MRC Laboratory for Molecular Cell Biology, and Anatomy and Developmental Biology Department, University College London, London, UK

Amrik Singh
Department of Pathology, Beth Israel Deaconess Medical Center, Harvard Medical School, Boston, Massachusetts

Usha Sivaprasad
Department of Molecular Biology and Immunology, University of North Texas Health Science Center, Fort Worth, Texas

Keli Song
Department of Pathology, Beth Israel Deaconess Medical Center and Harvard Medical School, Boston, Massachusetts

Zhou Songyang
Verna and Marrs McLean Department of Biochemistry and Molecular Biology, Baylor College of Medicine Houston, Texas

Emilie Tresse
Centre d'Immunologie INSERM-CNRS-Univ.Medit. de Marseille-Luminy, Marseille, France

Alan Tseng
Department of Pathology, Beth Israel Deaconess Medical Center, Harvard Medical School, Boston, Massachusetts

Vivian A. von Burstin
Department of Pharmacology and Institute for Translational Medicine and Therapeutics (ITMAT), University of Pennsylvania School of Medicine, Philadelphia, Pennsylvania

Loren D. Walensky
Department of Pediatric Oncology and the Program in Cancer Chemical Biology, Dana-Farber Cancer Institute, and the Division of Hematology/Oncology, Children's Hospital Boston, Harvard Medical School, Boston, Massachusetts

Danielle Weidman
Department of Physiology, Faculty of Medicine University of Manitoba, Winnipeg, Manitoba R2H2A6

Wenyi Wei
Department of Pathology, Beth Israel Deaconess Medical Center, Harvard Medical School, Boston, Massachusetts

Eileen White
Department of Molecular Biology and Biochemistry, Rutgers University, Piscataway, New Jersey, and University of Medicine and Dentistry of New Jersey, Robert Wood Johnson Medical School, Piscataway, New Jersey, and Center for Advanced Biotechnology and Medicine, Rutgers University, Piscataway, New Jersey, and The Cancer Institute of New Jersey, New Brunswick, New Jersey

Liqing Xiao
Department of Pharmacology and Institute for Translational Medicine and Therapeutics (ITMAT), University of Pennsylvania School of Medicine, Philadelphia, Pennsylvania

Elizabeth Yang
Departments of Pediatrics, Cancer Biology, Cell and Developmental Biology, Vanderbilt University School of Medicine, Nashville, Tennessee

Ling Yan
Departments of Pediatrics, Cancer Biology, Cell and Developmental Biology, Vanderbilt University School of Medicine, Nashville, Tennessee

Xiaolu Yang
Abramson Family Cancer Research Institute and Department of Cancer Biology, University of Pennsylvania School of Medicine, Philadelphia, Pennsylvania

Natalia Yurkova
Department of Physiology, Faculty of Medicine University of Manitoba, Winnipeg, Manitoba R2H2A6

Tong Zhang
Department of Physiology, Faculty of Medicine University of Manitoba, Winnipeg, Manitoba R2H2A6

Xuefeng Zhang
Department of Surgery, Beth Israel Deaconess Medical Center and Harvard Medical School, Boston, Massachusetts

Xiaoping Zhang
Department of Urology, Massachusetts General Hospital, Harvard Medical School, Boston, Massachusetts

PREFACE

The idea of assembling a *Methods in Enzymology* book on cell death at first struck us as not very different from assembling a *Methods* book on Biology. There are so many aspects to cover and such a diversity of techniques that it seemed impossible to put together a manual of direct laboratory use such as one might do by offering a volume on, for instance, the isolation and study of mitochondria. After some reflection, we realized that the complexity was the point: it would be useful to compile a work that would, first, clearly define the types of cell death; second, indicate that not all forms of death were apoptosis; third, indicate the major means of documenting apoptosis; and fourth, explain how to document forms of cell death that are not apoptosis. In this manner we would provide a guide for the junior investigator and define criteria in a manner that would help readers recognize and avoid the confusion that exists in the literature because of the lack of clear definitions.

There are many means of organizing these overlapping topics. We have chosen to group the more biologic and general topics into the first volume, with the more clinical and specific topics in the second volume. A particular reader will find it useful to pick and choose information, perhaps in several chapters. Nevertheless, readers can expect to find in the first volume: comparisons of the characteristics of apoptosis, necrosis, autophagy, and other forms of cell death (Chapters 1, 12, 13, 19, and 20); means of studying the often unusual forms of cell death in nonmammalian organisms (Chapters 21 and 22); more general and commonly used methods for studying apoptosis in mammalian cells (Chapters 2, 3, 10); and means for studying specific events or components of apoptosis: The CD95 DISC (Chapter 4); the TNF receptosome (Chapter 5); the apoptosome (Chapter 6); and four types of enzymes active in apoptosis and other forms of cell death: caspases (Chapter 7); lysosomal cathepsins (Chapter 8); granzymes (Chapter 9); and nucleases (Chapter 11). In addition, we include discussions of the metabolic components of apoptosis such as mitochondrial membrane permeabilization (Chapter 14); oxidative lipidomics (Chapter 15); endoplasmic reticulum stress-induced apoptosis (Chapter 16); membrane scrambling (Chapter 17); and glucose metabolism (Chapter 18). In the second volume, readers can find chapters on studying apoptosis in different model systems (Chapters 1, 2, 3, 4, 5), investigation of several of the key promoters and signaling pathways that regulate programmed cell death (Chapters 6, 7, 8, 9, 10,11, and 21), analysis of several mechanisms by which apoptotic pathways are regulated (Chapters 12, 13, 14), analysis of programmed cell death in different tissue and organ systems (Chapters 15, 16, and 17), methods for generation and analysis of death ligands (Chapters 18 and 19), methods for identifying caspase activity and substrates (Chapter 21), generation and analysis of BCL-2 stabilized peptides (Chapters 22 and 23), and, finally, methods for a genetic screening strategy that could be potentially be used in identifying novel factors regulating programmed cell death (Chapter 24).

Although each reader will undoubtedly identify particular items that he or she would have preferred to see, or a different order of presentation, we were very pleased that almost all of the contributors that we originally invited accepted our invitation and delivered to us manuscripts conforming to our goals. Thus, this work represents what we intended—a broad spectrum of approaches to apoptosis and programmed cell death by recognized leaders in their respective fields. Thus, we hope that it will contribute by serving as a guide and a means of helping researchers to ask the most meaningful questions and to design their experiments for the highest clarity.

METHODS IN ENZYMOLOGY

VOLUME 363. Recognition of Carbohydrates in Biological Systems (Part B)
Edited by YUAN C. LEE AND REIKO T. LEE

VOLUME 364. Nuclear Receptors
Edited by DAVID W. RUSSELL AND DAVID J. MANGELSDORF

VOLUME 365. Differentiation of Embryonic Stem Cells
Edited by PAUL M. WASSAUMAN AND GORDON M. KELLER

VOLUME 366. Protein Phosphatases
Edited by SUSANNE KLUMPP AND JOSEF KRIEGLSTEIN

VOLUME 367. Liposomes (Part A)
Edited by NEJAT DÜZGÜNEŞ

VOLUME 368. Macromolecular Crystallography (Part C)
Edited by CHARLES W. CARTER, JR., AND ROBERT M. SWEET

VOLUME 369. Combinational Chemistry (Part B)
Edited by GUILLERMO A. MORALES AND BARRY A. BUNIN

VOLUME 370. RNA Polymerases and Associated Factors (Part C)
Edited by SANKAR L. ADHYA AND SUSAN GARGES

VOLUME 371. RNA Polymerases and Associated Factors (Part D)
Edited by SANKAR L. ADHYA AND SUSAN GARGES

VOLUME 372. Liposomes (Part B)
Edited by NEJAT DÜZGÜNEŞ

VOLUME 373. Liposomes (Part C)
Edited by NEJAT DÜZGÜNEŞ

VOLUME 374. Macromolecular Crystallography (Part D)
Edited by CHARLES W. CARTER, JR., AND ROBERT W. SWEET

VOLUME 375. Chromatin and Chromatin Remodeling Enzymes (Part A)
Edited by C. DAVID ALLIS AND CARL WU

VOLUME 376. Chromatin and Chromatin Remodeling Enzymes (Part B)
Edited by C. DAVID ALLIS AND CARL WU

VOLUME 377. Chromatin and Chromatin Remodeling Enzymes (Part C)
Edited by C. DAVID ALLIS AND CARL WU

VOLUME 378. Quinones and Quinone Enzymes (Part A)
Edited by HELMUT SIES AND LESTER PACKER

VOLUME 379. Energetics of Biological Macromolecules (Part D)
Edited by JO M. HOLT, MICHAEL L. JOHNSON, AND GARY K. ACKERS

VOLUME 380. Energetics of Biological Macromolecules (Part E)
Edited by JO M. HOLT, MICHAEL L. JOHNSON, AND GARY K. ACKERS

VOLUME 381. Oxygen Sensing
Edited by CHANDAN K. SEN AND GREGG L. SEMENZA

VOLUME 382. Quinones and Quinone Enzymes (Part B)
Edited by HELMUT SIES AND LESTER PACKER

VOLUME 383. Numerical Computer Methods (Part D)
Edited by LUDWIG BRAND AND MICHAEL L. JOHNSON

VOLUME 384. Numerical Computer Methods (Part E)
Edited by LUDWIG BRAND AND MICHAEL L. JOHNSON

VOLUME 385. Imaging in Biological Research (Part A)
Edited by P. MICHAEL CONN

VOLUME 386. Imaging in Biological Research (Part B)
Edited by P. MICHAEL CONN

VOLUME 387. Liposomes (Part D)
Edited by NEJAT DÜZGÜNEŞ

VOLUME 388. Protein Engineering
Edited by DAN E. ROBERTSON AND JOSEPH P. NOEL

VOLUME 389. Regulators of G-Protein Signaling (Part A)
Edited by DAVID P. SIDEROVSKI

VOLUME 390. Regulators of G-Protein Signaling (Part B)
Edited by DAVID P. SIDEROVSKI

VOLUME 391. Liposomes (Part E)
Edited by NEJAT DÜZGÜNEŞ

VOLUME 392. RNA Interference
Edited by ENGELKE ROSSI

VOLUME 393. Circadian Rhythms
Edited by MICHAEL W. YOUNG

VOLUME 394. Nuclear Magnetic Resonance of Biological Macromolecules (Part C)
Edited by THOMAS L. JAMES

VOLUME 395. Producing the Biochemical Data (Part B)
Edited by ELIZABETH A. ZIMMER AND ERIC H. ROALSON

VOLUME 396. Nitric Oxide (Part E)
Edited by LESTER PACKER AND ENRIQUE CADENAS

VOLUME 397. Environmental Microbiology
Edited by JARED R. LEADBETTER

VOLUME 398. Ubiquitin and Protein Degradation (Part A)
Edited by RAYMOND J. DESHAIES

VOLUME 399. Ubiquitin and Protein Degradation (Part B)
Edited by RAYMOND J. DESHAIES

VOLUME 400. Phase II Conjugation Enzymes and Transport Systems
Edited by HELMUT SIES AND LESTER PACKER

CHAPTER ONE

ANALYSIS OF AUTOPHAGIC AND NECROTIC CELL DEATH IN *DICTYOSTELIUM*

Corinne Giusti, Artemis Kosta, David Lam, Emilie Tresse, Marie-Françoise Luciani, *and* Pierre Golstein

Contents

Abstract

Non-apoptotic cell death types can be conveniently studied in *Dictyostelium discoideum,* an exceptionally favorable model not only because of its well-known genetic and experimental advantages, but also because in *Dictyostelium* there is no apoptosis machinery that could interfere with non-apoptotic cell death. We show here how to conveniently demonstrate, assess, and study these non-apoptotic cell death types. These can be generated by use of modifications of the monolayer technique of Rob Kay *et al.*, and either wild-type HMX44A *Dictyostelium* cells, leading to autophagic cell death, or the corresponding *atg1*⁻ autophagy

Centre d'Immunologie INSERM-CNRS-Univ.Medit. de Marseille-Luminy, Marseille, France

Methods in Enzymology, Volume 446
ISSN 0076-6879, DOI: 10.1016/S0076-6879(08)01601-7

gene mutant cells, leading to necrotic cell death. Methods to follow these non-apoptotic cell death types qualitatively and quantitatively will be reported.

1. INTRODUCTION

Starvation-induced development in *Dictyostelium* ultimately leads to a fruiting body that includes stalk cells. These are highly vacuolated (de Chastellier and Ryter, 1977; George *et al.*, 1972; Maeda and Takeuchi, 1969; Quiviger *et al.*, 1980; Raper and Fennell, 1952; Schaap *et al.*, 1981) and dead as demonstrated by their inability to regrow in culture medium (Whittingham and Raper, 1960). *Dictyostelium* stalk cells may thus be considered one of the earliest known occurrences of developmental programmed cell death in eukaryote evolution. This cell death can be mimicked in *in vitro* tests, where depending on conditions (see later) it can be either autophagic or necrotic (Kosta *et al.*, 2004; Laporte *et al.*, 2007; Levraud *et al.*, 2003b). At least three main reasons make this model a very favorable one to study these non-apoptotic cell deaths: (1) marked genetic tractability, with in particular haploidy facilitating the search for causative molecules by random mutagenesis, (2) absence of the main families of molecules involved in apoptosis, limiting the risk of interference with the non-apoptotic mechanisms, and (3) relative molecular simplicity, decreasing the risk of functional redundancy.

Programmed cell death in *Dictyostelium* can be studied *in vitro* by use of conditions that are more amenable to microscopic observation and to further genetic manipulations than *in vivo* stalk cell death. Differentiation in monolayers (Kay, 1987a) of a *Dictyostelium* mutant strain called HM44 (Kopachik *et al.*, 1983) (derived from the original *Dictyostelium* strain V12M2, adapted for axenic growth as HMX44—J. G. Williams, University of Dundee—and subcloned in our laboratory as HMX44A) to stalk cells (i.e., to vacuolated, dead cells) seems to result from the sequential action of at least two main factors under starvation conditions: first, cyclic AMP (cAMP) and second, the dichlorinated hexanone differentiation-inducing factor DIF-1 (thereafter called DIF). DIF acts on starved cAMP-subjected cells and promotes their differentiation into stalk cells (Morris *et al.*, 1987; Sobolewski *et al.*, 1983; Town and Stanford, 1979; Town *et al.*, 1976). HM44 produces very little DIF but is sensitive to exogenous DIF (Kopachik *et al.*, 1983). On starvation and addition of exogenous DIF, HM44 differentiates into stalk cells; however without morphogenizing, HM44 differentiates into a sorocarp.

Dictyostelium HMX44A cells were subjected (Cornillon *et al.*, 1994; Kay, 1987a; Levraud *et al.*, 2003a) to incubation in starvation medium in the presence of cAMP, which does not lead in itself to cell death, followed by another period of incubation together with DIF, which triggers a cascade of events leading to cell death (Cornillon *et al.*, 1994; Levraud *et al.*, 2003a).

This autophagic vacuolar cell death is caspase- and paracaspase-independent (Olie *et al.*, 1998; Roisin-Bouffay *et al.*, 2004). Vacuolization, but not cell death itself, depends on an intact *atg1* autophagy gene (Kosta *et al.*, 2004). When *atg1* is mutated, under similar circumstances cell death is necrotic (Kosta *et al.*, 2004; Laporte *et al.*, 2007) and still does not require the paracaspase (D. Lam *et al.*, 2007). Although in this chapter we usually refer to cell death as studied in our laboratory, other groups have also studied cell death in *Dictyostelium*, often in non–developmental contexts (Arnoult *et al.*, 2001; Katoch and Begum, 2003; Kawli *et al.*, 2002; Li *et al.*, 2000; Schaap *et al.*, 1996; Tatischeff *et al.*, 2001).

In this chapter, a *Dictyostelium* cell will be considered dead when its membrane is altered either obviously or as reflected by an abnormally increased permeability to certain dyes, and/or when it has lost its ability to multiply as reflected by loss of clonogenicity. Some of the methods used have been described in detail in previous reviews (Kosta *et al.*, 2006; Levraud *et al.*, 2001), together with genetic approaches to the molecular mechanism of cell death (Levraud *et al.*, 2004) that will not be touched on here. A wealth of practical information can be obtained from the Dictybase Web site (http://dictybase.org/). In this chapter, we will consider how to induce and to study autophagic or necrotic cell death in this model.

2. Growing *Dictyostelium* Cells in Monolayer Cultures

Dictyostelium cells adapted for axenic growth can be grown as many eukaryotic cells in tissue culture plastic flasks (e.g., Falcon flasks) (Fig. 1.1). As an example, a 25-cm^2 Falcon flask would contain 10 ml of medium, with 10^6

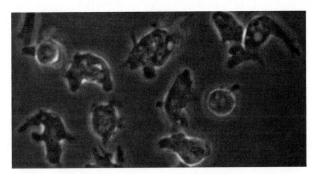

Figure 1.1 Vegetative cells growing in rich medium. HMX44A cells in exponential growing phase were seeded in Lab-Tek chamber in HL5 complete medium. Phase contrast microscopy, oil immersion ×100 objective.

to 10^7 cells. Saturating density is approximately 2×10^6 cells/ml of medium, and cells can be passed by 20 time dilutions every second or third day. Cell culture is at 22–24 °C (*Dictyostelium* cells die around 27 °C), in a temperature-regulated but normal atmosphere incubator. As culture medium we use HL-5, which includes 14.3 g/L bacteriologic peptone (OXOID LTD, Basingstoke, Hampshire-England, ref. L37); peptone from other sources can cause a dramatic decrease of *Dictyostelium* growth and impair development after starvation (our unpublished observations; see also Sussman, 1987); 7.15 g/L yeast extract (EZMix yeast extract, Sigma, ref. Y1626), 9 g/L maltose (Sigma, ref M5885), 0.93 g/L Na_2HPO_4, $7H_20$ (3.6 mM), 0.49 g/L KH_2PO_4 (3.6 mM), and water. To reduce variations in the quality of water, we have been routinely used Volvic source water instead of column-demineralized water. Source water contains a small amount of mineral salts and is therefore not strictly equivalent to fully demineralized water. Sterilize by autoclaving for 30 min. After cooling, filter through a 0.22-μm filter and store at 4 °C. This medium is stable for several months at 4 °C. Although we routinely grow *Dictyostelium* in sterile conditions in the absence of antibiotics, accidental bacteriologic contamination can usually be corrected by adding penicillin-streptomycin (100 units/mL and 100 mg/mL, respectively; Gibco-BRL, Grand Island, NY) to the culture medium. Fungal contaminations are more problematic, because *Dictyostelium* is killed by fungizone; nystatin or gentamycin may be of help.

3. INDUCING AUTOPHAGIC VACUOLAR CELL DEATH

Programmed cell death in *Dictyostelium* is the outcome of terminal differentiation of stalk cells. This can be obtained in two different ways: either by inducing normal development at an air–wet solid interface (stalk cells then constitute the stalks and basal disks of the resulting fruiting bodies), or by *in vitro* stalk cell differentiation in monolayers under submerged conditions. The first method is the "natural" one, but is less convenient than the second one for most applications because only ∼15% of the cells end up as stalk cells, the remainder differentiate into viable spores. Regrowth of stalk cells exclusively is thus difficult to score (although not impossible; see Whittingham and Raper, 1960). Also, microscopic observation of cells in a stalk is not easy; further manipulations are not only required to place the stalk on a microscope slide, but cell morphology is more difficult to assess, because cells are enclosed in the cellulose sheath tube (in addition to their own casing) and tightly packed. Furthermore, cells surviving in a stalk are difficult to isolate and manipulate.

The methods detailed here are meant for direct microscopic examination, usually of unfixed cells, mostly under the fluorescence microscope.

A technical problem linked with such examination is the fact that when wild-type *Dictyostelium* cells differentiate in monolayers, some adhere very tightly to the substrate, whereas others are found in suspension (most of them clustered). To get a representation of the total population, ideally both cell pools should be considered. If an inverted fluorescence microscope is available, differentiation may be conveniently carried out in plastic chambers on coverslips. These coverslips should be thin enough to allow microscopic examination by use of a $100\times$ oil-immersion objective. Lab-Tek chambers have proven very useful for this purpose (Ref # 155380, Nalge Nunc, Lab-Tek chambered Coverglass @1 German borosilicate sterile, 2 wells—in fact two plastic chambers, each with a surface area of 2×2 cm, set on a slide-size borosilicate coverslip). Cell manipulations, staining, washes, and microscopic examination can be performed directly in these chambers.

This protocol is applicable with most of the usual strains. However, the percentage of cells differentiating into stalk cells is strongly strain–dependent: cells of V12M2 origin (such as HMX44) differentiate more efficiently than cells of NC4 origin (such as AX-2). This is largely because of a difference in sensitivity of inhibition of the DIF-dependent step by cAMP (Berks and Kay, 1988), implying that an additional washing step is recommended for some strains.

The protocol detailed here, derived from the one described by Kay (Kay, 1987a), has been optimized for HMX44A. Collect vegetative cells growing in HL-5 medium in late log phase (i.e., no more than 2×10^6 cells/ml if cells are grown without stirring). Wash twice with SB buffer (Soerensen Buffer $50\times$ stock solution: 100 mM Na$_2$HPO$_4$, 735 mM KH$_2$PO$_4$, pH 6.0; filter-sterilize and store at $4°$. SB $1\times$ buffer is obtained by adding 10 ml of $50\times$ solution to 490 ml of autoclaved source water), count. In each of two Lab-Tek chambers add 3×10^5 cells in 1 ml of SB containing 3 mM cAMP (cyclic AMP; $3',5'$ adenosine cyclic monophosphate, sodium salt, Sigma A6885; make stock solution 60 mM in demineralized autoclaved water, filter sterilize, keep at $-20°$C in 1 mL aliquots). Incubate for 8 h at $22°$C. Most ($>80\%$) cells should adhere firmly to the bottom slide. Because of the high concentration of cAMP, almost no aggregation should be seen at this stage, and the cells should be randomly scattered and isolated. Carefully remove the liquid and wash once with 1 mL of SB. Replace with 1 mL of SB containing 100 nM DIF-1 in one chamber (DIF-1: differentiation-inducing-factor 1, 1-(3,5-dichloro-2, 6-dihydroxy-4-methoxyphenyl)-hexan-1-one; DN1000, Affiniti Research Products, Exeter, UK ; make stock solution 10 mM in absolute ethanol. Working stocks are diluted to 0.1 mM in absolute ethanol in 0.1 mL aliquots. Store at $-20°$C). In control chambers use 1 mL of SB with no DIF. Incubate for 24 h at $22°$C.

At this stage, in the DIF-containing chamber, most cells should be differentiated to "stalk" cells: highly vacuolated, cellulose-encased, non-refringent cells as seen by phase contrast microscopy (see, for instance,

Kay, 1987a). Almost no vacuolated cells should be seen in the control chamber if a cell line producing little endogenous DIF (e.g., HMX44) is used. Further incubation will increase the proportion of dead cells, up to 50% at 30 to 36 h as assessed by propidium iodide staining (see later and Cornillon *et al.*, 1994), and vacuolization will become progressively more prominent as the cytoplasm of dying/dead cells continues to shrink.

4. FURTHER EXPLORATION OF AUTOPHAGIC CELL DEATH

4.1. Phase contrast microscopy

Dictyostelium cells are small (approximately 10 μm across), thus an oil-immersion 100× objective is optimal. We use standard phase contrast (Nomarski DIC shows less details on these cells), mounted on an inverted Zeiss Axiovert 200 microscope, with an Axiocam MRC digital camera connected to a PC equipped with Axiovision (Zeiss). Time lapse microscopy is often of tremendous help, because, in particular, it allows us to easily rank in time the various steps of a given process (see for instance the movies in Levraud *et al.*, 2003a). In some cases, pictures were taken by use of a confocal microscope (Zeiss Axiovert 200, LSM510) with a 63× oil immersion objective, with the scanning module at 1024 × 1024 pixel resolution, 8 bits, by use of a scan average of 4.

HMX44A cells subjected to starvation only show typical flat morphology (Fig. 1.2A). The addition of DIF leads to paddle cells typically 12–16 h after DIF, then rounding, and vacuolization 20–24 h after DIF (Cornillon *et al.*, 1994; Levraud *et al.*, 2003a). All three aspects can be present at the intermediate time of 18 h (Fig. 1.2B).

4.2. Propidium iodide staining

Propidium iodide (PI) is a DNA intercalating dye that cannot cross cell membranes freely: thus, cells will fluoresce only if membranes have become permeable, a late sign of cell death. This implies that cells must be stained fresh (e.g. they cannot be fixed for this test). Add PI concentrated stock (propidium iodide, Sigma P4170, stock solution: 80 mM (53 mg/mL) in sterile water, keep at 4 °C protected from light. Gloves should be worn because PI is carcinogenic to cells in SB to reach a final concentration of 4 mM PI. Incubate for 10 min at room temperature away from light. Wash twice carefully with SB. View under the fluorescence microscope, observe red fluorescence by use of filter set 15 (Zeiss). Typically, this red fluorescence appears as a crescent between the vacuole and the cellulose shell of dead cells.

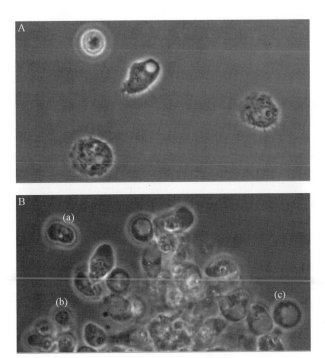

Figure 1.2 Autophagic vacuolar cell death. (A) Control cells starved with no DIF added, showing "flat" morphology. (B) After 8 h of incubation in starvation medium with cAMP, autophagic vacuolar cell death was induced by addition of DIF 10^{-7} M. HMX44A wild-type dying cells undergo a series of morphologic changes. First, they acquire a paddle shape (a), then transform into round cells (b) that vacuolize (c). Unfixed cells were directly visualized in Lab-Tek chambers by phase contrast microscopy, oil immersion ×100 objective.

4.3. Fluorescein diacetate staining

In contrast to PI, fluorescein diacetate (FDA) stains living cells. The non-fluorescent, hydrophobic compound freely enters the cell, where it is cleaved by cytoplasmic lipases of metabolically active cells into a green fluorophore unable to leave the cell if the membrane is intact. Again, cells must not have been fixed. Wash cells once with SB, add FDA (fluorescein diacetate, Sigma F7378, stock solution: 10 mg/mL in acetone; store at 4 °C away from light) concentrated stock to cells in SB to reach a final concentration of 0.05 mg FDA/mL, incubate for 10 min at room temperature away from light, wash twice with SB, view under the fluorescence microscope, observe green fluorescence with filter set 09 (Zeiss). Cells alive by this criterium appear uniformly bright green.

FDA and PI can be used for double staining by directly mixing these dyes.

4.4. Calcofluor staining

As they differentiate, stalk cells encase themselves in a cellulose shell that may be labeled with calcofluor. Positive staining does not constitute evidence of cell death but is nevertheless a useful differentiation marker. One should be aware that other cell types (e.g., spores and macrocysts) also secrete cellulose coats, but cell sizes are very different. A thin cellulose trail is also left on the substrate before the cell finally stops migrating and fully differentiates (Blanton, 1993). Cells may be fixed before staining. Wash cells with SB, add calcofluor concentrated stock (Calcofluor, Sigma F6259; now sold under catalog # F3397; also named "fluorescent brightener 28", "calcofluor white M2R", "C.I. (Colour Index) 40622", or "tinopal LPW"; stock solution 1% (w/v) in H_2O, keep at 4 °C protected from light) to cells in SB to reach a final calcofluor concentration of 0.1%, incubate for 5 min at room temperature protected from light, wash twice with SB, view under the fluorescence microscope, observe blue fluorescence by use of filter set 01 (Zeiss). Cellulose shells appear as very bright blue coats around dying or dead cells (Levraud *et al.*, 2003a).

4.5. Trypan blue staining

Trypan blue is a commonly used dye to discriminate between live and dead mammalian cells, because it stains only cells with a compromised plasma membrane. Staining with trypan blue or propidium iodide thus provides comparable information. Trypan blue–stained cells are less obvious than PI-labeled cells, but because fluorescence is not required, trypan blue is more convenient for routine quantification of live vs dead cells in a sample. Trypan blue (Sigma T8154) is sold as a 0.4% stock solution. Dilute four times with SB and store at room temperature under sterile conditions (may be frozen; 0.1% sodium azide can be added to prevent contamination). Gloves should be worn when manipulating trypan blue, which is teratogenic and may also include carcinogenic compounds. Trypan blue is added to cells at a final concentration of 0.1% to 0.03% for at least 10 min, and the sample directly examined by use of standard microscopy. The liquid layer has to be thin; for cells in 25-cm^2 flasks, 1 mL of trypan blue solution is suitable. It is important not to allow cells to dry, because the liquid tends to accumulate as a meniscus at the corners of the flask. Results will be optically better if the cells are placed between a microscopic slide and a coverslip. The use of a blue filter enhances contrast.

Other stains may be used as a function of specific questions being asked (e.g., to stain cell organelles such as mitochondria or structures such as cytoskeleton components, or to detect metabolites such as reactive oxygen species). These will not be described here.

4.6. Electron microscopy

Electron microscopy is a valuable tool that allows the structure of organelles and cellular subsystems to be characterized with unprecedented detail and reliability. The methods we use to investigate *Dictyostelium* cell death by electron microscopy have been described in detail elsewhere (Kosta *et al.*, 2006).

4.7. Regrowth assay

This assay is currently the only one quantifying surviving cells (if surviving means ability to multiply) after vacuolar autophagic cell death, other than tedious microscopy counting of percentages of vacuolized and/or FDA/PI-stained cells. Collect vegetative cells in late exponential growth phase. Wash twice with SB buffer, count (this test is sensitive to variations in initial density, so cells should be carefully counted before plating into flasks). Plate two Lab-Tek chambers each with 3×10^5 cells in 1 mL of SB containing 3 mM cAMP. Incubate for 8 h at 22 °C. Carefully remove the liquid, wash once with 1 ml of SB, and replace with 1 ml of SB + DIF $10^{-7}M$ in the first chamber, or 1 ml of SB in the second, control chamber. To be rigorous, a similar amount of absolute ethanol (the solvent of DIF) should be added to the control flask. However, we have never seen a significant effect of the addition of 0.1% ethanol in this assay. Incubate for 24 h at 22 °C. This leads as described previously to the death of most cells in the DIF-containing chamber. If cells are incubated in starvation medium with DIF for longer than 24 h, vacuolization may seem more complete, but other phenomena may interfere with the results. In particular, and unexpectedly, differentiation into what appears to be macrocysts may occur with HMX44A cells maintained in SB without DIF (unpublished observations), lowering the frequency of regrowing cells. From each chamber, carefully remove 0.5 mL of SB, and add 2 volumes (1 mL) of HL-5 to initiate regrowth of surviving cells. Incubate at 22 °C for 40 to 72 h. Detach cells by vigorous flushes with a pipette. Under an inverted microscope, check that all vegetative cells are detached; many stalk cells will still adhere, which is not a problem because they are not to be counted. Count amoeboid cells by use of a hemocytometer and phase contrast optics. The rare heavily vacuolated, nonrefringent stalk cells are easily distinguished and excluded. Trypan blue may be added to facilitate the identification of dead cells (see earlier). Calculate the ratio of the number of regrowing cells in the DIF chamber to the number of cells in the control chamber. Results of the test are collected after a period of exponential growth. Slight variations in culture conditions may thus significantly affect the results, although expression as a ratio prevents excessive departure from the usual values. For HMX44A, this should be approximately 0.15. This ratio expresses the percentage of cells surviving after DIF-induced cell death. Approximately 15% is the usual background of surviving HMX44A cells.

5. Inducing Necrotic Cell Death

Typically, as mentioned elsewhere (Laporte *et al.*, 2007) HMX44A. *atg1-1* (an HMX44A derivative mutated for the autophagy gene *atg1*, Kosta *et al.*, 2004) cells were routinely grown at 22 °C in HL5 medium (Sussman, 1987) as modified (Cornillon *et al.*, 1994) except for maltose that was 9 g/L. Blasticidin (10 μg/mL) was added to the HMX44A.*atg1-1* cultures. For experiments, vegetative cells in growth phase were washed once and resuspended in 10 mM MES (MES hydrate, Sigma M2933) prepared in demineralized water. The pH of this MES buffer was found to matter. PH < 6 increased, and pH > 6.6 decreased cell membrane rupture, perinuclear clustering and DC-FDA fluorescence (not shown). The pH was therefore adjusted at 6.4 with NaOH in all experiments, and drugs eventually added were checked not to alter this pH by more than 0.2 pH units. The cells in exponential growth at a maximum concentration of 10^6 cells/mL (20 mL in a 75 cm^2 flask) were adjusted at a concentration of 3×10^5 cells/mL. One mL of this cell suspension was distributed in each well of Lab-Tek culture chambers with addition of cAMP to a final concentration of 3 mM. Incubation proceeded for 16 h at 22 °C in MES buffer and cAMP (the previously used 8 h incubation period (Cornillon *et al.*, 1994; Kay, 1987b) was found limiting for *atg1-1* cells. A starvation period of sufficient duration was essential for efficient induction by DIF of necrotic cell death). Cells were then washed once by careful removal and addition of MES and incubated at 22 °C in either MES, or MES in the presence of the differentiation factor DIF at a final concentration of 10^{-7} M unless stated otherwise (varying the concentrations of DIF did not modify the type of death but changed it quantitatively). After incubation for the indicated period of time, cells in the Lab-Tek chambers were examined as indicated above by use of an inverted microscope. Approximately 20 min after addition of DIF, cell organelles, in particular mitochondria, clustered around the nucleus, giving the aspect shown in Fig. 1.3. Plasma membrane rupture ensued within approximately 150 min (Kosta *et al.*, 2004; Laporte *et al.*, 2007).

6. Further Exploration of Necrotic Cell Death

Tests assessing not cell death *per se* but mitochondrial or mitochondria-dependent events preceding it, such as measurement of DC-FDA fluorescence possibly revealing reactive oxygen species, of oxygen consumption or of ATP levels, have been described elsewhere (Laporte *et al.*, 2007) and will not be described in detail here.

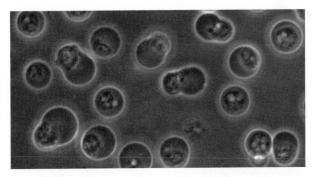

Figure 1.3 Morphology of cells dying in a necrotic manner. HMX44A *atg*1-1 mutant cells were subjected to starvation and cAMP for 16 h then to DIF 10^{-7} M. Prominent traits of necrotic cell death were, as soon as 20 min after DIF addition, perinuclear clustering of organelles, followed after approximately 150 min by plasma membrane rupture. There was no blebbing and only moderate cell swelling before membrane rupture. Unfixed cells were directly visualized in Lab-Tek chambers by phase contrast microscopy, oil immersion ×100 objective.

6.1. Lysosomal investigation

6.1.1. Acridin orange uptake assay

Acridin orange (AO), a metachromatic fluorophore and a lysosomotropic base, is retained in its charged form (AOH+) by proton trapping inside the acidic vacuolar compartment. Normal cells excited by blue light show highly concentrated lysosomal AO with intense red granular fluorescence, and only weak nuclei and cytosol diffuse green fluorescence. Lysosome alteration results in loss of lysosomal red fluorescence and appearance of bright green cytoplasmic and nucleolar fluorescence. Starved cells were stained at the indicated times with 5 μg/mL AO (A-6014; Sigma–Aldrich) for 10 min at 22 °C with protection from light, and washed with MES. Green fluorescence (FITC channel) of 10^4 cells per sample was determined by use of a BD FACScantoII flow cytometer (Becton Dickinson) equiped with a 488-nm argon laser. FlowJo software (Tree Star Inc.) was used for analysis. Such AO-labeled could also be examined by confocal microscopy (see later).

6.1.2. Lysosensor blue uptake assay

The acidotropic dye Lysosensor blue DND-167 (LB; L7533 ; Molecular Probes) appears to accumulate in acidic organelles as the result of protonation. It shows a pH-dependent decrease in fluorescence intensity on alkalinization. Cells in 12-well plates were labeled at the indicated times with LB 5 μM for 10 min then blue fluorescence was determined by flow cytometry (FACScantoII, Becton Dickinson) by use of a 405,407 excitation laser line and an am-cyan channel. FlowJo software (Tree Star Inc.)

was used for analysis. Alternately, starved cells (3×10^5 cells/ml) treated or not with DIF in Lab-Tek culture chambers as described previously, were AO-loaded (10 μg/ml) or LB-loaded (5 μM) for 10 min then examined by confocal microscopy (LSM 510 Zeiss, META UV) with a 63× oil immersion objective, with the scanning module at 1024×1024 pixel resolution, 8 bits, by use of a scan average of 4 with an argon/2 laser and a LP 650 for granular–AO fluorescence and a 405-nm excitation laser and a BP 420 to 480 for LB fluorescence.

6.1.3. Cathepsin B activity

For detection of cathepsin B and L activity, starved cells were treated with the corresponding substrate (0.5×) for 30 min by use of Magic Red Cathepsin Detection kit (Immunochemistry Technologies, LLC, Abc117-4053 and Abc117-4057) according to the manufacturer's instructions. Cells were then analyzed by confocal microscopy with an HeNe1 laser and a LP 650 filter. Objective and parameters of the scanning module were unchanged. Images were analyzed with LSM software by use of false colors.

6.1.4. Dextran endocytosis

Cells were starved in MES/cAMP buffer containing 0.5 mg/ml Texas red-dextran (Molecular Probes, D3329). After a 16 h starvation period, cells were washed and DIF was added. To visualize Texas Red-dextran by confocal microscopy, an HeNe2 laser with a LP 650 were used.

6.2. Quantifying cell death through plasma membrane rupture by flow cytometry

Flow cytometry is widely used to study the characteristics of cell death in mammalian cells. It allows quantitative measurement of fluorescence, size, and granularity of cells, which can be applied to statistically significant numbers of cells. However, flow cytometry can only be used with isolated cells in suspension. When they differentiate into stalk cells, wild-type *Dictyostelium* cells often adhere strongly to their substrate and form very tight cell clumps that are bound together with cellulose. This makes it difficult to analyze wild-type differentiating cells by flow cytometry. Vegetative cells, early wild-type differentiating cells, or round mutant $atg1^-$ cells that do not aggregate are, however, amenable to such analyses. Cytometry provides an easy, precise and objective quantification of necrotic $atg1$- cell death, which is particularly useful when testing the effect of modifiers (such as inhibitors) on cell death. Dead cells could be distinguished from live cells merely by morphologic criteria. Flow cytometry enables us to quantify the proportion of cells with a disrupted membrane.

Cell death is induced in monolayers by addition of DIF after a starvation period either in Lab-Tek culture chambers or in flasks. At the chosen time, the cells are resuspended and transferred in 5 mL round-bottom tube then directly analyzed. If one wants to use PI staining (which could be necessary if the treatment applied to cells affects their size or granulosity), two tubes for each sample are prepared: a control (without any staining) and a test tube for PI staining. PI is used at 1 μg/ml and incubation is done with minimum light exposure for 10 min at 22 °C, with no wash. Cells in Lab-Tek chambers were resuspended and transferred to tubes and data acquisition was performed on a FACScan cytometer or a FACScalibur cytometer with CELLQuest software (Becton Dickinson) for forward scatter (FSC) and side scatter (SSC) parameters that allowed to distinguish cells with a disrupted membrane from those with an intact one. To exclude debris, a FSC threshold was applied on the basis of the light properties of the cells populations in FSC and SSC modes. After set up the acquisition was made on 10^4 events per sample. Analysis of data was performed with either CELLQUEST or FlowJo software.

Altogether, in this privileged model organism with exceptional experimental and genetic advantages and absence of apoptotic machinery, both autophagic and necrotic cell death have been demonstrated and can now be further studied by use of a range of established methods.

REFERENCES

Arnoult, D., Tatischeff, I., Estaquier, J., Girard, M., Sureau, F., Tissier, J. P., Grodet, A., Dellinger, M., Traincard, F., Kahn, A., Ameisen, J. C., and Petit, P. X. (2001). On the evolutionary conservation of the cell death pathway: Mitochondrial release of an apoptosis-inducing factor during *Dictyostelium discoideum* cell death. *Mol. Biol. Cell.* **12,** 3016–3030.

Berks, M., and Kay, R. R. (1988). Cyclic AMP is an inhibitor of stalk cell differentiation in *Dictyostelium discoïdeum*. *Dev. Biol.* **126,** 108–114.

Blanton, R. L. (1993). Prestalk cells in monolayer cultures exhibit two distinct modes of cellulose synthesis during stalk cell differentiation in *Dictyostelium*. *Development* **119,** 703–710.

Cornillon, S., Foa, C., Davoust, J., Buonavista, N., Gross, J. D., and Golstein, P. (1994). Programmed cell death in *Dictyostelium*. *J.f Cell Sci.* **107,** 2691–2704.

de Chastellier, C., and Ryter, A. (1977). Changes of the cell surface and of the digestive apparatus of *Dictyostelium discoïdeum* during the starvation period triggering aggregation. *J. Cell Biol.* **75,** 218–236.

George, R. P., Hohl, H. R., and Raper, K. B. (1972). Ultrastructural development of stalk-producing cells in *Dictyostelium discoïdeum*, a cellular slime mould. *J. Gen. Microbiol.* **70,** 477–489.

Katoch, B., and Begum, R. (2003). Biochemical basis of the high resistance to oxidative stress in *Dictyostelium discoideum*. *J. Biosci.* **28,** 581–588.

Kawli, T., Venkatesh, B. R., Kennady, P. K., Pande, G., and Nanjundiah, V. (2002). Correlates of developmental cell death in *Dictyostelium discoideum*. *Differentiation* **70,** 272–281.

Kay, R. R. (1987a). Cell differentiation in monolayers and the investigation of slime mold morphogens. *Methods Cell Biol.* **28,** 433–448.

Kay, R. R. (1987b). Cell differentiation in monolayers and the investigation of slime mold morphogens. *In* "Methods in Cell Biology" (J. A. Spudich, Ed.), Vol. 28, pp. 433–448. Ac. Press, Orlando, FL.

Kopachik, W., Oohata, A., Dhokia, B., Brookman, J. J., and Kay, R. R. (1983). *Dictyostelium* mutants lacking DIF, a putative morphogen. *Cell* **33**, 397–403.

Kosta, A., Laporte, C., Lam, D., Tresse, E., Luciani, M. F., and Golstein, P. (2006). How to assess and study cell death in *Dictyostelium discoideum*. *Methods Mol. Biol.* **346**, 535–550.

Kosta, A., Roisin-Bouffay, C., Luciani, M. F., Otto, G. P., Kessin, R. H., and Golstein, P. (2004). Autophagy gene disruption reveals a non-vacuolar cell death pathway in *Dictyostelium. J. Biol. Chem.* **279**, 48404–48409.

Lam, D., Leuraud, J. P., Luciani, M. F., and Golstein, P. (2007). Autophagic or necrotic cell death in the absence of caspase and bcl-2 family members. *Biochem. Biophys. Res. Commun.* **363**, 536–541.

Laporte, C., Kosta, A., Klein, G., Aubry, L., Lam, D., Tresse, E., Luciani, M. F., and Golstein, P. (2007). A necrotic cell death model in a protist. *Cell Death Differ.* **14**, 266–274.

Levraud, J.-P., Adam, M., Cornillon, S., and Golstein, P. (2001). Methods to study cell death in *Dictyostelium discoideum*. *In* "Cell Death. Methods in Cell Biology" (L. M. Schwartz and J. Ashwell, Eds.), Vol. 66, pp. 469–497. Academic Press, San Diego.

Levraud, J.-P., Adam, M., Luciani, M.-F., Aubry, L., Klein, G., and Golstein, P. (2004). Cell death in *Dictyostelium*: Assessing a genetic approach. *In* "When Cells Die II" (R. A. Lockshin and Z. Zakeri, Eds.), pp. 59–77. John Wiley and Sons, Inc., New York.

Levraud, J.-P., Adam, M., Luciani, M.-F., De Chastellier, C., Blanton, R. L., and Golstein, P. (2003a). *Dictyostelium* cell death: Early emergence and demise of highly polarized paddle cells. *J. Cell Biol.* **160**, 1105–1114.

Levraud, J. P., Adam, M., Luciani, M. F., de Chastellier, C., Blanton, R. L., and Golstein, P. (2003b). *Dictyostelium* cell death: Early emergence and demise of highly polarized paddle cells. *J. Cell Biol.* **160**, 1105–1114.

Li, G., Alexander, H., Schneider, N., and Alexander, S. (2000). Molecular basis for resistance to the anticancer drug cisplatin in *Dictyostelium*. *Microbiol. UK* **146**, 2219–2227.

Maeda, Y., and Takeuchi, I. (1969). Cell differentiation and fine structures in the development of the cellular slime molds. *Dev. Growth Differen.* **11**, 232–245.

Morris, H. R., Taylor, G. W., Masento, M. S., Jermyn, K. A., and Kay, R. R. (1987). Chemical structure of the morphogen differentiation inducing factor from *Dictyostelium discoideum*. *Nature* **328**, 811–814.

Olie, R. A., Durrieu, F., Cornillon, S., Loughran, G., Gross, J., Earnshaw, W. C., and Golstein, P. (1998). Apparent caspase independence of programmed cell death in *Dictyostelium*. *Curr. Biol.* **8**, 955–958.

Quiviger, B., Benichou, J.-C., and Ryter, A. (1980). Comparative cytochemical localization of alkaline and acid phosphatases during starvation and differentiation of *Dictyostelium discoïdeum*. *Biol. Cellulaire* **37**, 241–250.

Raper, K. B., and Fennell, D. I. (1952). Stalk formation in *Dictyostelium*. *Bull. Torrey Botanical Club* **79**, 25–51.

Roisin-Bouffay, C., Luciani, M. F., Klein, G., Levraud, J. P., Adam, M., and Golstein, P. (2004). Developmental cell death in *Dictyostelium* does not require paracaspase. *J. Biol. Chem.* **279**, 11489–11494.

Schaap, P., Nebl, T., and Fisher, P. R. (1996). A slow sustained increase in cytosolic Ca^{2+} levels mediates stalk gene induction by differentiation inducing factor in *Dictyostelium*. *EMBO J.* **15**, 5177–5183.

Schaap, P., van der Molen, L., and Konijn, T. M. (1981). The vacuolar apparatus of the simple cellular slime mold *Dictyostelium minutum*. *Biol. Cell* **41**, 133–142.

Sobolewski, A., Neave, N., and Weeks, G. (1983). The induction of stalk cell differentiation in submerged monolayers of *Dictyostelium discoïdeum*. Characterization of the temporal sequence for the molecular requirements. *Differentiation* **25,** 93–100.

Sussman, M. (1987). Cultivation and synchronous morphogenesis of *Dictyostelium* under controlled experimental conditions. *In* "Methods in Cell Biology" (J. A. Spudich, Ed.), pp. 9–29. Harcourt Brace Jovanovich, New York.

Tatischeff, I., Petit, P. X., Grodet, A., Tissier, J. P., Duband-Goulet, I., and Ameisen, J. C. (2001). Inhibition of multicellular development switches cell death of *Dictyostelium discoideum* towards mammalian-like unicellular apoptosis. *Eur. J. Cell Biol.* **80,** 428–441.

Town, C., and Stanford, E. (1979). An oligosaccharide-containing factor that induces cell differentiation in *Dictyostelium discoïdeum*. *Proc. Natl. Acad. Sci. USA* **76,** 308–312.

Town, C. D., Gross, J. D., and Kay, R. R. (1976). Cell differentiation without morphogenesis in *Dictyostelium discoïdeum*. *Nature* **262,** 717–719.

Whittingham, W. F., and Raper, K. B. (1960). Non-viability of stalk cells in *Dictyostelium*. *Proc.Natl. Acad. Sci. USA* **46,** 642–649.

CHAPTER TWO

METHODS AND PROTOCOLS FOR STUDYING CELL DEATH IN *DROSOPHILA*

Donna Denton, Kathryn Mills, *and* Sharad Kumar

Contents

Abstract

Drosophila melanogaster is a highly amenable model system for examining programmed cell death during animal development, offering sophisticated genetic techniques and *in vivo* cell biological analyses. The reproducible pattern of apoptosis, as well as the apoptotic response to genotoxic stress, has been well characterized during *Drosophila* development. The main cellular components required for cell death are highly conserved throughout evolution. Central to the regulation of apoptosis is the caspase family of cysteine proteases, and studies in *Drosophila* have revealed insights into their regulation and function. This chapter describes protocols for detecting apoptotic cells

Hanson Institute, Institute of Medical and Veterinary Science, Adelaide, Australia

Methods in Enzymology, Volume 446
ISSN 0076-6879, DOI: 10.1016/S0076-6879(08)01602-9

during *Drosophila* development, as well as the use of *Drosophila* cell lines. Commonly used methods for detecting apoptosis are described, including TUNEL, acridine orange, and immunostaining with specific components of the apoptotic pathway such as active caspases. A crucial step in the induction of apoptosis is caspase activation and cleavage, which can be measured by use of fluorogenic peptide substrates or detection of cleaved protein products by immunoblotting, respectively. In addition, one of the advantages of the use of *Drosophila* as model is the ability to examine genetic interactions with various components of the cell death pathway.

1. INTRODUCTION

Programmed cell death (PCD) or apoptosis is an essential process for normal animal development and can be distinguished by morphologically distinct characteristics, including cellular shrinkage, membrane blebbing, and nuclear DNA fragmentation (reviewed in Baehrecke, 2002). This form of cell death is distinct from necrosis which, as a result of cellular injury, culminates in cell swelling and lysis. During development, the balance between cell proliferation, differentiation, and death is critical. Apoptosis is vital to remove unnecessary or excess cells during tissue pattern formation and also in maintaining adult homeostasis. In addition, apoptosis eliminates damaged or abnormal cells such as those subjected to DNA damage or those infected with pathogens. Misregulation of apoptosis can have severe effects on the organism and can lead to various developmental abnormalities or diseases including cancer.

1.1. *Drosophila* as a model system to study apoptosis

As the main components of the cell death machinery are highly evolutionarily conserved, *Drosophila* is an exceptional model system for examining apoptosis during animal development. In addition to its well-characterized developmental program and complete genome sequence, *Drosophila* offers sophisticated genetic techniques and *in vivo* cell biological analyses that are not possible in other systems (Adams *et al.*, 2000; Richardson and Kumar, 2000). The life cycle of *Drosophila* consists of a series of developmental stages: embryo, larvae (three larval instar stages), pupae, and adult flies. The steroid hormone ecdysone plays a significant role in controlling and coordinating progression of the animal through the major developmental stages. Increases in ecdysone titer initiate the onset of the larval molts, late larval wandering behavior, pupariation, pupation, and adult development (Fig. 2.1). Modulation of ecdysone can lead to severe developmental consequences arising from perturbations to apoptosis (reviewed in Baehrecke, 2000).

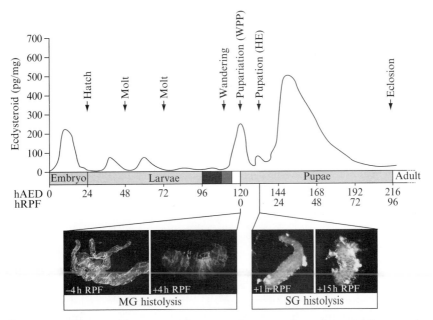

Figure 2.1 The steroid hormone ecdysone regulates morphogenic events. Increases in ecdysone titers initiate the onset of the larval molts, late larval wandering behavior, pupariation, pupation, and adult development. The late larval and prepupal ecdysone pulses correspond to histolysis of obsolete larval tissues. Midgut (MG) death can be detected at 120 h AED (0 h RPF), and 15 h later salivary gland (SG) death can be detected. Modified from Riddiford, 1993.

The reproducible pattern of apoptosis during *Drosophila* development has been well characterized (Abrams *et al.*, 1993). As early as 6 h after egg deposition (AED), apoptosis can be detected in a small number of cells. After this, apoptotic cells become more widespread, and at later stages of embryogenesis an increase in apoptotic cells can be identified in tissues such as the central nervous system. During metamorphosis, a dramatic increase in PCD is observed when obsolete larval tissues are removed in response to the developmental pulses of ecdysone (reviewed in Baehrecke, 2000; Kumar and Cakouros, 2004) (Fig. 2.1). The larval midgut is removed by apoptosis in response to the large ecdysone pulse during the late larval stage, and the adult gut begins to develop in its place. Approximately 12 h later, another ecdysone pulse signals the larval salivary glands to undergo histolysis. Correct patterning of the adult eye during the pupal stage also requires developmentally regulated PCD. The interommatidial cells of the pupal eye disc are rearranged to form a single layer around the photoreceptor clusters, with the excess interommatidial cells eliminated by apoptosis (reviewed in Brachmann and Cagan, 2003). During normal oocyte development, nurse cells undergo PCD after they deposit their cytoplasmic contents into the developing oocyte. Another role for apoptosis during oogenesis is to remove defective egg chambers unable to develop into fertile eggs (reviewed in Buszczak and Cooley, 2000). Unlike PCD during embryogenesis or metamorphosis, upstream components of the apoptotic pathway are not required for nurse cell death. Thus in *Drosophila*, the regulation of PCD can occur differently in specific tissues during development.

In addition to developmental PCD, apoptosis can also be induced in response to DNA damage induced by genotoxic stress such as ionizing radiation. This can be observed in embryos and in the proliferating larval imaginal disc tissue that undergoes cell death after irradiation. In wing imaginal discs an increase in cell death is observed 4 to 6 h after irradiation to remove damaged cells (Wichmann *et al.*, 2006).

1.2. Cell death pathway

Central to apoptosis is the highly conserved caspase family of cysteine proteases. These enzymes are produced as inactive zymogens and, upon activation after death-inducing signals, cleave multiple cellular proteins resulting in apoptosis (reviewed in Hay and Guo, 2006; Kumar, 2007). Caspases can be grouped into two classes: "initiator" caspases contain long prodomains and act as signal transducers that are required to cleave and activate the downstream "effector" caspases that subsequently orchestrate cleavage of cellular substrates and dismantling of the cell. The *Drosophila* genome encodes seven caspases, three initiator (Dronc, Dredd, and Strica), and four effector caspases (Drice, Dcp-1, Decay, and Damm) (Chen *et al.*, 1998; Dorstyn *et al.*, 1999a,b; Doumanis *et al.*, 2001; Fraser and Evan, 1997; Harvey *et al.*, 2001; Song *et al.*, 1997). Dronc is the only caspase recruitment domain (CARD) containing caspase in *Drosophila* and is essential for development, unlike the other initiator caspases. *dronc* null mutants are pupal lethal and display various cell death defects in embryos, larvae, and prepupae, as well as defects in response to stress-induced apoptosis (Chew *et al.*, 2004; Daish *et al.*, 2004). Although *strica* mutants are viable, *strica* seems to play a redundant role with *dronc* in PCD during oogenesis (Baum *et al.*, 2007). The other initiator caspase, *Dredd,* is not essential for developmental PCD but is required for innate immune response (Leulier *et al.*, 2000). The effector caspase *drice* is required for developmental and stress-induced apoptosis and mutants are pupal lethal with approximately 20% escapers (Kondo *et al.*, 2006; Muro *et al.*, 2006; Xu *et al.*, 2006). Dcp-1 plays a relatively minor role in PCD, required during mid-oogenesis, and overlaps with *drice* function (Kondo *et al.*, 2006; Muro *et al.*, 2006; Xu *et al.*, 2006). *decay* mutants are viable and fertile with normal developmental PCD (Kondo *et al.*, 2006), whereas a role for *Damm* in PCD has not been established, because no specific mutants are available. Thus, *Dronc* is believed to be the predominant initiator caspase

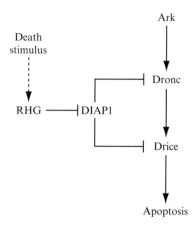

Figure 2.2 The main caspase activation pathway in *Drosophila*. On death stimulus the RHG family of proapoptotic proteins bind to DIAP1, which is then degraded by an ubiquitination-dependent mechanism. This allows released Dronc to become activated after binding with Ark and activate downstream effector caspase Drice.

and *Drice* the main effector caspase required for both developmental PCD and stress-induced apoptosis (Fig. 2.2).

Activation of Dronc requires association with the adaptor protein Ark (Apaf-1 related killer), and similar to *dronc, ark* is an essential gene with mutants showing severe defects in developmental PCD and stress-induced apoptosis (Mills *et al.*, 2006; Srivastava *et al.*, 2006). Until the appropriate death signal is received, Dronc is maintained in an inactive form by association with DIAP1, the *Drosophila* member of the inhibitor of apoptosis protein (IAP) family (reviewed in Kumar, 2007; Mills *et al.*, 2005) (Fig. 2.2). A loss-of-function mutation in the gene encoding DIAP1, *thread*, is embryonic lethal as a result of an increase in cell death during embryogenesis (Hay *et al.*, 1995). In *Drosophila*, initiation of cell death requires the products of the *reaper (rpr), grim* and *head involution defect (hid/Wrinkled)* genes, as deletion of all three genes shows an absence of cell death during embryogenesis (White *et al.*, 1994). In addition to Rpr, Grim, and Hid, other cell death regulators include Sickle and Jafrac2 and are collectively referred to as the RHG proteins (Fig. 2.2). This group of proteins acts to initiate cell death by binding to and sequestering DIAP1 from interacting with caspases (Hay and Gao, 2006; Kumar, 2007).

1.3. Protocols described

This chapter describes methods used to examine apoptosis in *Drosophila*. The first section describes protocols to examine cell death during development and includes methods for detecting apoptosis *in situ* by DNA fragmentation with TUNEL and by use of the vital dye acridine orange (AO) in live tissue.

Another *in situ* approach includes immunostaining with specific markers of apoptosis such as antibodies to active caspase-3. The second section details biochemical approaches. A crucial step in the induction of apoptosis is caspase activation and cleavage, which can be measured by use of synthetic peptides or detection of cleaved substrates by immunoblotting. In addition to detecting apoptosis in whole animals, a variety of *Drosophila* cell lines can also be used. The third section illustrates one of the great advantages of the use of *Drosophila* as model to examine apoptosis by the ability to test genetic interactions by expression of components of the cell death pathway in the animal.

2. CELL DEATH ANALYSIS DURING DEVELOPMENT

The cellular basis of *Drosophila* development, including the reproducible patterns of PCD, is well described (Abrams *et al.*, 1993). In addition, differential regulation of PCD has been characterized in the embryo, larvae, and adult (ovaries). The protocols described in this section include TUNEL, acridine orange staining, and immunostaining requiring either fixed or live samples. Different tissues have different requirements for fixation, as described in the following.

2.1. Fixation of samples

2.1.1. Collection and fixation of embryos

Flies can be set to lay in "lay tubes" on top of grape juice agar plates (see below) with a small amount of yeast paste. We use plastic tubing sealed at one end with fine wire mesh as lay tubes. The embryos are then harvested from the agar plate by washing with water and brushing with a small paint brush into a collection basket (made by placing fine wire mesh over a cutoff 1.5-ml Microfuge tube or by use of cell strainers available from BD Falcon). The chorion is then removed by use of a pipette to continually rinse embryos with a 50% bleach solution (domestic brand) for 3 min. The dechorionated embryos are washed thoroughly with water and transferred to a tube containing a two-phase mix of equal parts 4% formaldehyde in 1 × HEN buffer (10 × HEN buffer: 1 M HEPES, 0.5 M EGTA, 0.1% NP-40, pH to 6.9, and filter sterilize) and heptane. The sample is then shaken for 20 min on an orbital platform such that the interface between the liquid phases is disrupted and the embryos are in an emulsion. After fixation, the embryos will settle on the interface and the bottom (aqueous) phase can be removed and replaced with an equal volume of methanol. Shake vigorously for 30 s to remove the vitelline membrane and let stand for 1 min. Devitellinized embryos will sink from the interface and can be collected from the bottom by use of a cut-off pipette tip (discard embryos with vitelline membrane as they will not stain). Embryos are then rinsed in methanol at which point they can be stored at $-20\,°C$ in ethanol. Before staining, embryos are rehydrated by serial incubation in 75, 50, and 25% (v/v) methanol in PBT for 5 min each and then washed with PBT.

Note: As the strength of bleach can vary, dechorionation can be monitored by use of a dissection microscope until 80% of the dorsal appendages have dissolved. It is important not to overfix embryos, because this results in decreased efficiency of devitellinization. If cloudiness or precipitation is present in the $10 \times$ HEN solution, the NP-40 can be omitted. The addition of heptane in the fixative is essential to create holes in the vitelline membrane. Overnight storage in methanol often improves image quality, however, extended periods of storage (more than a month) can result in loss of image quality.

2.1.1.1. Grape juice agar plates

0.3% agar, 25% grape juice, 0.3% sucrose, 0.03% tegosept (10% *para*-hydroxybenzoate in ethanol). Combine agar, grape juice, sucrose, and water together and boil by heating in a microwave. Mix and to cool to $60\,°C$ before adding tegosept and pouring into petri dishes (appropriately sized to fit the lay tube). Once the plates have set, they can be stored at $4\,°C$.

2.1.2. Staging and fixation of larval tissues

To stage larvae for analysis of midgut (MG) and salivary gland (SG) histolysis, larvae can be propagated on food containing 0.05% bromophenol blue. As larvae commence the wandering behavior, they cease feeding and begin emptying their gut contents. Therefore, their age relative to pupariation, when the cuticle hardens and forms a pupal case, can be estimated on the basis of gut clearance (Andres and Thummel, 1994; Maroni and Stamey, 1983). Wandering larvae are collected from the side of the vial with a wet paint brush and transferred to a petri dish lined with wet Whatmann paper. Animals can be monitored regularly (every 15, 30, or 60 min) for gut clearance or pupariation. Larvae with dark blue guts represent -24 to -12 h relative to pupariation (RPF), light blue guts -12 to -5 h RPF, and clear guts -6 to -1 h RPF. Immediately before pupariation, the larvae shorten, evert spiracles, and stop moving. During the first 1 h after pupariation, the cuticle appears white, after which the cuticle tans. This coloring can be used to stage animals at 0 to $+1$ h RPF and is also referred to as white prepupae (WPP) (Andres and Thummel, 1994; Maroni and Stamey, 1983) (see Fig. 2.1). At the desired stage, animals are transferred to another petri dish lined with wet Whatmann paper and aged until the desired time for experimental analysis. For analysis of apoptosis in the MG, we use clear gut larvae (-6 to -1 h RPF) and WPP. For SG analysis we age animals to 14 h RPF that is equivalent to 2 h after head eversion (AHE). Head eversion marks the transition between prepupal and pupal stages. During this stage, abdominal muscles contract, pushing an air bubble from the posterior to the anterior end, resulting in a space for the head to evert. This process occurs rapidly in approximately 10 min. Dissect tissue in 1× PBS and fix in 4% formaldehyde in PBT_W (0.1% Tween-20 in PBS) for 20 min at room temperature. Wash twice with PBT_X (0.1% TritonX-100 in PBS).

2.1.3. Fixation of ovaries

To ensure ovaries are well developed and all stages represented, females are aged at 25 °C for 3 to 7 days in uncrowded conditions on food supplemented with fresh yeast in the presence of an equal number of males to stimulate egg production. Dissect females in PBS in a depression plate by use of fine forceps to hold the submerged fly between the thorax and abdomen and a second pair of forceps to hold the external genitalia and pull to release ovaries into PBS. Remove any debris from the ovary and tease apart ovarioles by use of a fine needle. Transfer ovarioles into a 1.5-ml tube containing a two-phase fix solution of equal parts heptane (400 μl) and 4% formaldehyde in PBS (400 μl) and shake for 25 min at room temperature. Remove heptane/fix and wash twice with PBT, ensuring removal of all heptane.

Note: Sylgard® 184 Silicone Elastomer Kit (Dow Corning) can be used according to manufacturers instructions to make dissection dishes with a soft silicone base, which are useful to prevent dissecting forceps and needles from damage.

2.2. Detection of apoptotic cells with TUNEL

A frequently used method for detecting apoptosis by DNA fragmentation is terminal deoxynucleotidyl transferase (TdT)–mediated dUTP nick-end labeling (TUNEL) staining and has been used routinely on whole-mount tissue samples (White *et al.*, 1994). This technique specifically detects apoptotic cells by preferentially labeling DNA strand breaks. Several commercial kits are available and should be used according to manufacturers protocols. Both fluorescent and color detection kits are available (extra labeling steps are required for color detection) and offer reliable results.

After fixation (appropriate for your tissue of interest), wash twice with PBT_X (0.1% TritonX-100) then twice with PBT_{X5} (0.5% TritonX-100) and permeabilize in 100 mM sodium citrate in PBT_X at 65 °C for 30 min. Wash three times with PBT_{X5} before addition of TUNEL mix according to the manufactures instructions and incubate at 37 °C for 3 h on a rotating platform. Wash three times with PBT_X and store in 80% glycerol in PBS. To mount samples place a strip of double-sided tape across each end of the slide, placing samples between the tape, dissect the midgut or tissue of interest away from remaining tissue. Gently place a coverslip resting on the tape over samples. Examine samples by epifluorescence or confocal microscopy.

Whilst TUNEL staining produces highly reproducible results, it is advisable to set up both positive and negative controls. For TUNEL staining, a DNase-treated sample will serve as a positive control, and negative control can be performed by the addition of water instead of the active enzyme TdT.

Note: Ensure the sodium citrate solution is made up fresh. If difficulties in labeling are encountered, an alternative permeabilization to increase penetrance can be achieved by proteinase K treatment (10 μg/ml in PBS for 3 to 5 min at room temperature) (Arama and Stellar, 2006; McCall and Peterson, 2004). Antibody staining can be performed sequentially in combination with TUNEL staining. Samples can also be co-stained to detect nuclei, by use of Hoechst 33258 at 4 μg/ml in PBS for 1 min before storage in 80% glycerol in PBS. When assaying for cell death after irradiation, it is preferable to use acridine orange instead of TUNEL, because TUNEL can (in principle) detect the DNA damage induced by the treatment.

2.3. Acridine orange staining

The vital dye acridine orange (AO) can be used to observe apoptosis in live tissue (Arama and Steller, 2006). The advantage of AO compared with TUNEL staining is the speed of the staining. However, because it is carried out on live tissue, multiple labeling cannot be performed, and the tissue must be examined immediately. A stock solution of AO can be made up in ethanol (1 mM stock: 1.85 mg AO in 5 ml ethanol) and stored in the dark at room temperature for several months. The AO stock solution is then diluted to the final concentration in PBS immediately before use.

2.3.1. Embryo staining
After egg collection and dechorionation, transfer embryos to a tube containing equal volume of heptane and 5 μg/ml AO in PBS. Shake vigorously for 5 min at room temperature to generate a fine emulsion. Allow the liquid phases to separate and transfer the embryos from the interface onto a glass microscope slide in a drop of PBS.

2.3.2. Staining larval tissue
From appropriately aged larvae dissect tissue of interest away from other tissue. Incubate the dissected tissue in a drop (20 to 30 μl) of 5 μg/ml AO solution for 5 to 15 min. Wash briefly by transferring tissue to a fresh drop of PBS. Mount in a drop of PBS on a glass microscope slide.

2.3.3. Ovary staining
Transfer dissected ovarioles into a 1.5-ml tube containing equal volumes of heptane and 10 μg/ml AO in PBS. Mix gently and rotate for 5 min. Allow ovarioles to sink to bottom and replace AO solution with PBS to rinse briefly. Transfer ovarioles to a slide and gently spread out.

To prevent damage and squashing of samples, double-sided tape can be positioned across each end of the microscope slide to support the coverslip. Examine samples immediately by epifluorescence/confocal microscopy. The best sensitivity can be observed by use of the green channel (522 nm); however, the red channel (568 nm) can be used to provide less background (Arama and Steller, 2006).

In addition to the test samples, positive and negative control samples should be examined. Comparisons can be made between tissues from wild-type animals at different stages of development, depending on the level of apoptosis at that specific stage.

Note: Ensure no detergent is present in dissection and AO staining solutions, because this will abolish staining. Additional protocols for AO and TUNEL staining in embryos and testis can be found in Arama and Steller (2006). Halocarbon oil can be used as an alternative mounting solution if PBS evaporates too quickly.

2.4. Immunostaining: Caspase-3 and cytochrome *c*

Antibodies that specifically recognize active caspases can be used to detect apoptotic cells *in vivo*. Various tissue and developmental stages can be analyzed and require fixation (as described previously), before incubation with a primary antibody. Of particular use is the antibody against the active form of caspase-3, which has been shown to cross-react with *Drosophila* effector caspases (anti-active caspase-3, Cell Signaling Technology). In addition, the anti-cytochrome *c* antibody that recognizes an altered configuration of the protein can be used to specifically label dying cells (Varkey *et al.*, 1999). However, it should be noted that, unlike in mammalian cells, cytochrome *c* is not generally released during apoptotic signaling in *Drosophila* cells and is not required for caspase activation (Dorstyn *et al.*, 2002; 2004).

After fixation, block sample in 5 to 10% normal goat sera for 1 h at room temperature in PBT then incubate in primary antibody diluted in blocking solution overnight at 4 °C with gentle shaking. Remove the primary antibody and wash three times with PBT over an hour. Longer washes and more rinses will result in less background. Add fluorescently labeled secondary antibody, also diluted in blocking solution, and incubate at room temperature for 1 h. Once the fluorescent secondary has been added, samples should be kept in the dark. Remove the secondary antibody and wash in PBT as described previously. Store samples in 80% glycerol in PBS before mounting for microscopy.

Note: Alternative blocking solutions such as 1% BSA in PBT can be used, depending on the primary antibody. Samples can also be co-stained to detect nuclei (as described). Alexa Fluor–conjugated secondary antibodies are available in a wide range of fluorescence emission wavelengths (we commonly use Alexa Fluor 488 and 568) and offer increased intensity and photostability compared with others products. If photobleaching occurs, commercially available anti-fade products can be used as the mounting medium.

2.5. γ-Irradiation–induced apoptosis

Induction of apoptosis can be observed after genotoxic stress in proliferating larval imaginal tissue. Wandering third-instar larvae are collected and exposed to varying doses of γ-irradiation, usually ranging from 20 up to 40 Gy, by use of a ^{137}Cs source (Kondo *et al.*, 2006). After irradiation, larvae are allowed to recover at 25 °C for an appropriate length of time before dissection and staining of tissue. A dose response should be determined and a nonirradiated control included.

Note: Alternative sources such as ultraviolet irradiation and X-irradiation may also be used, however, suitable doses and recovery times should be optimized to suit the equipment used.

3. Biochemical Analysis

The protocols described in this section include the use of cell lines to analyze the apoptotic response to death stimuli, measurement of caspase activity by cleavage of synthetic peptide substrates, and detection of caspase processing by immunoblotting with specific antibodies.

3.1. Cell culture

Drosophila cell culture can be useful to complement animal studies and, consistent with *Drosophila* development, central death activators induce cell death in cultured cells. Various *Drosophila* cell lines are available from the *Drosophila* Genomics Resource Center (DGRC), and general protocols can be found on their web site (https://dgrc.cgb.indiana.edu/). Commonly maintained lines include SL2, mbn2, BG2 and Kc cells.

3.1.1. Maintaining cell lines

Drosophila cell lines are grown in Schneider's cell medium supplemented with penicillin and streptomycin, 10% fetal bovine serum, and 1% glutamine. Some cell lines require additional growth factors, for example BG2 cells are supplemented with insulin.

The advantage of SL2 and BG2 cells is that they are adherent, fast growing, relatively robust, and good for localization studies. Whereas other cells such as mbn2 are responsive to ecdysone and are, therefore, useful for analyzing ecdysone responsive cell death pathways. Several different vectors are available for inducible or constitutive expression and can be obtained from the DGRC or other Biotech companies. The inducible vector, pRmHa-3, contains a metallothionein promoter that is induced by the addition of $CuSO_4$ to the cell culture medium. Vectors for constitutive expression are also available. We routinely use pIE (Novagen), however, others that also contain epitope tags are available.

3.1.2. Transient transfections

Transient transfections can be performed with Cellfectin® (Invitrogen) transfection reagent according to manufacturers guidelines with a resultant transfection efficiency of approximately 20%. Plate exponentially growing cells at 6×10^5 cells/ml in six-well plates (or 35-mm plates) in Schneider's medium and leave for 1 h to overnight for cells to adhere. Prepare the transfection solutions, by diluting 1 to 2 μg DNA into 100 μl of media without serum and antibiotics and dilute 1.5 to 9 μl of Cellfectin® into 100 μl of media without serum and antibiotics. Combine the two solutions, mix gently, and incubate at room temperature for 20 to 30 min. Remove the media from the plated cells and wash once with media without serum and antibiotics. To the transfection mix add 0.8 ml serum and antibiotic-free media, mix gently, and add to the cells after the removal of the wash media. Incubate cells at 27 °C for 5 h. Add 2 ml of complete media containing serum and antibiotics. Incubate cells at 27 °C for the desired experimental time, usually 24 to 48 h. If an inducible expression system is used, after 24 to 48 h incubation, add 0.7 mM $CuSO_4$ to induce expression and incubate for the desired experimental time. Be sure to include a sample that has not been treated with $CuSO_4$ as noninduced control. Cells are harvested either by scraping by use of a commercial sterile cell scraper or by pipetting.

3.1.3. Viability assays

Cell viability can be assessed by use of the Trypan blue exclusion assay. Trypan blue is taken up by dying cells, and the percentage viable cells can be estimated by counting the number of Trypan blue–negative (unstained) cells compared with Trypan blue–positive (blue) cells. Alternately, cells can be stained with DAPI (2 μg/ml in methanol) and scored on the basis of nuclear morphology. The nuclei of apoptotic cells appear condensed, and at later stages apoptotic nuclear membrane blebbing is visible. To score transfected cells for cell death, cells can be co-transfected with a β-gal or GFP reporter vector and morphology of β-gal or GFP-positive transfected cells scored by microscopy.

Note: Drosophila cells can be susceptible to low cell density and variations in batches of media. The same transfection protocol can be used to generate stable cell lines by use of vectors with appropriate selection.

3.2. Measurement of caspase activity by use of synthetic peptide substrates

A frequently used method to detect caspase activity after the induction of apoptosis is measuring the cleavage of synthetic substrates that have been incubated with cell lysates. The technique was originally described by Pennington and Thornberry (1994) and has been further developed with a variety of synthetic substrates becoming commercially available. The synthetic peptides are conjugated to a fluorochrome, such as 7-amino-4-methylcoumarin (AMC) (other fluorometric or colorimetric substrates are available). The basis of the assay is that the peptide substrate is cleaved by caspases that recognize the substrate cleavage site, releasing the fluorochome. The intensity of the fluorescent signal measured by a spectrophotometer is proportional to the amount of cleaved substrate, which depends on caspase activity from the apoptotic cells present in the cell population.

3.2.1. Preparation of protein lysates

Protein lysates can be prepared from appropriately staged whole animals, dissected tissue samples, or tissue culture cell lysates. The sample is homogenized in lysis buffer (below) by use of a pestle in a 1.5-ml Microfuge tube and subjected to three rounds of freezing in liquid nitrogen and thawing. Debris is removed by spinning the extracts at 13,000 rpm for 20 min at 4 °C then transferring the supernatant to a fresh tube. After determining protein concentration, the lysates can be used in caspase activity assays as described below or can be stored at -70 °C in aliquots for several months.

Note: Samples can be snap frozen and stored at -70 °C before preparation of lysates. The volume of lysis buffer used can vary, depending on the protein yield of the sample. As a guide, we use 10 μl per whole animal or 5 μl per dissected tissue.

3.2.1.1. Lysis buffer 50 mM HEPES, pH 7.5; 100 mM NaCl; 1 mM EDTA; 0.1% CHAPS; 10% sucrose; 5 mM DTT; 0.5% TritonX-100; 4% glycerol; 1$\times$ protease inhibitor cocktail (Complete Roche).

3.2.2. Measurement of caspase activity

Cleavage of the caspase peptide substrates VDVAD–AMC and DEVD–AMC can be used to determine enzyme activity, because they show the highest level of activity for *Drosophila* caspases. Caspase activity assays should be set up in triplicate with 20 to 50 μg of protein lysate incubated with 100 μM of VDVAD–AMC or DEVD–AMC in a final volume of 100 μl made up with lysis buffer. The assay should also include positive and negative control samples, as well as a fluorescence calibration standard. Fluorescence is quantified by use of a spectrophotometer (excitation, 385 nm; emission, 460 nm) over a time course of up to 5 h at 37 °C. The fluorescence standard curve is obtained by plotting the relative fluorescence of the calibration standard over the concentration, and the slope of the line is used as the conversion factor (1/slope) when performing the caspase assay. The relative fluorescence units for each sample can be calculated by plotting the linear region of the change in fluorescence over time, whereby the slope of the line can be determined. This is then used with the conversion factor to calculate the activity expressed as pmol/min (Fig. 2.3).

Note: This method of detecting caspase activity is widely used, because it is quantitative, relatively quick, and sensitive. Even though different peptide substrates are available to distinguish caspase activities, there is redundancy in the sequence specificity of caspases for substrates. That is, several caspases will cleave the same substrates, although with different efficiency, so the peptide sequences should be viewed as preferred substrate motifs and rather than absolute (Berger *et al.*, 2006).

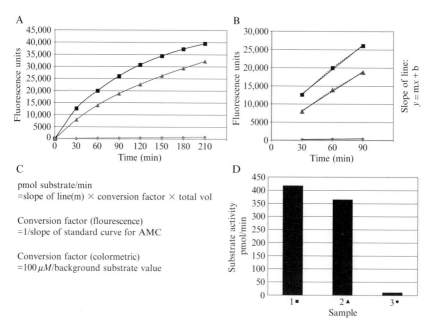

Figure 2.3 Determining the specific activity for cleavage of the synthetic fluorogenic substrate. (A) Example of raw data of the substrate cleavage activity expressed as fluorescence units and the change over time. (B) Plot of the linear region of the change in fluorescence over time and use an appropriate linear regression program to obtain the slope of the line (m). (C) Calculations used to determine the activity of the samples expressed as pmol substrate cleaved per min. (D) Plot of the substrate cleavage activity is expressed in pmol/min.

3.3. Immunoblotting cleaved substrates

The activation of caspases coincides with the cleavage of a caspase precursor, therefore, the appearance of large and small subunits can be used as a marker for processing and activation of individual caspases. Cell extracts can be analyzed by immunoblotting with antibodies to detect cleaved caspase subunits. The standard protocols for immunoblot analysis can be used (Sambrook and Russell, 2001). The protein lysates (as made previously for caspase assay) can be prepared for electrophoresis by mixing with an equal volume of 2× protein loading buffer. The proteins are separated by sodium dodecyl sulfate-polyacrylamide gel electrophoresis (SDS-PAGE) and transferred to a membrane (nitrocellulose or polyvinylidene difluoride, PVDF). The membrane is blocked with 5% dry milk powder dissolved in phosphate-buffered saline (PBS) and probed with a primary antibody against the caspase of interest. Useful antibodies include anti-Dronc (Dorstyn *et al.*, 2002; Quinn *et al.*, 2000) and anti-Drice (Dorstyn *et al.*, 2002). The primary antibody can be detected by use of horseradish peroxidase (HRP) or alkaline phosphatase (AP)-conjugated secondary antibodies and visualized by enhanced chemiluminescence (ECF) substrate or enhanced chemifluorescence ECF substrates.

4. GENETIC ANALYSIS

4.1. Genetic dissection of cell death pathway

Genetic interaction screens have been commonly used to identify genes that modify a phenotype caused by misexpression of a gene of interest in the *Drosophila* eye. This type of screen is also particularly useful in determining the order of new components within a biological pathway. The following section is designed to give an introduction into the principles of examining genetic interactions in the *Drosophila* adult eye.

The *Drosophila* adult compound eye is an excellent tissue to examine genetic interactions, because it is not essential for viability and is easy to examine (reviewed in Thomas and Wassarman, 1999). The compound eye consists of approximately 800 units called ommatidia in a precise array and is derived from a single layer epithelium, the eye imaginal disc, that is set aside during embryogenesis and proliferates during larval development increasing several fold in size (Wolff and Ready, 1991a). Differentiation of the eye disc initiates during the third–instar larval stage and requires cell death as part of its normal development, and perturbations to this process are visible in the adult eye (Wolff and Ready, 1991b). Several components of the cell death pathway, when expressed in the eye imaginal disc, perturb eye development and result in a disorganized or rough adult eye (see below).

The GAL4/UAS system is a powerful tool in *Drosophila* for driving expression of a gene of interest by specific promoters (Brand and Perrimon, 1993; reviewed in Duffy, 2002). It is a two-component system where *GAL4* expression is driven by tissue-specific enhancers, and the presence of GAL4 results in transcriptional activation from the GAL4 binding sites, *UAS*, driving a gene of interest (Fig. 2.4). The *GMR* (glass multimer reporter) promoter drives expression of *GAL4* in the posterior region of the eye imaginal disc (Hay *et al.*, 1994) (*GMRGAL4* is available from Bloomington *Drosophila* Stock Center). The expression of several apoptotic regulators in the eye has been characterized such as the baculovirus caspase inhibitor p35, in which *GMRp35* results in adults with a rough eye because of the survival of cells (supernumerary retinal cells) that would normally be lost (Hay *et al.*, 1994). Conversely, expression of the apoptotic activator *hid* in the eye,

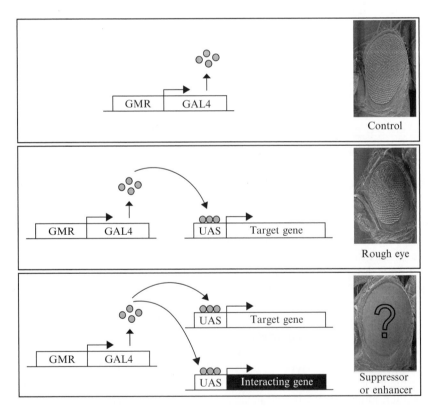

Figure 2.4 The GAL4/UAS system used to examine genetic interactions in the adult eye. The GMR enhancer is used to drive eye-specific expression of GAL4. The target line containing the UAS sites is crossed to flies containing GMRGAL4, enabling control of target gene expression, producing a rough eye phenotype. Testing for genetic interaction is achieved by co-expression from the UAS sites present in both the target and the test interactor gene.

GMRhid, results in rough eyes because of an increase in cell death (Hay *et al.*, 1994). A number of fly strains have been generated with cell death genes under the control of UAS sites that can be used for ectopic expression in the eye and some of these are shown in Table 2.1.

To uncover a genetic interaction between your candidate gene and known components of the cell death pathway several approaches can be undertaken. These include (1) gene activation screen, where the gene is overexpressed; (2) dominant modifier screen by use of single gene disruption alleles; and (3) target gene inactivation screen undertaken by the knockdown of gene expression. These are all based on the ability of the gene expression, either increased or decreased, to modify a sensitized rough eye phenotype.

To achieve overexpression of a gene in the eye, the GAL4/UAS system is used such that the gene of interest is inserted downstream of UAS. This can be achieved by direct cloning into the *P*-element vector, pUAST, that contains UAS sites (available from *Drosophila* Genomics Resource Center https://dgrc.cgb.indiana.edu/) and then generating transgenic flies by *P*-element mediated transformation (by standard procedures; Ashburner *et al.*, 2005). Alternately, a collection of misexpression *Drosophila* lines (*P{EP}*) are available that contain a *P*-element with UAS sites, that when inserted near the 5′ end of gene in the correct orientation can be activated by GAL4 to drive expression of the downstream gene (Rørth, 1996; Rørth *et al.*, 1998). Approximately 2300 *P{EP}* lines are available from the various

Table 2.1 Cell death transgenes available for expression in the eye

Transgene	Effect on cell death in the eye	Source
GMRhid★	Increase	Bloomington
GMRgrim★	Increase	Bloomington
UASreaper	Increase	Bloomington
UASprodronc	Increase	Meier *et al.*, 2000; Quinn *et al.*, 2000
GMRdrice	Increase	Song *et al.*, 2000
GMRDcp-1	Increase	Song *et al.*, 2000
UASp35	Decrease	Bloomington
UASDIAP1	Decrease	Bloomington
UASdroncDN	Decrease	Meier *et al.*, 2000; Quinn *et al.*, 2000
UASRNAi	Increase/decrease	Available from VDRC

★ If *hid* or *grim* overexpression is used, direct *GMRhid* and *GMRgrim* lines will need to be used as *GMRGAL4*, *UAShid* (or *grim*) is lethal. *GMRGAL4* will still be required to co-express an interacting gene with UAS sites.
A full description of the transgene can be found on FlyBase: http://flybase.bio.indiana.edu/Bloomington *Drosophila* Stock Center at Indiana University: http://flystocks.bio.indiana.edu/The Vienna Drosophila RNAi Center (VDRC): www.vdrc.at

stock centers, and a list of *Drosophila* stock centers is available at FlyBase (http://flybase.bio.indiana.edu/) (reviewed in Adams, 2002).

Traditionally, genetic screens have required the generation of mutations in genes resulting in decreased function, and many thousands of mapped gene disruption lines are available (Bellen *et al.*, 2004). Another approach is to knockdown a gene of interest by use of RNA-mediated gene interference (RNAi). RNAi can be triggered in *Drosophila* by the presence of long double-stranded hairpin RNA that can be expressed from a transgene. UAS-RNAi transgenic lines are available from Vienna *Drosophila* RNAi Center (VDRC, www.vdrc.at; Dietzl *et al.*, 2007) and the *Drosophila* Genetic Resource Center (www.dgrc.kit.ac.jp/en/) in Japan or can be generated by construction of inverted repeats by use of a vectors such as pRISE and pWIZ, available from the *Drosophila* Genomics Resource Center (https://dgrc.cgb.indiana.edu/).

The availability of such overexpression and knockdown fly lines makes it feasible to examine genetic interactions with components of the cell death pathway. This can be done by screening for modification, that is, either suppression (more mild) or enhancement (more disrupted) of the eye-specific phenotype from expression of a cell death regulator (e.g., *GMRhid* or *GMRp35*). To achieve this, the gene of interest (e.g., P{*EP*} or RNAi) line can be crossed to the sensitized cell death eye phenotype line (Fig. 2.4).

The use of loss-of-function mutations can also be examined in this manner, where the effect of a reduction in gene dose on eye phenotype can be determined. A prerequisite for this type of analysis is that a reduction in gene dose by half must be sufficient to modify the phenotype, therefore enabling the dissection of the contribution of your gene of interest to the cell death pathway.

A rough eye phenotype associated with either increased or decreased cell death (Table 2.1) can be used in crosses to examine the effect of your transgene of interest. Crosses with *GMRGAL4* are routinely set up at 25 °C, because the activity of GAL4 is temperature dependent (Duffy, 2002). If a severe effect is observed, reducing the temperature to 18 °C may enable more subtle differences to become evident. It is also necessary to keep in mind that *GMRGAL4* alone induces cell death in the eye (Kramer and Staveley, 2003) and so must always be included as a control by crossing to a wild-type line (does not contain UAS transgene).

The phenotype of the eye can be examined by use of a dissecting microscope comparing the gene of interest to an appropriate control. Photographic images can be taken under the dissecting microscope, or samples can be dehydrated in serial dilutions of acetone (25, 50, 75 and 100% acetone) for scanning electron microscopy. After dehydration, flies can be air-dried, mounted onto EM studs, and viewed without coating. Alternately, they can be dried by critical point drying and coated before electron microscopy.

Note: The exact requirements for setting up crosses will vary between experiments, therefore, detailed instructions cannot be given within the scope of this section. Useful resources include Ashburner *et al.* (2005), Greenspan (2004), and Sullivan *et al.* (2000). The UAS constructs can be used with alternative tissue-specific or ubiquitous GAL4 drivers to express at different stages of development to induce or inhibit apoptosis. A *P*-element is a type of modified transposable element commonly used in generating transgenic flies as well as insertional mutagenesis.

REFERENCES

Abrams, J. M., White, K., Fessler, L. I., and Steller, H. (1993). Programmed cell death during *Drosophila* embryogenesis. *Development* **117**, 29–43.

Adams, M. D., and Sekelsky, J. J. (2002). From sequence to phenotype: Reverse genetics in *Drosophila melanogaster*. *Nat. Rev. Genet.* **3**(3), 189–198.

Adams, M. D., *et al.* (2000). The genome sequence of *Drosophila melanogaster*. *Science* **287**, 2185–2195.

Andres, A. J., and Thummel, C. S. (1994). Methods for quantitative analysis of transcription in larvae and prepupae. *Methods Cell Biol.* **44**, 565–573.

Arama, E., and Steller, H. (2006). Detection of apoptosis by terminal deoxynucleotidyl transferase-mediated dUTP nick-end labeling and acridine orange in *Drosophila* embryos and adult male gonads. *Nat Protoc.* **1**, 1725–1731.

Ashburner, M., Golic, K. G., and Hawley, R. S. (2005). *Drosophila*: A Laboratory Handbook 2nd Ed. Cold Spring Harbor Laboratory Press, Cold Spring Harbour, NY.

Baehrecke, E. H. (2000). Steroid regulation of programmed cell death during *Drosophila* development. *Cell Death Differ.* **7**, 1057–1062.

Baehrecke, E. H. (2002). How death shapes life during development. *Nat. Rev. Mol. Cell Biol.* **3**, 779–787.

Baum, J. S., Arama, E., Steller, H., and McCall, K. (2007). The *Drosophila* caspases Strica and Dronc function redundantly in programmed cell death during oogenesis. *Cell Death Differ.* **14**, 1508–1517.

Bellen, H. J., *et al.* (2004). The BDGP gene disruption project: Single transposon insertions associated with 40% of *Drosophila* genes. *Genetics* **167**, 761–781.

Berger, A. B., Sexton, K. B., and Bogyo, M. (2006). Commonly used caspase inhibitors designed based on substrate specificity profiles lack selectivity. *Cell Res.* **16**, 961–963.

Brachmann, C. B., and Cagan, R. L. (2003). Patterning the fly eye: The role of apoptosis. *Trends Genet.* **19**(2), 91–96.

Brand, A. H., and Perrimon, N. (1993). Targeted gene expression as a means of altering cell fates and generating dominant phenotypes. *Development* **118**, 401–415.

Buszczak, M., and Cooley, L. (2000). Eggs to die for: Cell death during *Drosophila* oogenesis. *Cell Death Differ.* **11**, 1071–1074.

Chen, P., Rodriguez, A., Erskine, R., Thach, T., and Abrams, J. M. (1998). Dredd, a novel effector of the apoptosis activators *reaper, grim* and *hid* in *Drosophila*. *Dev. Biol.* **201**, 202–216.

Chew, S. K., Akdemir, F., Chen, P., Lu, W. J., Mills, K., Daish, T., Kumar, S., Rodriguez, A., and Abrams, J. M. (2004). The apical caspase *dronc* governs programmed and unprogrammed cell death in *Drosophila*. *Dev. Cell* **7**, 897–907.

Daish, T. J., Mills, K., and Kumar, S. (2004). *Drosophila* caspase DRONC is required for specific developmental cell death pathways and stress-induced apoptosis. *Dev. Cell* **7**, 909–915.

Dietzl, G., *et al.* (2007). A genome-wide transgenic RNAi library for conditional gene inactivation in *Drosophila*. *Nature* **448,** 151–156.

Doumanis, J., Quinn, L., Richardson, H., and Kumar, S. (2001). STRICA, a novel *Drosophila* caspase with an unusual serine/threonine–rich prodomain, interacts with DIAP1 and DIAP2. *Cell Death Differ.* **8,** 387–394.

Dorstyn, L., Colussi, P. A., Quinn, L. M., Richardson, H., and Kumar, S. (1999a). DRONC, a novel ecdysone-inducible *Drosophila* caspase. *Proc. Natl. Acad. Sci. USA* **96,** 4307–4312.

Dorstyn, L., Mills, K., Lazebnik, Y., and Kumar, S. (2004). The two cytochrome *c* species, DC3 and DC4, are not required for caspase activation and apoptosis in *Drosophila* cells. *J. Cell Biol.* **167,** 405–410.

Dorstyn, L., Read, S., Cakouros, D., Huh, J. R., Hay, B. A., and Kumar, S. (2002). The role of cytochrome *c* in caspase activation in *Drosophila melanogaster* cells. *J. Cell Biol.* **18,** 1089–1098.

Dorstyn, L., Read, S. H., Quinn, L. M., Richardson, H., and Kumar, S. (1999b). DECAY, a novel *Drosophila* caspase related to mammalian caspase-3 and caspase-7. *J. Biol. Chem.* **274,** 30778–30783.

Duffy, J. B. (2002). GAL4 system in *Drosophila*: A fly geneticist's Swiss army knife. *Genesis* **34,** 1–15.

Fraser, A. G., and Evan, G. I. (1997). Identification of a *Drosophila melanogaster* ICE/CED-3-related protease, drICE. *EMBO J.* **16,** 2805–2813.

Greenspan, R. J. (2004). Fly Pushing: The Theory and Practice of *Drosophila* Genetics. Cold Spring Harbor Laboratory Press, Cold Spring Harbour, NY.

Harvey, N. L., Daish, T., Mills, K., Dorstyn, L., Quinn, L. M., Read, S. H., Richardson, H., and Kumar, S. (2001). Characterization of the *Drosophila* caspase, DAMM. *J. Biol. Chem.* **276,** 25342–25350.

Hay, B. A., and Guo, M. (2006). Caspase-dependent cell death in *Drosophila. Annu. Rev. Cell Dev. Biol.* **22,** 623–650.

Hay, B. A., Wassarman, D. A., and Rubin, G. M. (1995). *Drosophila* homologs of baculovirus inhibitor of apoptosis proteins function to block cell death. *Cell* **83,** 1253–1262.

Hay, B. A., Wolff, T., and Rubin, G. M. (1994). Expression of baculovirus P35 prevents cell death in *Drosophila. Development* **120,** 2121–2129.

Kondo, S., Senoo-Matsuda, N., Hiromi, Y., and Miura, M. (2006). DRONC coordinates cell death and compensatory proliferation. *Mol. Cell. Biol.* **26,** 7258–7268.

Kramer, J. M., and Staveley, B. E. (2003). GAL4 causes developmental defects and apoptosis when expressed in the developing eye of *Drosophila melanogaster. Genet. Mol. Res.* **31,** 43–47.

Kumar, S. (2007). Caspase function in programmed cell death. *Cell Death Differ.* **14,** 32–43.

Kumar, S., and Cakouros, D. (2004). Transcriptional control of the core cell-death machinery. *Trends Biochem Sci.* **29,** 193–199.

Leulier, F., Rodriguez, A., Khush, R. S., Abrams, J. M., and Lemaitre, B. (2000). The *Drosophila* caspase Dredd is required to resist gram-negative bacterial infection. *EMBO Rep.* **1,** 353–358.

Maroni, G., and Stamey, S. C. (1983). *Drosophila Information Service* **59,** 142.

McCall, K., and Peterson, J. S. (2004). Detection of apoptosis in *Drosophila. Methods Mol. Biol.* **282,** 191–205.

Meier, P., Silke, J., Leevers, S. J., and Evan, G. I. (2000). The *Drosophila* caspase DRONC is regulated by DIAP1. *EMBO J.* **19,** 598–611.

Mills, K., Daish, T., and Kumar, S. (2005). The function of the *Drosophila* caspase DRONC in cell death and development. *Cell Cycle* **4,** 744–746.

Mills, K., Daish, T., Harvey, K. F., Pfleger, C. M., Hariharan, I. K., and Kumar, S. (2006). The *Drosophila melanogaster* Apaf-1 homologue ARK is required for most, but not all, programmed cell death. *J. Cell Biol.* **172,** 809–815.

Muro, I., Berry, D. L., Huh, J. R., Chen, C. H., Huang, H., Yoo, S. J., Guo., M., Baehrecke, E. H., and Hay, B. A. (2006). The *Drosophila* caspase Ice is important for many apoptotic cell deaths and for spermatid individualization, a nonapoptotic process. *Development* **133,** 3305–3315.

Pennington, M. W., and Thornberry, N. A. (1994). Synthesis of a fluorogenic interleukin-1β converting enzyme substrate based on resonance energy transfer. *Pept. Res.* **7,** 72–76.

Quinn, L. M., Dorstyn, L., Mills, K., Colussi, P. A., Chen, P., Coombe, M., Abrams, J., Kumar, S., and Richardson, H. (2000). An essential role for the caspase dronc in developmentally programmed cell death in *Drosophila. J Biol. Chem.* **275,** 40416–40424.

Richardson, H., and Kumar, S. (2002). Death to flies: *Drosophila* as a model system to study programmed cell death. *J. Immunol. Methods* **265,** 21–38.

Riddiford, L. M. (1993). Hormones and *Drosophila* development. *In* "The Development of *Drosophila melanogaster*." (M. Bate and A. A. Martinez Arias, eds.), Cold Spring Harbor Laboratory Press, Cold Spring Harbor, NY.

Rørth, P. (1996). A modular misexpression screen in *Drosophila* detecting tissue-specific phenotypes. *Proc. Natl. Acad. Sci. USA* **93,** 12418–12422.

Rørth, P., Szabo, K., Bailey, A., Laverty, T., Rehm, J., Rubin, G. M., Weigmann, K., Milan, M., Benes, V., Ansorge, W., and Cohen, S. M. (1998). Systematic gain-of-function genetics in *Drosophila. Development* **125,** 1049–1057.

Sambrook, J., and Russell, D. W. (2001). "Molecular Cloning: A Laboratory Manual." 3rd Ed. Cold Spring Harbor Laboratory PressCold Spring Harbor, NY.

Song, Z., McCall, K., and Steller, H. (1997). DCP-1, a *Drosophila* cell death protease essential for development. *Science* **275,** 536–540.

Song, Z., Guan, B., Bergman, A., Nicholson, D. W., Thornberry, N. A., Peterson, E. P., and Steller, H. (2000). Biochemical and genetic interactions between *Drosophila* caspases and the proapoptotic genes *rpr, hid,* and *grim. Mol. Cell. Biol.* **20,** 2907–2914.

Srivastava, M., Scherr, H., Lackey, M., Xu, D., Chen, Z., Lu, J., and Bergmann, A. (2006). ARK, the Apaf-1 related killer in *Drosophila*, requires diverse domains for its apoptotic activity. *Cell Death Differ.* **14,** 92–102.

Sullivan, W., Ashburner, M., and Hawley, R. S., Eds. (2000). *In* "*Drosophila* Protocols." Cold Spring Harbor Laboratory Press, Cold Spring Harbor, NY.

Thomas, B. J., and Wassarman, D. A. (1999). A fly's eye view of biology. *Trends Genet.* **15,** 184–190.

Varkey, J., Chen, P., Jemmerson, R., and Abrams, J. M. (1999). Altered cytochrome c display precedes apoptotic cell death in *Drosophila. J. Cell Biol.* **144,** 701–710.

White, K., Grether, M. E., Abrams, J. M., Young, L., Farrell, K., and Steller, H. (1994). Genetic control of programmed cell death in *Drosophila. Science* **264,** 677–683.

Wichmann, A., Jaklevic, B., and Su, T. T. (2006). Ionizing radiation induces caspase-dependent but Chk2- and p53-independent cell death in *Drosophila melanogaster. Proc. Natl. Acad Sci. USA* **103,** 9952–9957.

Wolff, T., and Ready, D. F. (1991a). The beginning of pattern formation in the *Drosophila* compound eye: The morphogenetic furrow and the second mitotic wave. *Development* **113,** 841–850.

Wolff, T., and Ready, D. F. (1991b). Cell death in normal and rough eye mutants of *Drosophila. Development* **113,** 825–839.

Xu, D., Wang, Y., Willecke, R., Chen, Z., Ding, T., and Bergmann, A. (2006). The effector caspases drICE and dcp-1 have partially overlapping functions in the apoptotic pathway in *Drosophila. Cell Death Differ.* **13,** 1697–1706.

CHAPTER THREE

IN VIVO AND IN VITRO METHODS FOR STUDYING APOPTOTIC CELL ENGULFMENT IN DROSOPHILA

Elizabeth A. Silva,* Jemima Burden,* *and* Nathalie C. Franc*

Contents

Abstract

Proper development of all multicellular organisms involves programmed apoptosis. Completion of this process requires removal of the resulting cell corpses through phagocytosis by their neighbors or by macrophages. Studies in *C. elegans* have been fruitful in the genetic dissection of key pathways, but

* Medical Research Council Cell Biology Unit, MRC Laboratory for Molecular Cell Biology, and Anatomy and Developmental Biology Department, University College London, London, UK

Methods in Enzymology, Volume 446
ISSN 0076-6879, DOI: 10.1016/S0076-6879(08)01603-0

they lack the professional immune system of higher organisms. Mammalian studies have identified a plethora of factors that are required for engulfment, but redundancy in the pathways has made it difficult to explain the genetic hierarchy of these factors. Thus, *Drosophila* has proven to be a useful evolutionary intermediate in which to examine this phenomenon. Here we describe methods used for dissecting the mechanisms and pathways involved in the engulfment of apoptotic cells by *Drosophila* phagocytes. Included are methods to be used for *in vivo* studies in the early embryo that can be used to examine engulfment of dying cells at various stages of embryogenesis. We also describe *in vitro* techniques for the use of *Drosophila* cell culture, including cell engulfment assays, that can be used for general phenotypic analysis, as well as live cell studies. We provide advice on imaging, including the preparation of samples for high-resolution microscopy and quantification of potential engulfment phenotypes for both *in vivo* and *in vitro* methods.

1. INTRODUCTION

Programmed cell death is required for the normal development of all multicellular organisms (Raff *et al.*, 1994; Wyllie *et al.*, 1980). Integral to this process is the clearance of the cell corpses that are produced as a consequence (deCathelineau and Henson, 2003). Failure to clear these cells can result in the release of toxins and inflammation of the surrounding environment, and even contribute to the development of autoimmune disorders such as systemic lupus erythematosus (Maderna and Godson, 2003; Nagata, 2006). Engulfment is even required for completion of the apoptotic program in some contexts; in the absence of phagocytosis, cells can recover from an initiated program or even elude apoptosis altogether (Li and Baker, 2007; Reddien *et al.*, 2001).

Undoubtedly, some mechanisms are common to all forms of phagocytosis, whether the engulfment is of pathogens, apoptotic cells, or even synthetic factors such as latex beads. However, there must necessarily be some distinction between apoptotic cell engulfment and microbial engulfment, because they elicit distinct responses (Erwig and Henson, 2007). For example, apoptotic cell engulfment triggers neither an adaptive immune response nor an accumulation of histamines (Majai *et al.*, 2006). Efforts to identify mechanisms that are both shared and distinct to each type of engulfment event are fundamental to our overall understanding of the immune system.

Genetic studies in *C. elegans* identified two major cassettes required for apoptotic cell engulfment (Kinchen *et al.*, 2005; Liu and Hengartner, 1999; Zhou *et al.*, 2004). However, the absence of a professional immune system may limit the extent to which these data can be applied to higher organisms. Studies in mammalian systems are complicated by redundancy of the pathways involved (Fadok and Chimini, 2001; Franc, 2002; Savill *et al.*, 2002). Thus, we have found *Drosophila* to be a useful intermediate model for studying these processes within professional phagocytes. As well as its amenity for *in vivo* and genetic studies, work in *Drosophila* can be expanded to cell culture–based models, where established cell lines are known to exhibit all macrophage characteristics (Abrams *et al.*, 1992; Gateff, 1978; Schneider, 1972).

Here we provide a thorough description of the methods used for examining phagocytosis of apoptotic cells *in situ* in the *Drosophila* embryo, including conditions for collecting and culturing the embryos, visualizing the apoptotic cells and the macrophages, imaging, and quantification of potential engulfment defects. We also describe cell culture–based methods, in particular an *in vitro* engulfment assay that takes advantage of the phagocytic properties of established *Drosophila* cell lines.

2. Culture Conditions

2.1. Collecting the embryos

To obtain an efficient collection, we use only males and females that have eclosed within the previous 5 days. It often takes 2 to 3 days before the females start to lay their maximum brood size, so it is advisable to maintain your cage for a couple of days before collecting.

Plates are made by combining 30 g agar, 150 ml of molasses, and topping up to 1 L with water. Autoclave the mixture; once cooled, add 1 ml of tegosept (from a stock solution of 10% in 95% ethanol) and 10 ml of ethyl acetate to prevent fungal growth. Pour into petri dishes of the appropriate size for your cage.

When ready to collect, wipe any debris or accumulated embryos from the rim of the cage and add a fresh molasses plate with a smear of yeast paste (baker's yeast mixed with water to a creamy consistency). It is important to do a "pre-collection" of at least 1 h first thing in the morning. This will allow females to extrude any eggs that may have been retained in the abdomen during the night; discard this collection. We then take collections of 2 h each at 25 °C.

2.2. Staging the embryos

For the most part, we examine embryos at stage 13. At this stage, the macrophages are fully functional and should have migrated throughout the head, the nervous system, and into the posterior (Tepass *et al.*, 1994). There are also high levels of developmentally regulated cell death (Abrams *et al.*, 1993), resulting in robust conditions in which to analyze potential engulfment defects. Embryos at this stage will have completed germ band retraction but not yet have initiated dorsal closure (Fig. 3.1).

We usually age a 2-h collection for a further 18 h at 18 °C to reach this stage. However, it is occasionally necessary to culture flies at 25 °C, or even 29 °C, particularly when analyzing phenotypes of transgenic lines. It is important to note that a change in temperature often results in a difference in the amount of engulfment, which is at least partially because of the levels of cell death. For example, at 29 °C we see slightly elevated levels of cell death, as well as engulfment; therefore, wild-type collections are always taken in parallel, particularly when doing statistical analysis.

It is also common to examine cell death and engulfment in the developing nervous system, usually at stage 16. This method is most commonly used for analyzing the engulfment of apoptotic corpses by glial cells. For this analysis, embryos collected over the course of 3 h are aged for a further 13 h at 25 °C, or 26 h at 18 °C (Freeman *et al.*, 2003; Sears *et al.*, 2003).

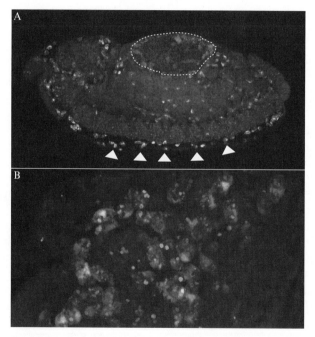

Figure 3.1 Stage 13 embryo labeled for macrophages and apoptotic corpses. Macrophages are labeled with the *Crq-eGFP* transgene (green) and the Crq antibody (blue); apoptotic corpses are labeled with 7-AAD (brighter red). (A) This stage is characterized by the completion of germ-band retraction and the initiation of dorsal closure, at which point the amnioserosa is fully exposed on the dorsal side (outlined in white). Macrophages in the CNS along the ventral side of the embryo tend to be smaller and engulf fewer corpses (arrowheads) than in the head and tail regions. Anterior is to the left. (B) Macrophages in the head region engulf between three and seven corpses each (also see Fig. 3.3). (See color insert.)

2.3. General fixing protocol

Appropriately staged embryos are collected by washing them off the plate and into a 70-μm nylon cell strainer (BD Falcon, cat: 352350). Rinse with water to remove all yeast paste. De-chorionate by washing for 2 to 3 min in freshly prepared 50% bleach, then rinse thoroughly in water. Properly de-chorionated embryos will appear shiny and tend to cluster together at the surface. At this stage it is important to avoid letting the embryos dry out. To fix, transfer to an Eppendorf tube containing one volume of heptane and one volume of 4% paraformaldehyde (EM grade, TAAB Laboratories, cat: F006) in PBS (phosphate buffered saline) and shake for 20 min at room temperature.

To de-vitellinize, remove the paraformaldehyde (bottom, aqueous layer) and add one volume of methanol. Vortex or shake vigorously immediately and for at least 1 min. All de-vitellinized embryos will sink to the bottom; remove the heptane, methanol, and any embryos still floating. From this point on the embryos will settle to the bottom of the Eppendorf tube when given sufficient time (they will sink faster in alcohols than aqueous solutions so expect to allow one or 2 min in the latter case). To change the solutions, it is, therefore, best to simply let the embryos sink, then decant the solution with a pipette. Rinse twice in 100% methanol. Embryos can be stored indefinitely at $-20\,^{\circ}$C or rehydrated for use in immunostaining as follows. Note that paraformaldehyde is corrosive and a possible carcinogen, and methanol and heptane are damaging to the environment. All waste solutions must, therefore, be collected separately and disposed of according to local regulations.

To rehydrate the embryos, the following washes are done for 5 min each, at room temperature on a rotating platform: twice in 100% ethanol; once in 3:1 ethanol to PBS-Triton (PBS with 0.1% Triton-100X); once in 1:1 ethanol to PBS-Triton; once in 1:3 ethanol to PBS-Triton; twice in PBS-Triton. The embryos are now ready for immunostaining.

3. Labeling of Apoptotic Cells

3.1. Assessing the levels of cell death

Essential to the analysis of any engulfment phenotype are methods for labeling the apoptotic corpses. First, it is important to ensure that any change in amount of engulfment is not simply a result of altered levels of cell death. For example, fewer corpses will result in a reduced engulfment, whereas more corpses will result in enhanced engulfment. Second, the pattern of the corpses themselves can indicate the presence of an engulfment defect: clustering of apoptotic cells in grapelike bunches indicates successful engulfment where each bunch represents several corpses within a single macrophage; failure to cluster is indicative of a defect (Fig. 3.2).

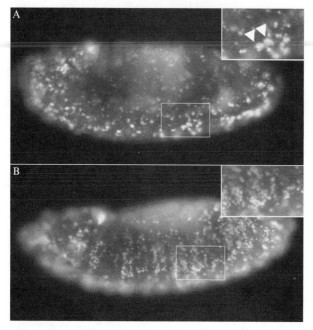

Figure 3.2 Efficient engulfment results in clustering of apoptotic cells. Wild-type (A) and engulfment-defective (B) embryos labeled with acridine orange (AO) to mark apoptotic cells. In wild-type embryos, apoptotic cells appear in clusters (white arrowheads). In embryos deficient for a region covering an as yet unidentified gene required for phagocytosis of apoptotic cells, this clustering is absent despite comparable levels of apoptosis.

Many effective methods can be used for marking apoptotic cells in the *Drosophila* embryo that have been widely described. We have generally relied on TUNEL labeling to assess the levels of cell death and acridine orange (AO) staining as a rapid method to look for patterns of clustering among the corpses, both of which are described here.

3.1.1. TUNEL labeling

The terminal deoxynucleotidyl transferase biotin-dUTP nick end labeling technique (TUNEL) takes advantage of the DNA cleavage that is typically found in apoptotic cells (Gavrieli *et al.*, 1992). The free DNA ends are targeted by terminal deoxytransferase (TdT), incorporating digoxigenin-labeled nucleotides.

We use ApopTag Peroxidase *In Situ* Apoptosis Detection Kit from Chemicon International (cat: S7100). Appropriately staged embryos are fixed, de-vitellinized, and rehydrated as described previously. They are then treated with 10 μg/ml of proteinase K in PBS for 4 to 5 min at room temperature. Wash twice in PBS-Tween (phosphate-buffered saline with 0.1% Tween 20) and post-fix in 4% paraformaldehyde in PBS for 20 min at room temperature. Wash five times for 10 min in PBS-Tween. Incubate in the kit-supplied equilibration buffer for 1 h at room temperature followed by overnight incubation at 37 °C in the kit-supplied reaction buffer containing TdT enzyme (2:1) and Tween20 to 0.3%. The following day, the reaction is stopped by use of the stop/wash buffer (1:34 in water) for 3 h at 37 °C. After three 10-min washes in PBS-Tween, embryos are blocked for 1 h at room temperature in PBS-Tween containing 2 mg/ml bovine serum albumin (BSA) and 5% normal goat serum (NGS). Blocking is followed by incubation with fluorescein-coupled anti-dioxygen in antibody at a 1/100 dilution for 1 h at room temperature. Three washes of 20 min in PBS-Tween are performed at room temperature before mounting in Vectashield (Vector Laboratories, H-1000).

3.1.2. Acridine orange

The molecular basis of AO staining is its selective retention in dying cells. This allows one to observe the pattern of apoptosis in living embryos or freshly dissected larval tissues (Gavrieli *et al.*, 1992). Stock solutions of 10 mg/ml are kept in the dark at 4 °C (Molecular Probes, cat: A3568). AO is excited by use of the 488 Argon laser, resulting in emission largely in the green wavelengths with overlap into the red. It is important to note that AO is a mutagen and should be handled with gloves.

Begin by de-chorionating the embryos as described previously; do not fix. All traces of bleach must be eliminated by extensive washes in dH$_2$O or the AO staining will not work. After de-chorionation, the embryos are still surrounded by a hydrophobic membrane. By shaking embryos hard at the interface of heptane and the aqueous dye, some of the dye is driven through this membrane, and retained by the apoptotic cells. Transfer the embryos into 1 ml of heptane in a glass tube. Add 1 ml of 2.5 μg/ml AO in 0.1 *M* phosphate buffer, pH 7.4 (this diluted AO solution should be kept in the dark and is stable for months at room temperature. In fact, aged solutions seem to give lower background than fresh ones). Shake vigorously or vortex for 2 min. Place the tube in an Eppendorf shaker for 5 min or continue to shake by hand for an additional 5 min. Use a glass Pasteur pipette that has been rinsed in heptane, remove the embryos from the interface, being careful to take as little of the aqueous (lower) phase as possible. Mount the embryos in mineral oil, and visualize within 20 min, because the staining will rapidly fade over time.

3.2. Engulfed apoptotic cells

Here we describe two methods for visualizing engulfed apoptotic corpses. We generally rely on the first method, 7-amino actinomycin D (7-AAD). 7-AAD will stain all apoptotic corpses, but we include it here because we find it particularly reliable for analysis when used in conjunction with labeling of the macrophages themselves. We then analyze the embryos by confocal microscopy to distinguish between engulfed and unengulfed corpses.

The second method, which involves labeling of fragmented DNA in dICAD mutants, is a method that has been described for visualizing only those corpses that are undergoing lysosomal degradation. When this method is used, there is no need to visualize the macrophages to assess any engulfment defect.

3.2.1. 7-Amino actinomycin-D

To visualize apoptotic cells that have been engulfed *in vivo* we use 7–AAD (Molecular probes/Invitrogen, cat: A1310). It is a fluorescent DNA inter-calator that can be used on fixed tissues, where it will label all nuclei with moderate intensity, but the condensed DNA of apoptotic nuclei more brightly. DNA/7–AAD complexes are maximally excited at 543 nm and emit at 655 nm (Gill *et al.*, 1975). Stock solutions of 1 mg/ml of 7–AAD in water can be kept indefinitely at −20 °C. It is important to note that 7–AAD is carcinogenic, and solutions must be disposed of according to local regulations.

To label fixed embryos with 7–AAD, first wash for 10 min in PBS to remove any detergent. Replace with fresh PBS, to which is added the appropriate volume of 7–AAD stock solution for a final concentration of 5 μg/ml. Incubate for 20 to 30 min at room temperature on a rocking platform in the dark. Wash briefly in PBS before mounting; extensive washing will result in leaching of the dye from the cells. If 7–AAD is to be used in conjunction with other protocols such as immunolabeling, 7–AAD staining must be performed as the final step.

3.2.2. dICAD mutants

In embryos deficient for dICAD, an absence of DNA fragmentation nor-mally occurs in apoptotic cells. Thus, any DNA fragmentation in these embryos should be the consequence of lysosomal degradation after the engulfment of the corpse (Manaka *et al.*, 2004). DNA fragmentation can then be detected by any of a number of methods, such as TUNEL or *in situ* nick translation (ISNT). One of the major drawbacks of this method is that it requires the incorporation of dICAD mutations into the background of your gene of interest, thus we have not used this technique ourselves.

4. VISUALIZING THE MACROPHAGES

A number of methods are available for visualizing the macrophages. Here we describe publicly available transgenic lines that can be used to express eGFP, as well as a number of previously described antibodies.

4.1. Genetic lines

The use of genetic lines has the advantages of availability, consistency, and flexibility. For example, there is no reliance on possibly limited quantities of polyclonal antibodies, and expression of a marker protein is generally consistent and can even be dose-sensitive. In addition, such lines allow for analysis of both live and fixed samples and, with the wide availability of various anti-GFP antibodies, can be used in conjunction with immunodetection of any other molecule of interest as described later, or in conjunction with other staining methods, including 7–AAD staining of apoptotic corpses as described previously.

The lines described here rely on the Gal4–UAS system (Brand and Perrimon, 1993). Enhanced green fluorescent protein (eGFP) expression is driven by binding of the yeast-derived transcriptional co-activator, Gal4, to the upstream activating sequence (UAS). Cell type- or tissue-specific Gal4 expression is driven by the promoter of choice, in this case by the promoter regions of genes expressed in macrophages. Various transgenic lines expressing eGFP under control of UAS are available from the Bloomington *Drosophila* Stock Center, Indiana University; http://flystocks.bio.indiana.edu/.

We usually rely on Gal4 expression from the *croquemort* (*crq*) promoter. Crq is a scavenger receptor and ortholog of CD36 that is expressed specifically in the *Drosophila* macrophages during embryogenesis (Franc *et al.*, 1996). Embryos carrying the Crq-Gal4, UAS-eGFP transgenic combination exhibit eGFP expression throughout the cytoplasm of all macrophages (generated by H. Agaisse and N. Perrimon) (Figs. 3.1 and 3.3). Furthermore, expression of eGFP in this line is dose-dependent, making it particularly useful as a marker for genotyping. There are lines carrying the Crq-Gal4 driver alone on the X chromosome and in combination with the UAS-eGFP on the third chromosome.

Another widely used and publicly available transgenic line, *srp*Hemo-Gal4, relies on Gal4 expression from selected regulatory regions of the *serpent* (*srp*) gene (Bruckner *et al.*, 2004). Serpent is a GATA factor that is essential for the development of all *Drosophila* embryonic blood cells. Expression of eGFP from the *srp*Hemo-Gal4 line is visible as early as stage 9 of embryonic development, in the procephalic mesoderm, and in hemocytes/macrophages thereafter. This transgene is inserted on the second chromosome, and we are not aware of the existence of a chromosome carrying both *srp*Hemo-Gal4 and UAS-eGFP.

Hemolectin (Hml) was first identified as a protein secreted from the *Drosophila* cell line Kc167 (Goto *et al.*, 2001) and was subsequently characterized for its role *in vivo*. By use of a transgenic Gal4 expression line it was revealed to be expressed in a subset of plasmatocytes and crystal cells, from embryonic stages through to late third instar, but not in the lamellocytes. Hml itself seems to function primarily in wound healing rather than the characterized innate immunity pathways, but when combined with UAS-eGFP is a useful marker for the hemocytes (Goto *et al.*, 2003).

Peroxidasin (Pxn) is produced early in hemocyte development and is essential for macrophage maturation; it is a component of the extracellular matrix but in cell culture is secreted into the medium (Abrams *et al.*, 1993; Nelson *et al.*, 1994). A transgenic line expressing Gal4 expression from the *peroxidasin* promoter was generated for studying wound healing in the embryo (Stramer *et al.*, 2005).

4.2. Immunostaining of macrophages

A number of published antibodies also either label or are specific for macrophages. The advantage of the use of antibodies over transgenic lines is, of course, that there is no need to incorporate a specific genetic background when analyzing your gene of choice. Here we describe several antigens for which there are antibodies available, followed by a general immunostaining protocol.

We generally rely on Crq, the CD36 ortholog and putative scavenger receptor, for immunostaining of macrophages. Crq is required for the engulfment of apoptotic corpses in the embryo and most strongly localizes to the site of engulfment and to the membrane of vesicles enclosing engulfed corpses (Franc *et al.*, 1999). During embryogenesis, Crq is exclusively expressed in macrophages, and the CRQ antibody is the only reagent, as far as we are aware, to label macrophages to the exclusion of all other embryonic tissues (Figs. 3.1 and 3.3). Of note is that whereas the CRQ antibody also labels larval and adult macrophages (also known as plasmatocytes), it is not exclusive to these cells at these stages.

Pxn and macrophage derived proteoglycan-1 (Mdp-1) are two other proteins that are strongly expressed in macrophages and for which there are available antibodies. Mdp-1, like Pxn as described earlier, is an extracellular matrix protein that is expressed in all hemocytes in the embryo (Hortsch *et al.*, 1998). However, unlike Pxn, the precise role of Mdp-1 is not known.

Two other antigens that have been used to visualize macrophages are Serpent (Srp), as described previously, and the putative *Drosophila* engulfment receptor, Draper (Drpr). It is again important to note that Srp expression is not limited to the hemocytes. As revealed both by immunostaining and *in situ* hybridization, Srp is also expressed in the gut and the procephalic mesoderm, as well as in the adult ovaries (Bruckner *et al.*, 2004). Drpr has proven useful as a marker for engulfing cells, where its expression becomes elevated. However, like Srp, Drpr is not specific for macrophages and in *Drosophila* was first identified for its role in glial cells (Freeman *et al.*, 2003). Immunostaining against Drpr is frequently used in combination with Repo to distinguish glia from macrophages in the developing nervous system (Freeman *et al.*, 2003).

4.3. Immunostaining protocol

For immunostaining against any of the preceding molecules, we rely on a basic protocol that works for most antibodies. Any modifications to this protocol that may be required for the antibodies described here can be found in Table 3.1.

Collect, stage, fix, and rehydrate the embryos as described earlier. Immediately after rehydration, block the embryos for 30 min at room temperature in PBS with 0.1% Triton-X100 (PBS-Triton) with 10% bovine serum albumin (w/v, BSA). Dilute the appropriate primary antibody in PBS-Triton with 1% BSA and incubate on a rocking platform overnight at 4 °C. Wash three times for 20 min in PBS-Triton and follow with the appropriate secondary antibody, also diluted in PBS-Triton with 1% BSA. We generally rely on fluorophore-conjugated secondary antibodies from Jackson Labs at a dilution of 1/500 to 1/1000. Wash three times for 10 min in PBS-Triton. If desired, follow with 7-AAD staining as described previously before mounting in Vectashield. Images are usually collected by confocal microscopy, which is discussed in the following section.

Table 3.1 Antibodies that detect macrophage-derived antigens

Antigen	Dilution	Source/Reference	Notes
Croquemort	1/500	(Franc et al., 1996, 1999)	none
Draper	1/500	(Freeman *et al.*, 2003)	none
Green Fluorescent Protein	1/1000	Roche (cat: 11814460001)	none
Macrophage derived proteoglycan-1	1/500	(Hortsch *et al.*, 1998)	Devitellinize using ethanol in place of methanol
Peroxidasin	1/1000	(Abrams *et al.*, 1993; Nelson *et al.*, 1994)	Block in PBS, 0.0125% saponin, 1% bovine serum albumin, and 4% normal goat serum (PSN)
Serpent	1/500	(Han and Olson, 2005; Petersen *et al.*, 1999)	none

5. Quantification and Statistics

Unless there is a complete absence of engulfment, analysis of any mutant phenotype will inevitably require some sort of quantification. This quantification is referred to as the phagocytic index and, *in vivo*, is simply a count of the average number of engulfed corpses per macrophage.

In practice, a statistically significant measurement will require counts of 150 to 400 macrophages, because the phagocytic index can vary greatly. As discussed earlier, this variation can depend on the amount of cell death, which in turn depends on culturing conditions. It is, therefore, important to measure the phagocytic index of macrophages in the wild-type embryo collections that are taken in parallel to each experiment. We also ensure that measurements are taken from at least five different embryos of any given genotype. In the stage 13 embryo, the amount of engulfment also depends on the location of the macrophages; those in the central nervous system (CNS), along the ventral axis of the embryo, often engulf fewer corpses than those in the head and tail regions. We, therefore, ensure that we have taken images of macrophages in all three regions (head, tail, and CNS) for calculating the phagocytic index.

For *in vivo* quantification we take confocal Z-stacks of embryos that have been labeled for apoptotic corpses by use of 7-AAD and for macrophages by use of either the crq-Gal4, UAS-eGFP transgenes or the Crq antibody. By taking a stack of images through a labeled macrophage it is possible to count the number of 7-AAD–stained corpses (Fig. 3.3). The interval between each stack should be no more than 1 μm. Note that this method will not provide an absolute count of the number of corpses per macrophage, because those that are almost fully degraded may no longer be positive for 7-AAD or may be too small to be caught in the stack. The important measure is the relative amount of engulfment when comparing mutants to wild type.

5.1. High-resolution imaging

Once we have established the presence of an engulfment-defective phenotype, we often turn to high-resolution analysis by transmission electron microscopy. By use of this method we are able to obtain a more accurate impression of where in the engulfment process the defect occurs. For example, a phenotype could result from a failure to recognize the corpse, in phagosome maturation and degradation of the engulfed body, or at any stage in between. To achieve this, embryos are fixed by high-pressure freezing and freeze substitution. We find that this method is best for preserving the membrane ultrastructure, which is necessary for visualizing events such as membrane ruffling. It is also possible to calculate the phagocytic index

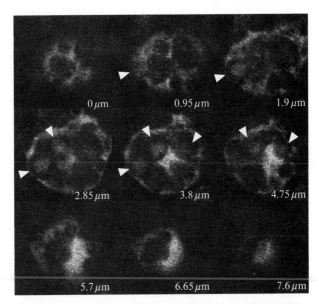

Figure 3.3 Confocal stack through a single wild-type macrophage. Z-stack of nine images of a single macrophage taken at intervals of 0.95 μm. The macrophage is marked by the cytoplasmic expression of a UAS-eGFP transgene under the control of the *Crq-Gal4* driver (green) and the Crq antibody (blue); the apoptotic corpses are labeled with 7-AAD (brighter red). Each of three apoptotic corpses engulfed by this macrophage become visible at different levels within the stack (arrowheads). The localization of Crq itself is to the membrane of vesicles surrounding the engulfed corpses. (See color insert.)

by examining transmission electron micrographs. This method is far less likely to miss even the smallest engulfed corpses and will, therefore, result in a higher phagocytic index. However, it is a very time-consuming method for gathering sufficient numbers of macrophages to obtain a statistically significant phagocytic index and requires serial sectioning.

Staged embryos are placed in 1.2-mm diameter flat specimen carriers, covered with a yeast paste (a thick mixture of baker's yeast and water) and transferred to an EMPACT high-pressure freezer (Leica Microsystems UK) to be frozen as described by Studer *et al.* (2001). The specimens are then transferred under liquid nitrogen to a Leica AFS freeze substitution unit and precooled to −90 °C. The samples are freeze-substituted in 1% osmium tetroxide (crystals, TAAB), 0.4 % uranyl acetate in extra dry acetone (Fisher Scientific) for 5 h at −90 °C. The temperature is then raised to 0 °C at a rate of 5 °C per hour. The samples are then incubated in extra dry acetone for 1 h at 0 °C, before being warmed to 20 °C over 1 h, and incubated for a further hour in extra dry acetone. After two 10-min incubations in propylene oxide, the samples are infiltrated in Epon 812:propylene oxide (1:1) for 1 h, followed by two further incubations in Epon 812, for 2 h each at room temperature. At this stage, the embryos are carefully teased out of the carriers and separated from the yeast to orient them for embedding: for transverse sectioning we orient and embed the embryos in coffins; for longitudinal sectioning we orient and embed the embryos in mesh-bottomed capsules (Leica Microsystems UK). All samples are then baked overnight at 60 °C. Prepared samples are trimmed and then cut into ultrathin sections by use of an UltraCut UCT ultramicrotome (Leica Microsystems UK). The sections are then post-stained with lead citrate before imaging.

We have not combined this method with immunostaining; however, there is no reason that it could not be. The combination of osmium tetroxide, uranyl acetate, and lead citrate are sufficient for visualizing macrophages, as well as darkly stained apoptotic corpses (Fig. 3.4).

6. ENGULFMENT IN CELL CULTURE

Drosophila cell culture is particularly amenable to engulfment studies, because the most commonly used cell lines have been shown to be phago-cytic (Abrams *et al.*, 1992; Gateff, 1978; Schneider, 1972). We have found analysis of engulfment phenotypes with cell culture to be useful in a number of contexts, for example in conjunction with RNAi when no mutant allele is available for the gene of interest. These cells will engulf bacteria, apoptotic corpses, latex beads, and even small fungi (Abrams *et al.*, 1992; Ramet *et al.*, 2001); we have seen full engulfment and degradation of labeled corpses take place as quickly as 30 min, making live studies quite feasible. Here we describe two such cell lines, as well as methods for *in vitro* studies of engulfment processes.

6.1. Macrophage-like cell lines

One of the most commonly used *Drosophila* cell lines, and the most frequently used for engulfment studies, is the Schneider 2 (S2) cell line (ATCC, http://www.lgcpromochem-atcc.com/). This cell line was originally established from a mixed embryonic primary culture.

The culture medium is revised Schneider's medium with L-glutamine (Invitrogen, cat: 21720-024) containing 10% fetal bovine serum (Invitrogen, heat-inactivated, cat: 10108-165), 50 units/ml penicillin, and 50 μg/ml streptomycin (Sigma, cat: P0781). S2 cells are semiadherent; they will adhere to the culture vessel until they approach confluence, at which point they can detach and grow in suspension and will also grow in suspension if cultured in a rotating vessel. For protocols such as for stable

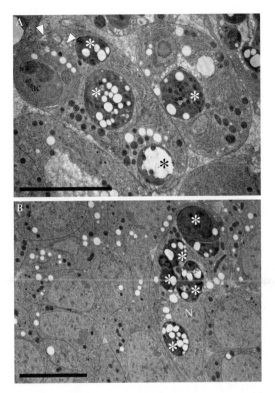

Figure 3.4 Transmitting electron microscopy section of a wild-type embryonic macrophage. (A) Two wild-type (*yw*) macrophages containing a total of four apoptotic corpses at varying stages of degradation, which are easily identified by their dark bodies against the lighter background of the live cells (marked with *). In the macrophage on the left, there is an additional large, recently engulfed corpse (ac); the surrounding vesicular membrane is clearly visible and contains some hemolymph captured during engulfment (arrowhead). (B) A wild-type (yw) macrophage containing five apoptotic corpses (*). In this image the macrophage is easily distinguished from the surrounding cells. "N" denotes the nucleus of the macrophage; scale bars are 5 μm. Images were taken with a Tecnai 120 Spirit (FEI Eindhooven, The Netherlands) and a Morada digital camera (Olympus-SIS).

or transient transfection they should be treated as cells in suspension. One important caveat when S2 cells are used in the assays described in the following sections is that their engulfment efficiency may significantly decrease when they have been passaged 25 to 30 times; it is, therefore, advisable to freeze aliquots of cells that can later be thawed as necessary.

Another cell line that has been used in engulfment assays is the l(2)mbn cell line (Manaka *et al.*, 2004). l(2)mbn cells are derived from tumorous larval hemocytes and share many of the properties of S2 cells (Gateff, 1978). However, for l(2)mbn cells to be fully functional as hemocytes they require treatment with 1 μM 20-hydroxyecdysone (Sigma) for 48 h (Dimarcq *et al.*, 1997). Cultures are maintained in Schneider's *Drosophila* revised medium (as described earlier).

6.2. Methods for inducing apoptosis

Any study of apoptotic cell engulfment will require a population of corpses that can be fed to the phagocytes. Here we describe several alternate methods that can be used to induce cell death in a population of S2 cells, which can subsequently be harvested and added to a live culture.

6.2.1. Chemical induction of apoptosis

A number of methods are available to choose from when generating apoptotic cells. The most commonly used chemical methods are to induce death by the addition of actinomycin-D or cycloheximide. Both chemicals have been demonstrated to induce apoptosis in S2 cells, as characterized by exposure of phosphatidylserine on the surface of the dying cells, activation of the caspase drICE, condensation of the chromatin and fragmentation of the DNA (Nagano *et al.*, 2000; Wright *et al.*, 2005).

We have found the use of actinomycin-D to be effective and reliable for induction of apoptosis in S2 cell culture. However, we find that the efficiency of engulfment of these cells can vary from batch to batch, a phenomenon for which we currently have no explanation. Thus, as described here, we induce cell death in a large culture, then fix a quantity that is sufficiently large for the number of experiments required, including all controls.

We induce death in cultures that are aged 5 to 6 days after splitting. At this stage they will have extended beyond confluence and be growing in suspension. Add 7.5 μl of actinomycin-D (0.5 mg/ml stock, Invitrogen, cat: 118050017) to each 12-ml culture, which has grown in a 75 cm^2 flask. Leave at 25 °C for approximately 18 h. To induce cell death by use of cycloheximide, incubate for 24 h in 1.5 μg/ml (Sigma, cat: C4859). Transfer the cells and the media to falcon tubes and centrifuge at 3000 rpm in a benchtop centrifuge; note that the apoptotic cells are spun at a higher velocity than live cells, because apoptotic cells tend to be smaller. Remove the supernatant and resuspend the cells in 10% formaldehyde (EM grade, TAAB laboratories, cat: F006) that is diluted in serum-free media. Use 1 ml fixative for each 12-ml culture and incubate at room temperature for 20 min. If desired, cells may be transferred to Eppendorf tubes at this time. Wash the cells twice in serum-free media with spins at 6500 rpm in a microcentrifuge, or 3000 rpm in a benchtop centrifuge. Each 12-ml culture should be resuspended in 1 ml serum-free medium. The cells can be stored at 4 °C for 2 to 3 weeks. As mentioned previously, the efficiency with which each batch of cells is engulfed does vary, so before initiating any large-scale or laborious experiments such as RNAi knockdown, check for engulfment in untreated S2 cells and use the same batch for each complete experiment.

6.2.2. Induction by UV irradiation

Induction of apoptosis by UV irradiation has several key advantages. The most significant is the absence of any chemicals in the media, which allows one to use the conditioned media for assaying the release of cell death signals from the apoptotic corpses themselves. In addition, it is relatively safe, because it precludes the need to handle potentially toxic or mutagenic substances. To induce cell death by UV irradiation, simply place the flask directly on a UV transilluminator. We expose the cells for 5 min on a GVM20 transilluminator, part of the Geneflash Gel documentation system (Syngene, 302 nm, 100% intensity). If you wish to keep the cells to be used for engulfment assays, fix as described for chemical treatments. Alternately, spin the cells as described previously and harvest the supernatant to test the effect of the conditioned medium.

6.3. Engulfment assay

Once the apoptotic cells have been generated, they are simply added to a live, macrophage-derived culture; no further treatment is required for engulfment to occur. However, here we describe a few techniques that we have found to optimize the visualization of this process.

To ensure a healthy culture and high efficiency of engulfment, live cells should be split 1 to 2 days before the experiment and be subconfluent. To visualize the apoptotic corpses we surface-label them with fluorescein (FITC, isomer 1, Molecular Probes/Invitrogen, cat: F-1906). The fluorescein–isomer suspension is 1 mg/ml in DMSO, which we divide into 25-μl aliquots, dry in a speed-vac, and store at $-20\,°C$. To label 1 ml of apoptotic corpses, resuspend a single 25-μl aliquot in DMSO and add it to the cells. Incubate on a rocking platform for 1 h at room temperature. Remember that FITC is light sensitive so the cell/FITC mixture must be kept in the dark, for example by covering with aluminium foil. Perform two washes in serum-free media (SFM, Invitrogen, cat: 10797) as described in the protocol for fixation. Resuspend the cells in 1 ml SFM. This labeling must be done immediately before use, because the staining will not keep.

For a robust incidence of engulfment, we aim to add apoptotic cells to live culture in a ratio of 10:1 (it is difficult to get an accurate count of apoptotic cells because they are small, variable in size, and tend to cluster; as such, the apoptotic cell count used to calculate this ratio is made on the basis of the original live culture, before induction of cell death). This ratio translates to approximately 50μL of prepared, labelled corpses to each well of live cells in a 24-well plate. It is important to dissociate the aggregates of corpses before adding them, which we find can be effectively accomplished mechanically by pipetting the cells repeatedly with a fine-bore tip. Incubate the culture for the desired amount of time. As mentioned previously, it is possible to visualize engulfment and full degradation of a live culture in the first 30 min. After 4 to 5 h at 25 °C, approximately 40% of the cells will contain at least one engulfed corpse. Should you wish to fix the samples and counterstain to analyze the engulfment process, we would recommend doing so at the latter time. This will ensure that the sample includes the cells at all stages of engulfment. This time is also where we would recommend fixing samples for taking the phagocytic index.

The phagocytic index *in vitro* is usually calculated differently from what we have already described *in vivo*. In culture it is difficult to accurately discern the number of corpses engulfed by each cell, thus the index is usually calculated as the percentage of cells containing at least one corpse. As mentioned, a healthy wild-type S2 cell culture should have a phagocytic index of approximately 40% after 4 to 5 h of engulfment.

6.4. Primary hemocytic cell culture

Methods have been described that allow for the isolation of live hemocytes, and even separate specific subsets, from *Drosophila* lymph glands and hemolymph (Tirouvanziam *et al.*, 2004). Rather strikingly, these methods can be effectively applied to single animals.

Third instar larvae are first collected, rinsed in PBS, then transferred to a pool of ice-cold Schneider's medium with 1× complete protease inhibitor cocktail (Roche Applied Science) to prevent melanization, clump formation, and autolysis. Hemolymph from the larvae is collected by simply rupturing the cuticle with a pair of fine forceps and transferring the surrounding medium to a fresh Eppendorf tube; lymph glands are dissected out carefully, rinsed, then ruptured by repeated pipetting with a siliconized tip. Collect the cells by spinning at 1500 rpm in a low-speed centrifuge at 4 °C.

6.4.1. Isolating the hemocytes

To discriminate live hemocytes from dead cells and other contaminating debris, the isolated larval cells can be incubated with either 20 μM monochlorobimane (MCB, Molecular Probes) or 1.0 μ dihydrorhodamine 123 (DHR, Sigma) in 1 ml of staining medium (Schneider's medium, pH 6.5, with 2.5 mM probenecid (Sigma)). Follow with one wash in 10 ml of cold staining medium, then resuspend in 100 μl staining medium with 2 μg/ml propidium iodide (PI, Sigma) to label the dead cells. MCB enters cells to react with intracellular reduced glutathione (GSH), which is a well-conserved antioxidant present in all immune cells, to produce glutathione-S-bimane (GSB); GSB is excited by a UV laser (maximal excitation at 380 nm) and emits at 461 nm. DHR reacts with reactive oxygen species and localizes to active mitochondria, both of which are indicators of immune cell function. The advantage of DHR over MCB is that it can be excited by use of an argon 488 laser, which can be used in conjunction with excitation of PI. DHR is maximally excited at 507 nm and emits at 529 nm. Cells should be subjected to two to three rounds of sorting; live hemocytes will be reserved in the fraction that is negative for PI and has high levels of either GSB or DHR.

6.4.2. Sorting of hemocyte subsets

With minor modification to these methods, Tirouvanziam and colleagues described various reporters, antibodies, and metabolic indicators that can be used to discriminate between subsets of hemocytes (Tirouvanziam *et al.*, 2004). Because of space constraints and because they have already been well described, we do not include these methods here.

ACKNOWLEDGMENTS

We would like to thank Emeline van Goethem for details of the TUNEL protocol, and Herve Agaisse and Norbert Perrimon for the Crq-Gal4, UAS-eGFP line.

REFERENCES

Abrams, J. M., Lux, A., Steller, H., and Krieger, M. (1992). Macrophages in *Drosophila* embryos and L2 cells exhibit scavenger receptor-mediated endocytosis. *Proc. Natl. Acad. Sci. USA* **89,** 10375–10379.

Abrams, J. M., White, K., Fessler, L. I., and Steller, H. (1993). Programmed cell death during *Drosophila* embryogenesis. *Development* **117,** 29–43.

Brand, A. H., and Perrimon, N. (1993). Targeted gene expression as a means of altering cell fates and generating dominant phenotypes. *Development* **118,** 401–15.

Bruckner, K., Kockel, L., Duchek, P., Luque, C. M., Rorth, P., and Perrimon, N. (2004). The PDGF/VEGF receptor controls blood cell survival in *Drosophila*. *Dev. Cell* **7,** 73–84.

deCathelineau, A. M., and Henson, P. M. (2003). The final step in programmed cell death: Phagocytes carry apoptotic cells to the grave. *Essays Biochem.* **39,** 105–17.

Dimarcq, J. L., Imler, J. L., Lanot, R., Ezekowitz, R. A., Hoffmann, J. A., Janeway, C. A., and Lagueux, M. (1997). Treatment of l(2)mbn *Drosophila* tumorous blood cells with the steroid hormone ecdysone amplifies the inducibility of antimicrobial peptide gene expression. *Insect Biochem. Mol. Biol.* **27,** 877–86.

Erwig, L. P., and Henson, P. M. (2007). Immunological consequences of apoptotic cell phagocytosis. *Am. J. Pathol.* **171,** 2–8.

Fadok, V. A., and Chimini, G. (2001). The phagocytosis of apoptotic cells. *Semin Immunol.* **13,** 365–72.

Franc, N. C. (2002). Phagocytosis of apoptotic cells in mammals, caenorhabditis elegans and *Drosophila melanogaster*: Molecular mechanisms and physiological consequences. *Front Biosci.* **7,** d1298–313.

Franc, N. C., Dimarcq, J. L., Lagueux, M., Hoffmann, J., and Ezekowitz, R. A. (1996). Croquemort, a novel *Drosophila* hemocyte/macrophage receptor that recognizes apoptotic cells. *Immunity* **4,** 431–443.

Franc, N. C., Heitzler, P., Ezekowitz, R. A., and White, K. (1999). Requirement for croquemort in phagocytosis of apoptotic cells in *Drosophila*. *Science* **284,** 1991–1994.

Freeman, M. R., Delrow, J., Kim, J., Johnson, E., and Doe, C. Q. (2003). Unwrapping glial biology: Gcm target genes regulating glial development, diversification, and function. *Neuron* **38,** 567–580.

Gateff, E. (1978). Malignant neoplasms of genetic origin in *Drosophila melanogaster*. *Science* **200,** 1448–1459.

Gavrieli, Y., Sherman, Y., and Ben-Sasson, S. A. (1992). Identification of programmed cell death *in situ* via specific labeling of nuclear DNA fragmentation. *J. Cell Biol.* **119,** 493–501.

Gill, J. E., Jotz, M. M., Young, S. G., Modest, E. J., and Sengupta, S. K. (1975). 7-Amino-actinomycin D as a cytochemical probe. *I.* Spectral properties. *J. Histochem. Cytochem.* **23,** 793–799.

Goto, A., Kadowaki, T., and Kitagawa, Y. (2003). *Drosophila* hemolectin gene is expressed in embryonic and larval hemocytes and its knock down causes bleeding defects. *Dev. Biol.* **264,** 582–591.

Goto, A., Kumagai, T., Kumagai, C., Hirose, J., Narita, H., Mori, H., Kadowaki, T., Beck, K., and Kitagawa, Y. (2001). A *Drosophila* haemocyte-specific protein, hemolectin, similar to human von Willebrand factor. *Biochem. J.* **359,** 99–108.

Han, Z., and Olson, E. N. (2005). Hand is a direct target of Tinman and GATA factors during *Drosophila* cardiogenesis and hematopoiesis. *Development* **132,** 3525–3536.

Hortsch, M., Olson, A., Fishman, S., Soneral, S. N., Marikar, Y., Dong, R., and Jacobs, J. R. (1998). The expression of MDP-1, a component of *Drosophila* embryonic basement membranes, is modulated by apoptotic cell death. *Int. J. Dev. Biol.* **42,** 33–42.

Kinchen, J. M., Cabello, J., Klingele, D., Wong, K., Feichtinger, R., Schnabel, H., Schnabel, R., and Hengartner, M. O. (2005). Two pathways converge at CED-10 to mediate actin rearrangement and corpse removal in *C. elegans*. *Nature* **434,** 93–99.

Li, W., and Baker, N. E. (2007). Engulfment is required for cell competition. *Cell* **129,** 1215–1225.

Liu, Q. A., and Hengartner, M. O. (1999). The molecular mechanism of programmed cell death in *C. elegans*. *Ann. NY Acad. Sci.* **887,** 92–104.

Maderna, P., and Godson, C. (2003). Phagocytosis of apoptotic cells and the resolution of inflammation. *Biochim. Biophys. Acta* **1639,** 141–151.

Majai, G., Petrovski, G., and Fesus, L. (2006). Inflammation and the apopto-phagocytic system. *Immunol. Lett.* **104,** 94–101.

Manaka, J., Kuraishi, T., Shiratsuchi, A., Nakai, Y., Higashida, H., Henson, P., and Nakanishi, Y. (2004). Draper-mediated and phosphatidylserine-independent phagocytosis of apoptotic cells by *Drosophila* hemocytes/macrophages. *J. Biol. Chem.* **279,** 48466–48476.

Nagano, M., Ui-Tei, K., Suzuki, H., Piao, Z., and Miyata, Y. (2000). CDK inhibitors suppress apoptosis induced by chemicals and by excessive expression of a cell death gene, reaper, in *Drosophila* cells. *Apoptosis* **5,** 543–550.

Nagata, S. (2006). Apoptosis and autoimmune diseases. *IUBMB Life* **58,** 358–362.

Nelson, R. E., Fessler, L. I., Takagi, Y., Blumberg, B., Keene, D. R., Olson, P. F., Parker, C. G., and Fessler, J. H. (1994). Peroxidasin: A novel enzyme-matrix protein of *Drosophila* development. *EMBO J.* **13,** 3438–3447.

Petersen, U. M., Kadalayil, L., Rehorn, K. P., Hoshizaki, D. K., Reuter, R., and Engstrom, Y. (1999). Serpent regulates *Drosophila* immunity genes in the larval fat body through an essential GATA motif. *EMBO J.* **18,** 4013–4022.

Raff, M. C., Barres, B. A., Burne, J. F., Coles, H. S., Ishizaki, Y., and Jacobson, M. D. (1994). Programmed cell death and the control of cell survival. *Philos. Trans. R Soc. Lond B Biol. Sci.* **345,** 265–268.

Ramet, M., Pearson, A., Manfruelli, P., Li, X., Koziel, H., Gobel, V., Chung, E., Krieger, M., and Ezekowitz, R. A. (2001). *Drosophila* scavenger receptor CI is a pattern recognition receptor for bacteria. *Immunity* **15**, 1027–1038.

Reddien, P. W., Cameron, S., and Horvitz, H. R. (2001). Phagocytosis promotes programmed cell death in *C. elegans*. *Nature* **412**, 198–202.

Savill, J., Dransfield, I., Gregory, C., and Haslett, C. (2002). A blast from the past: Clearance of apoptotic cells regulates immune responses. *Nat. Rev. Immunol.* **2**, 965–75.

Schneider, I. (1972). Cell lines derived from late embryonic stages of *Drosophila melanogaster*. *J. Embryol. Exp. Morphol.* **27**, 353–365.

Sears, H. C., Kennedy, C. J., and Garrity, P. A. (2003). Macrophage-mediated corpse engulfment is required for normal *Drosophila* CNS morphogenesis. *Development* **130**, 3557–3565.

Stramer, B., Wood, W., Galko, M. J., Redd, M. J., Jacinto, A., Parkhurst, S. M., and Martin, P. (2005). Live imaging of wound inflammation in *Drosophila* embryos reveals key roles for small GTPases during *in vivo* cell migration. *J. Cell Biol.* **168**, 567–573.

Studer, D., Graber, W., Al-Amoudi, A., and Eggli, P. (2001). A new approach for cryofixation by high-pressure freezing. *J. Microsc.* **203**, 285–294.

Tepass, U., Fessler, L. I., Aziz, A., and Hartenstein, V. (1994). Embryonic origin of hemocytes and their relationship to cell death in *Drosophila*. *Development* **120**, 1829–1837.

Tirouvanziam, R., Davidson, C. J., Lipsick, J. S., and Herzenberg, L. A. (2004). Fluorescence-activated cell sorting (FACS) of *Drosophila* hemocytes reveals important functional similarities to mammalian leukocytes. *Proc. Natl. Acad. Sci. USA* **101**, 2912–2917.

Wright, C. W., Means, J. C., Penabaz, T., and Clem, R. J. (2005). The baculovirus anti-apoptotic protein Op-IAP does not inhibit *Drosophila* caspases or apoptosis in *Drosophila* S2 cells and instead sensitizes S2 cells to virus-induced apoptosis. *Virology* **335**, 61–71.

Wyllie, A. H., Kerr, J. F., and Currie, A. R. (1980). Cell death: The significance of apoptosis. *Int. Rev. Cytol.* **68**, 251–306.

Zhou, Z., Mangahas, P. M., and Yu, X. (2004). The genetics of hiding the corpse: Engulfment and degradation of apoptotic cells in *C. elegans* and *D. melanogaster*. *Curr. Top. Dev. Biol.* **63**, 91–143.

CHAPTER FOUR

A Mouse Mammary Epithelial Cell Model to Identify Molecular Mechanisms Regulating Breast Cancer Progression

Vassiliki Karantza-Wadsworth[*,†] and Eileen White[†,‡,§,¶]

Contents

Abstract

Breast cancer, like any other human cancer, results from the accumulation of mutations that deregulate critical cellular processes, such as cell proliferation and death. Activation of oncogenes and inactivation of tumor suppressor genes are common events during cancer initiation and progression and often determine treatment responsiveness. Thus, recapitulating these events in mouse cancer models is critical for unraveling the molecular mechanisms involved in tumorigenesis and for interrogating their possible impact on response to

[*] Division of Medical Oncology, Department of Medicine, University of Medicine and Dentistry of New Jersey, Robert Wood Johnson Medical School; Piscataway, New Jersey
[†] The Cancer Institute of New Jersey, New Brunswick, New Jersey
[‡] Center for Advanced Biotechnology and Medicine, Rutgers University, Piscataway, New Jersey
[§] University of Medicine and Dentistry of New Jersey, Robert Wood Johnson Medical School, Piscataway, New Jersey
[¶] Department of Molecular Biology and Biochemistry, Rutgers University, Piscataway, New Jersey

Methods in Enzymology, Volume 446
ISSN 0076-6879, DOI: 10.1016/S0076-6879(08)01604-2

anticancer drugs. We have developed a novel mouse mammary epithelial cell model, which replicates the steps of epithelial tumor progression and takes advantage of the power of mouse genetics and the ability to assess three-dimensional morphogenesis in the presence of extracellular matrix to model human breast cancer.

1. INTRODUCTION

Mouse models have played an essential role in the study of human breast cancer. The first generation of such models involved studies on inbred mice and resulted in the discovery that MMTV-induced mammary tumorigenesis resulted from insertional inactivation of specific genes in the *wnt* and *fgf* families (Callahan and Smith, 2000). In addition, use of chemical carcinogens and hormones on inbred mice revealed the contribution of these factors to mammary tumorigenesis and provided model systems for prevention studies (Medina, 2006). The second generation of mouse breast cancer models involved constitutive overexpression of oncogenes, such as *c-myc* (Stewart *et al.*, 1984), polyoma *mt* (Guy *et al.*, 1992a), *neu* (Guy *et al.*, 1992b), mutant *p53* (Li *et al.*, 1997), and *cyclin D1* (Wang *et al.*, 1994) targeted to the mammary gland. These models have provided valuable information on the role of single genes in mammary tumorigenesis but are limited by the leakiness and multi-tissue expression of the MMTV promoter and the pregnancy-type hormonal dependence of the whey acidic protein (WAP) and β-lactoglobulin promoters. There is also difficulty in controlling the level of oncogene overexpression, which is often activated constitutively throughout development, not likely representing the physiologic situation in human breast cancer. The third, and most recent, generation of mouse breast cancer models involves mammary gland–specific gene deletion and activation by use of technology such as the Cre-lox system and tetracycline-responsive transactivation to induce tissue-specific gene mutations in adult somatic tissue and, as such, provides the opportunity to more authentically model human cancer (Furth, 1997; Liu *et al.*, 2007; Moody *et al.*, 2002; Wijnhoven *et al.*, 2005).

Mouse models have provided valuable information on the genetic events contributing to mammary tumorigenesis but are not easily amenable to the investigation of the biochemical and cell biological pathways involved in tumor formation. As an alternative to the use of epithelial cells cultured as monolayers on tissue culture plastic for such studies, three-dimensional (3D) culture systems have been developed and have proven very useful for interrogating the effects of oncogenes on glandular architecture and the role of epithelial-stroma interactions in mammary tumorigenesis (Bissell, 2007; Debnath and Brugge, 2005). A system extensively used in 3D-morphogenesis assays involves culture of the immortalized, nontransformed human mammary epithelial cell line MCF-10A on a reconstituted basement membrane (Debnath *et al.*, 2003). In 3D-culture, MCF-10A cells form polarized acinar structures that resemble glandular epithelium *in vivo* and have been successfully used to investigate mechanisms involved in tumor initiation and progression ((Debnath and Brugge, 2005) and to perform functional screens for proteins implicated in breast cancer (Witt *et al.*, 2006). However, MCF-10A cells are negative for estrogen receptor-alpha (ER-α), are cytogenetically abnormal with *myc* amplification and deletion of the locus containing *p16* and *p14ARF* (Debnath *et al.*, 2003), have been immortalized by events not well defined, and may not be adequately representative of the spectrum of premalignant human breast disease.

2. A NOVEL MOUSE MAMMARY EPITHELIAL CELL MODEL

2.1. Generation and characterization of the model

We have developed a novel mouse mammary epithelial cell model that complements and expands the applications of the mouse and human breast cancer models described earlier (Karantza-Wadsworth *et al.*, 2007). Our model involves isolation and immortalization of primary mouse mammary epithelial cells by inactivation of the retinoblastoma and p53 pathways (Fig. 4.1), as previously described for mouse kidney epithelial cells (Degenhardt and White, 2006; Degenhardt *et al.*, 2002a,b; Mathew *et al.*, 2008).

Immortalized mouse mammary epithelial cells (iMMECs) from wild-type mice exhibit typical cuboidal epithelial morphology in two-dimensional (2D) culture, form tight junctions, and express ER-α and luminal epithelial cell markers (Fig. 4.1) (Karantza-Wadsworth *et al.*, 2007). In 3D-morphogenesis assays, iMMECs form polarized acini that generate lumens through apoptosis (Fig.4.1), similarly to MCF-10A cells (Debnath *et al.*, 2002; 2003), and secrete β-casein into the acinar lumen upon lactogenic stimulation (Karantza-Wadsworth *et al.*, 2007). iMMECs are not tumorigenic, forming clonal adenocarcinomas with long latency, only after acquisition of secondary genetic or epigenetic changes (Fig. 4.1) (Karantza-Wadsworth *et al.*, 2007). Expression of different oncogenes in iMMECs affects 3D morphogenesis (Fig. 4.1) and *in vivo* tumorigenicity in an oncogene-dependent manner (Karantza-Wadsworth *et al.*, 2007). Therefore, wild-type iMMECs represent an early stage in mammary tumorigenesis and provide a facile platform for studying the role of oncogenes (expressed individually or in combination) in breast cancer progression and treatment. iMMECs can also be used for investigating the role of tumor suppressors by knocking down gene expression by RNAi technology. Alternatively, the molecular mechanisms by which oncogenes and tumor

Identification of mammary epithelial tumor promoting functions

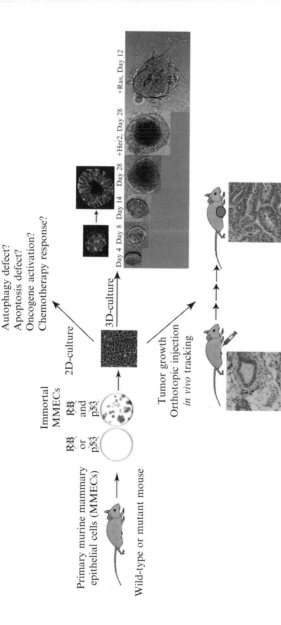

Figure 4.1 Identification of mammary epithelial tumor–promoting functions. A novel mouse mammary epithelial cell model for investigating the impact of oncogene activation and tumor suppressor inactivation on mammary tumorigenesis is summarized here. Primary mouse mammary epithelial cells (MMECs) are isolated from wild-type or mutant mice and immortalized (iMMECs) by concurrent inactivation of the Rb and p53 pathways. iMMECs can be grown and studied in standard two–dimensional (2D) *in vitro* culture. The functional properties of iMMECs can also be interrogated in a 3D morphogenesis assay, where iMMECs from wild-type mice initially form a solid acinus, as seen by β–catenin staining (green) to outline the cell periphery and Dapi counterstaining (blue) for DNA. With time (14 to 21 days), cell death in the acinar center creates a hollow lumen, closely mimicking duct formation in mammary epithelium *in vivo*. Oncogenic events perturb 3D morphogenesis, as illustrated in the phase–contrast images of the structures generated by Her2/neu-expressing iMMECs (larger, but hollow, acini) and by activated Ras-expressing iMMECs (nonacinar structures invading the basement membrane). iMMECs from wild-type mice are poorly, clonally tumorigenic on orthotopic transplantation into the mammary fat pad. Tumors formed by wild-type iMMECs are distinctly adenocarcinomas (right) compared with normal mouse mammary tissue (left), as indicated by histologic analysis of H&E–stained sections.

suppressor genes contribute to mammary tumorigenesis and impact response to anticancer drugs can be studied by generating iMMECs from mutant mice of interest and further genetically manipulating them *in vitro* as needed. Furthermore, iMMECs can be used in coculture with wild-type or mutant mouse embryonic fibroblasts (MEFs) to investigate the role of epithelial–stroma interactions in breast cancer.

2.2. Studying the role of autophagy in mammary tumorigenesis

We recently reported the application of our model in the study of the role of autophagy in mammary tumorigenesis (Karantza-Wadsworth *et al.*, 2007). iMMECs were generated from *beclin1*$^{+/+}$ and *beclin1*$^{+/-}$ mice and studied in 2D culture, 3D morphogenesis, and *in vivo* tumorigenicity assays. We found that allelic loss of *beclin1* compromises the autophagy potential of iMMECs and results in increased susceptibility of iMMECs to metabolic stress and accelerated lumen formation in mammary acini. Defective autophagy also activates the DNA damage response in iMMECs and mammary acini *in vitro* and in mammary tumors *in vivo*. Furthermore, we showed that monoallelic deletion of *beclin1* promotes gene amplification *in vitro* and accelerates mammary tumorigenesis, supporting the hypothesis that autophagy limits metabolic stress to protect the genome, whereas autophagy defects increase DNA damage and genomic instability that may ultimately facilitate breast cancer progression (Karantza-Wadsworth and White, 2007; Karantza-Wadsworth *et al.*, 2007). Similar results were obtained with *beclin1*$^{+/-}$ and *atg5*$^{-/-}$ kidney epithelial cells (Degenhardt *et al.*, 2006; Mathew *et al.*, 2007b), indicating that defective autophagy can promote genomic instability and epithelial tumor progression independent of cell type or means of autophagy inactivation. Thus, use of our novel mouse mammary epithelial cell model has provided valuable insight into the role of autophagy as a tumor suppressive mechanism in mammary tumorigenesis and, together with parallel studies on kidney epithelial cells (Degenhardt *et al.*, 2006; Mathew *et al.*, 2007b), has set the foundation for understanding the complex interplay between autophagy, metabolic stress management, and cancer progression and for identifying ways to manipulate autophagy for maximum therapeutic benefit (Mathew *et al.*, 2007a).

3. PROTOCOLS

This chapter provides detailed protocols for iMMEC generation and culture, 3D-morphogenesis assays, and *in vivo* tumorigenicity studies and will hopefully serve as a valuable addition to the technical repertoire used in breast cancer research.

3.1. Mouse mammary epithelial cell isolation and immortalization

3.1.1. Mouse mammary gland harvesting

Six- to 8-week-old female mice (wild-type or mutant, virgin or pregnant) are euthanized as required by the local Institutional Animal Care and Use Committee. Each mouse is then placed on its back on a corkboard, pinned in place through the feet, and swabbed with 70% ethanol. A ventral midline incision through the skin is made to expose the five pairs of subcutaneous mammary glands (Rasmussen et al., 2000). Additional incisions from the midline down each rear leg and below each ear facilitate the removal of the No. 4 and No. 1 fat pads, respectively. The skin flaps, with mammary glands attached, are carefully separated from the peritoneum with a blunt-edged instrument. The free edge of each skin flap is pinned to the corkboard, thereby exposing the adherent mammary glands, which can then be excised from the skin flap. Removal of the thoracic glands Nos. 2 and 3 should be performed with additional care to leave lymph nodes and muscle behind. The No. 1 fat pad may not be removed, if one is unable to distinguish it from the salivary gland, which is darker in color than the mammary tissue. Expected mammary tissue yield is 0.3 to 0.5 g for a virgin mouse and 1.0 to 1.2 g for a pregnant mouse. The mammary tissue is collected in phosphate-buffered saline (PBS) with antibiotics (pen/strep) on ice.

3.1.2. Mammary gland digestion and epithelial cell recovery

The mammary tissue pieces that float are transferred to a 10-cm sterile petri dish, whereas any tissue piece that precipitates is discarded. Mammary tissue is mechanically minced with two scalpels inside the petri dish under sterile conditions until uniform and oily in appearance (0.5- to 1-mm pieces), and then transferred to a small flask containing PBS (10 ml/g tissue) and antibiotics. Appropriate volume of 1% collagenase A stock Table 4.1 is added to achieve a final collagenase concentration of 0.05 to 0.1%, and tissue is stirred at 37° for 30 to 90 min, depending on collagenase activity. Stirring should be fast enough to cause thorough mixing without splashing of medium and tissue onto flask walls. The completeness of digestion can be checked starting at 30 min by aseptically removing small aliquots for low-power microscopic examination ($40\times$ total magnification) with an inverted stage, phase microscope. The desired endpoint is an epithelial preparation with no visible tissue pieces remaining and more than 80% of epithelial organoids free of adhering stromal tissue. If cells have lysed and DNA has been released producing a "stringy" clumping of organoids, DNAse stock solution Table 4.1 can be added until all clumping is cleared. On completion of digestion, 20 ml of F12 plus 5% fetal bovine serum (FBS) are added to the tissue digest and clumps are allowed to settle for 2 min.

Table 4.1 Reagents for iMMEC culture

Reagent	Source	Stock	Storage
Collagenase A	Roche	1% in H_2O	-20°
DNAse	Sigma	0.4 mg/ml in PBS	-20°
EGF	Sigma	10 μg/ml in PBS	4°
Fetuin	Sigma	2 mg/ml in 20% FBS	4°
F12	Invitrogen		4°
FBS	Invitrogen		-20°
Hydrocortisone	Sigma	1 mg/ml in EtOH	-20°
Insulin	Sigma	10 mg/ml in 25 mM HEPES, pH 8.2	4°
Matrigel, growth factor reduced	BD		-20°
Pen/Strep	Invitrogen	10,000 units	-20°
Prolactin	Sigma		4°
Prolong Antifade	Molecular Probes		-20°
Trypsin, regular strength	Invitrogen	0.05%	-20°
Trypsin, lower strength	Clonetics	0.025%	-20°

Epithelial organoids (mouse mammary epithelial cells, MMECs) are collected from the supernatant by centrifugation at 1500 rpm for 5 min and are then washed three times with F12 plus 5% FBS. Percoll gradient purification of epithelial cells is not routinely performed, because a highly enriched epithelial cell population is usually obtained.

3.1.3. Primary MMEC culture

For primary culture, MMECs from three mice can be pooled during washing, pelleted, resuspended in $2\times$ hormone plating medium (F12, 10 μg/ml insulin, 2 μg/ml hydrocortisone, 10 ng/ml EGF) Table 4.2 and plated in two 6-cm or one 10-cm plate (4 or 8 ml of resuspended cells per plate, respectively) that have been precoated with equal volume of 100 μl/cm^2 fetuin-serum solution (2 mg/ml fetuin in 20% FBS) Table 4.1 for 4 to 5 h at 37° (Rijnkels and Rosen, 2001). Cells are not routinely counted before plating, because single-cell suspensions required for accurate cell counting negatively impact viability of MMECs in primary culture. MMECs are allowed to settle for 36 to 48 h in the resultant $1\times$ plating medium with 10% FBS before switching to reduced-FBS growth medium (F12, 5% FBS, 5 μg/ml insulin, 1 μg/ml hydrocortisone, 5 ng/ml EGF) Table 4.2 to minimize fibroblast growth and are kept in a humidified incubator with 8.5% CO_2 at 37°. Medium changes are performed every 4 days.

Table 4.2 Medium recipes for iMMECs

Component	2× hormone plating medium	Reduced-FBS growth medium	Regular growth medium	Cloning medium	Differentiation medium
F12	250 ml	500 ml	500 ml	50 ml	500 ml
FBS	No FBS! (already in fetuin-serum solution)	25 ml (5% final)	50 ml (10% final)	10 ml (20% final)	
EGF (10 μg/ml stock in PBS)	250 μl (10 ng/ml final)	250 μl (5 ng/ml final)	250 μl (5 ng/ml final)	25 μl (5 ng/ml final)	
Hydrocortisone (1 mg/ml stock in EtOH)	500 μl (2 μg/ml final)	500 μl (1 μg/ml final)	500 μl (1 μg/ml final)	50 μl (1 μg/ml final)	500 μl (1 μg/ml final)
Insulin (10 mg/ml stock)	250 μl (10 μg/ml final)	250 μl (5 μg/ml final)	250 μl (5 μg/ml final)	25 μl (5 μg/ml final)	250 μl (5 μg/ml final)
Prolactin					1.5 mg (3 μg/ml final)
Pen/Strep (10,000 units)	5 ml	5 ml	5 ml	0.5	5 ml

Primary MMECs do not tolerate trypsinization well but remain viable for approximately 10 to 14 days in culture without cell passage.

3.1.4. Primary MMEC transfection and plating

MMECs from the mammary glands of a single mouse are resuspended in 250 μl F12 plus 5% FBS and electroporated in the presence of 10 μg ScaI-linearized cytomegalovirus (CMV)-driven adenovirus type 5 E1A plasmid (pCMVE1A [White et al., 1991]), 10 μg ScaI-linearized dominant negative mouse p53 plasmid (p53DD [Shaulian et al., 1992]), and 100 μg salmon sperm carrier DNA, as previously described for baby mouse kidney (BMK) epithelial cells (Degenhardt and White, 2006; Degenhardt et al., 2002b). During electroporation, the cells are pulsed at 220 V and 950 μF. MMECs from one electroporation are resuspended in 16 ml of 2× hormone plating medium Table 4.2 and plated in four 6-cm or two 10-cm fetuin-coated plates Table 4.1, as described previously. Again, MMECs are allowed to settle for 36 to 48 h in 1× plating medium with 10% FBS before switching to reduced-FBS growth medium Table 4.2 and are kept in a humidified incubator with 8.5% CO_2 at 37°. Medium changes are performed every 4 days without cell passage.

3.1.5. Cloning and expansion of immortalized MMEC colonies

Over a period of 4 to 6 weeks, primary (nontransfected) or singly transfected MMECs and any contaminating fibroblasts undergo cell death, whereas colonies of doubly transfected, immortalized MMECs (iMMECs) arise and start growing. iMMEC colonies first appear at 2 to 3 weeks after initial plating (typically, 1 to 5 colonies per 10-cm plate) and are composed of tightly packed, cuboidal cells with typical epithelial morphology (Fig. 4.1) (Karantza-Wadsworth et al., 2007). When colonies reach 0.6 to 0.8 cm in diameter, the medium is removed, and the plate is washed once with PBS. Individual colonies are surrounded by glass cloning cylinders (6 × 8 mm to 10 × 10 mm, depending on colony size; Bellco Biotechnology) secured in place by autoclaved vacuum grease (VWR). iMMECs are recovered by limited trypsin digestion with lower strength trypsin (0.025%) Table 4.1 for 5 to 10 sec and are transferred to 24-well plates in coloning medium F12, 20% FBS, 5 μg/ml insulin, 1 μg/ml hydrocortisone, 5 ng/ml EGF, Table 4.2 (0.5 ml medium per well, no need for fetuin-precoating). Colonies isolated from different plates are truly independent. However, as many colonies as possible are recovered from every plate, because colony survival after transfer to 24-well plates is approximately 50 to 70%. iMMECs are subsequently expanded in regular growth medium (F12, 10% FBS, 5 μg/ml insulin, 1 μg/ml hydrocortisone, 5 ng/ml EGF) Table 4.2 and can be frozen in 92% FBS–8% DMSO from 10-cm plates at approximately 70% confluency for long-term storage in vapor phase nitrogen. During regular tissue culture, iMMECs require refeeding every 3 days and passage every 3 to 5 days, usually at 1:5 split.

3.1.6. iMMEC characterization by WB and IF

Once multiple iMMEC colonies have been generated from a particular mouse strain, iMMECs are examined by Western blotting (WB) for expression of E1A and p53DD (immortalizing proteins), ER-α and epithelial cell markers, such as cytokeratin (CK) 5/6, CK8, CK14, E-cadherin, β-catenin, Ep-CAM, vimentin Table 4.3. For immunofluorescence (IF), iMMECs can be grown on glass coverslips to 70% confluency and then fixed with 1:1 methanol/acetone at −20° for 10 min. Coverslips are washed with PBS three times for 5 min each time (5 min × 3), and then incubated with primary antibody (at 1:100-1:400 dilution, depending on antibody) in 5% bovine serum albumin (BSA, Sigma), PBS, 0.1% Tween-20 (PBST) for 1 h at 37°. After PBST washes (5 min × 3), coverslips are incubated with fluorescein- or rhodamine-conjugated secondary antibody (at 1:100 dilution) for 40 min at room temperature (RT). Coverslips are again washed with PBST (5 min × 3), incubated with 0.5 ng of DAPI (4′, 6′-diamidino-2-phenylindole, Sigma) for 15 min at RT, and finally washed with PBS for 5 min, before being mounted with the antifade agent Prolong (Molecular Probes).

3.2. Lactogenic stimulation

3.2.1. In 2D culture

Cells are grown on plastic culture dishes until confluent and are subsequently induced with differentiation medium (F12, 5 μg/ml insulin, 1 μg/ml hydrocortisone, 3 μg/ml prolactin) Table 4.2 ±2% Matrigel (Streuli et al., 1995) for 6 days, with medium changes every 2 days.

Table 4.3 Useful antibodies and fluorescent reagents for iMMEC analysis

Antibody or stain	Purpose	Source
Activated (cleaved) caspase-3	Apoptosis marker	Cell signaling
β-casein	Milk protein	Santa Cruz
β-catenin	Cell–cell junctions	Zymed
Cytokeratin 5/6	Myoepithelial cell marker	Covance
Cytokeratin 8	Luminal cell marker	Abcam
Cytokeratin 14	Myoepithelial cell marker	Covance
DAPI	Nuclear counterstain	Sigma
E1A	Immortalization marker	Oncogene
E-cadherin	Cell–cell junctions	RDI
Ep-CAM	Epithelial cell marker	Santa Cruz
ER-α	Hormone receptor	Santa Cruz
Occludin	Cell–cell junctions	Zymed
p53 (for p53DD)	Immortalization marker	Oncogene
Smooth muscle actin (SMA)	Myoepithelial cell marker	Sigma
Vimentin	Myoepithelial cell marker	Santa Cruz
ZO-1	Cell–cell junctions	Zymed

3.2.2. In 3D culture

Mammary acini are grown on Matrigel for 12 days as described section 3.4.1 and then induced with differentiation medium containing 2% Matrigel for 2 additional days. Mammary acini are subsequently fixed and processed for IF, as described section 3.4.2.

3.3. Generation of stable cell lines

Proteins of interest can be easily expressed or down regulated in iMMECs, so that the impact of oncogene activation and tumor suppressor inactivation on mammary tumorigenesis and treatment responsiveness can be readily investigated. So far, iMMECs have been engineered to express human Bcl-2, H-Ras^{V12}, *myr*-Akt, wild-type human HER2/neu, and the vector control Table 4.5 by electroporation (as described in section 3.1.4) with pcDNA3.1hBcl-2, pcDNA3.1H-RasV12, pcDNA3.1Myr-Akt, pcDNA3.1wtHER2/neu, and pcDNA3.1 vector (Invitrogen), respectively, followed by selection with geneticin (Karantza–Wadsworth *et al.*, 2007) Table 4.5. For studying the autophagy potential of iMMECs generated from *beclin1*$^{+/+}$ and *beclin1*$^{+/-}$ mice, stable expression of EGFP-LC3 is performed by electroporation with pcDNA3.EGFP-LC3, followed by selection with geneticin for apoptosis-competent iMMECs, and by electroporation with pcDNA3.EGFP-LC3 and pcDNA3.1zeo (Invitrogen), followed by double selection with geneticin and zeocin for Bcl-2 expressing iMMECs (Karantza–Wadsworth *et al.*, 2007) Table 4.5. Geneticin and zeocin are used at 300 and 100 μg/ml respectively Table 4.4 for 10 to 14 days. Individual drug-resistant colonies are isolated and expanded to stable cell lines, as described in section 3.1.5. The drug used for selection is routinely kept in the regular growth medium thereafter.

3.4. Three-dimensional (3D) morphogenesis

Three-dimensional culture of iMMECs on a reconstituted basement membrane is performed according to a modified version of the protocol previously described for the immortalized, nontransformed human mammary epithelial cell line MCF-10A (Debnath *et al.*, 2003).

Table 4.4 Drug selection for stable transfection of iMMECs

Drug	Source	Final concentration
Geneticin	Invitrogen	300 μg/ml
Zeocin	Invitrogen	100 μg/ml

Table 4.5   Mouse (C57BL/6, Wild-type or mutant) mammary epithelial cell lines, E1A and p53DD-derived (Karantza-Wadsworth *et al.*, 2007)

Genotype	Transgene	iMMEC cell line name
Wild-type		WTA (or 21), WTB, WTC, WTD, WT3, WT5
Wild-type	pcDNA3.1 vector	WTA.V, WT3.V
Wild-type	Bcl-2 (human)	WTA.B1, WTA.B4, WT3.B1, WT3.B2, WT3.B3, WT3.B8
Wild-type	EGFP-LC3	WTA-LC3, WT3-LC3
Wild-type	Bcl-2, EGFP-LC3	WTA.B4-LC3, WT3.B3-LC3
Wild-type	HER2/neu (human, wild-type)	WTA.H2, WTA.H3
Wild-type	*myr*-Akt	WTA.A5, WTA.A7
Wild-type	H-Ras^{V12}	WTA.R3, WTA.R5
beclin1$^{+/-}$		BLN2 (or 2.1), BLN4
beclin1$^{+/-}$	pcDNA3.1 vector	BLN2.V
beclin1$^{+/-}$	Bcl-2	BLN2.B2, BLN2.B4, BLN2.B5, BLN2.B8
beclin1$^{+/-}$	EGFP-LC3	BLN2-LC3
beclin1$^{+/-}$	Bcl-2, EGFP-LC3	BLN2.B4-LC3.5

3.4.1. iMMEC trypsinization and plating on matrigel

Eight-well RS glass slides (BD Falcon) are coated with 40 μl per well of growth factor-reduced Matrigel thawed overnight at 4°. Trypsin (1 ml) is added to a 10-cm confluent plate of iMMECs, swirled around, and immediately aspirated to leave a thin film behind, which prevents cell clumping and ensures a single-cell suspension. After 15 to 20 min in a 8.5% CO_2 humidified incubator at 37°, iMMECs get dislodged and are collected in 2 ml growth medium with repeated pipetting to break up cell clumps, and are then counted. iMMECs are pelleted, resuspended in growth medium at a concentration of 25,000 cells/ml, and mixed 1:1 with growth medium containing 4% Matrigel; 400 μl of the resultant solution (12,500 cells/ml in growth medium with 2% Matrigel) is plated in each well. Growth medium containing 2% Matrigel is replaced every 4 days.

3.4.2. Fixation and immunofluorescence

Mammary acini are fixed in 4% formalin for 25 min at room temperature. Fixed structures are washed with PBS-glycine (PBS, 100 mM glycine) three times for 15 min each time. The structures are then blocked with IF buffer (PBS, 0.1% BSA, 0.2% Triton X-100, 0.05% Tween-20) plus 10% goat serum for 30 min at 37°, followed by secondary block [IF buffer containing

10% goat serum and 20 μg of goat anti-mouse F(ab')$_2$/ml] for 10 min at 37°, and then incubation with primary antibody (in secondary block solution, usually at 1:100 dilution) for 90 min at 37 degrees. Structures are then washed three times in IF buffer for 15 min each. Anti-mouse or anti-rabbit secondary antibodies coupled with fluorescein or rhodamine are diluted in IF buffer containing 10% goat serum (usually at 1:100 dilution), followed by incubation for 40 min at RT. After three washes with IF buffer for 15 min each, structures are incubated with 0.5 ng of DAPI for 15 min. Structures are washed with PBS for 5 min before being mounted with the antifade agent Prolong (Molecular Probes). Confocal laser scanning microscopy can be done with a Zeiss LSM510-META confocal microscope system. The percentage of acini with lumen formation is the mean of two independent experiments (for each experiment, 100 acini are scored for each cell line at each time point).

3.4.3. Histology
Mammary acini are grown on Matrigel for 12 days as described previously, fixed in 10% neutral buffered formalin, scraped from the glass slide with a razor blade, pelleted, embedded in paraffin, and processed for H&E staining.

3.4.4. Electron microscopy
Mammary acini grown for 12 days on Matrigel as described previously are fixed with electron microscopy fixative (1.2% paraformaldehyde/2.5% glutaraldehyde/0.03% picric acid) in 100 mM cacodylate buffer for 1 h at RT and then overnight at 4°. Fixed acini are scraped from the coverslip with a razor blade, pelleted, and processed for EM with standard procedure.

3.5. Orthotopic tumor growth

Cells are harvested by trypsinization, washed, and resuspended in PBS (10^7 cells/ml). Orthotopic mammary gland implantation of iMMECs is performed with IACUC-approved protocol; 5- to 8-week-old NCR nude female mice are anesthetized with ketamine (100 mg/kg intraperitoneally, IP) and xylazine (10 mg/kg IP). A small incision is made to reveal the right second or third mammary gland, and 10^6 cells are injected into the mammary fat pad. The incision is closed with surgical clips that are removed 10 days later. Tumor outgrowth is monitored by weekly measurements of tumor length (L) and width (W). Tumor volume is calculated as $\pi LW^2/6$. At the time of animal euthanasia and mammary tumor dissection, the left second or third mammary gland is collected as a normal control. Clonal mammary tumors generated by wild-type iMMECs after acquisition of secondary genetic or epigenetic changes appear at 3 to 4 months with 60% penetrance (in 3 of 5 mice), whereas highly tumorigenic iMMECs expressing activated H–Ras or wild–type Her2/neu form mammary tumors in 3 to 5 weeks with 100% penetrance (in 5 of 5 mice).

4. Concluding Remarks

The mouse mammary epithelial cell model presented takes advantage of the strength of mouse genetics in combination with 3D morphogenesis and orthotopic tumor growth assays for the study of oncogene and tumor suppressor functions as they pertain to mammary tumorigenesis, and by extension to human breast cancer. Generation of iMMECs from $beclin1^{+/+}$ and $beclin1^{+/-}$ mice and assessment of their properties in 2D culture, 3D morphogenesis, and *in vivo* tumorigenicity (Karantza-Wadsworth *et al.*, 2007) has provided valuable insight into the role of autophagy in mammary tumorigenesis and has set the stage for future investigations focused on the intriguing relationship between autophagy, metabolism, stress response, and cancer progression and treatment (Karantza-Wadsworth and White, 2007; Mathew *et al.*, 2007a). The *in vitro* and *in vivo* protocols described here will hopefully be widely applicable in breast cancer research involving mouse modeling and will provide new tools for successfully investigating the molecular mechanisms implicated in breast cancer progression and treatment responsiveness.

REFERENCES

Bissell, M. J. (2007). Modelling molecular mechanisms of breast cancer and invasion: Lessons from the normal gland. *Biochem. Soc. Trans.* **35**, 18–22.

Callahan, R., and Smith, G. H. (2000). MMTV-induced mammary tumorigenesis: Gene discovery, progression to malignancy and cellular pathways. *Oncogene.* **19**, 992–1001.

Debnath, J., and Brugge, J. S. (2005). Modelling glandular epithelial cancers in three-dimensional cultures. *Nat. Rev. Cancer* **5**, 675–688.

Debnath, J., Mills, K. R., Collins, N. L., Reginato, M. J., Muthuswamy, S. K., and Brugge, J. S. (2002). The role of apoptosis in creating and maintaining luminal space within normal and oncogene-expressing mammary acini. *Cell* **111**, 29–40.

Debnath, J., Muthuswamy, S. K., and Brugge, J. S. (2003). Morphogenesis and oncogenesis of MCF-10A mammary epithelial acini grown in three-dimensional basement membrane cultures. *Methods* **30**, 256–268.

Degenhardt, K., Chen, G., Lindsten, T., and White, E. (2002a). BAX and BAK mediate p53-independent suppression of tumorigenesis. *Cancer Cell* **2**, 193–203.

Degenhardt, K., and White, E. (2006). Autophagy promotes tumor cell survival and restricts necrosis, inflammation, and tumorigenesis. *Cancer Cell* **10**, 51–64.

Degenhardt, K., Sundararajan, R., Lindsten, T., Thompson, C., and White, E. (2002b). Bax and Bak independently promote cytochrome C release from mitochondria. *J. Biol. Chem.* **277**, 14127–14134.

Degenhardt, K., and White, E. (2006). A mouse model system to genetically dissect the molecular mechanisms regulating tumorigenesis. *Clin. Cancer Res.* **12**, 5298–5304.

Furth, P. A. (1997). Conditional control of gene expression in the mammary gland. *J. Mammary Gland Biol. Neoplasia* **2,** 373–383.

Guy, C. T., Cardiff, R. D., and Muller, W. J. (1992a). Induction of mammary tumors by expression of polyomavirus middle T oncogene: A transgenic mouse model for metastatic disease. *Mol. Cell. Biol.* **12,** 954–961.

Guy, C. T., Webster, M. A., Schaller, M., Parsons, T. J., Cardiff, R. D., and Muller, W. J. (1992b). Expression of the neu protooncogene in the mammary epithelium of transgenic mice induces metastatic disease. *Proc. Natl. Acad. Sci. USA* **89,** 10578–10582.

Karantza-Wadsworth, V., and White, E. (2007). Autophagy mitigates metabolic stress and genome damage in mammary tumorigenesis. *Genes Dev.* **21,** 1621–1635.

Karantza-Wadsworth, V., and White, E. (2007). Role of autophagy in breast cancer. *Autophagy* **3,** 610–613.

Li, B., Rosen, J. M., McMenamin-Balano, J., Muller, W. J., and Perkins, A. S. (1997). neu/ERBB2 cooperates with p53-172H during mammary tumorigenesis in transgenic mice. *Mol. Cell. Biol.* **17,** 3155–3163.

Liu, X., Holstege, H., van der Gulden, H., Treur-Mulder, M., Zevenhoven, J., Velds, A., Kerkhoven, R. M., van Vliet, M. H., Wessels, L. F., Peterse, J. L., Berns, A., and Jonkers, J. (2007). Somatic loss of BRCA1 and p53 in mice induces mammary tumors with features of human BRCA1-mutated basal-like breast cancer. *Proc. Natl. Acad. Sci. USA* **104,** 12111–12116.

Mathew, R., Karantza-Wadsworth, V., and White, E. (2007a). Role of autophagy in cancer. *Nat. Rev. Cancer* **7,** 961–967.

Mathew, R., Kongara, S., Beaudoin, B., Karp, C. M., Bray, K., Degenhardt, K., Chen, G., Jin, S., and White, E. (2007b). Autophagy suppresses tumor progression by limiting chromosomal instability. *Genes Dev.* **21,** 1367–1381.

Mathew, R., Karp, C. M., and White, E. (2008). A mouse epithelial cell model to study the role of apoptosis and autophagy in cancer. *Methods Enzymol.*

Medina, D. (2006). Chemical carcinogenesis of rat and mouse mammary glands. *Breast Dis.* **28,** 63–68.

Moody, S. E., Sarkisian, C. J., Hahn, K. T., Gunther, E. J., Pickup, S., Dugan, K. D., Innocent, N., Cardiff, R. D., Schnall, M. D., and Chodosh, L. A. (2002). Conditional activation of Neu in the mammary epithelium of transgenic mice results in reversible pulmonary metastasis. *Cancer Cell* **2,** 451–461.

Rasmussen, S. B., Young, L. J. T., and Smith, G. H. (2000). Preparing mammary gland whole mounts from mice. *In* "Methods in Mammary Gland Biology and Breast Cancer Research." (M. M. Ip. and B. B. Asch, Eds.), pp. 75–85. Kluwer Academic/Plenum Publishers, New York.

Rijnkels, M., and Rosen, J. M. (2001). Adenovirus-Cre–mediated recombination in mammary epithelial early progenitor cells. *J. Cell. Sci.* **114,** 3147–3153.

Shaulian, E., Zauberman, A., Ginsberg, D., and Oren, M. (1992). Identification of a minimal transforming domain of p53: Negative dominance through abrogation of sequence-specific DNA binding. *Mol. Cell. Biol.* **12,** 5581–5592.

Stewart, T. A., Pattengale, P. K., and Leder, P. (1984). Spontaneous mammary adenocarcinomas in transgenic mice that carry and express MTV/myc fusion genes *Cell* **38,** 627–637.

Streuli, C. H., Schmidhauser, C., Bailey, N., Yurchenco, P., Skubitz, A. P., Roskelley, C., and Bissell, M. J. (1995). Laminin mediates tissue-specific gene expression in mammary epithelia. *J. Cell. Biol.* **129,** 591–603.

Wang, T. C., Cardiff, R. D., Zukerberg, L., Lees, E., Arnold, A., and Schmidt, E. V. (1994). Mammary hyperplasia and carcinoma in MMTV-cyclin D1 transgenic mice. *Nature* **369,** 669–671.

White, E., Cipriani, R., Sabbatini, P., and Denton, A. (1991). Adenovirus E1B 19-kilodalton protein overcomes the cytotoxicity of E1A proteins. *J. Virol.* **65,** 2968–2978.

Wijnhoven, S. W., Zwart, E., Speksnijder, E. N., Beems, R. B., Olive, K. P., Tuveson, D. A., Jonkers, J., Schaap, M. M., van den Berg, J., Jacks, T., van Steeg, H., and de Vries, A. (2005). Mice expressing a mammary gland-specific R270H mutation in the p53 tumor suppressor gene mimic human breast cancer development. *Cancer Res.* **65,** 8166–8173.

Witt, A. E., Hines, L. M., Collins, N. L., Hu, Y., Gunawardane, R. N., Moreira, D., Raphael, J., Jepson, D., Koundinya, M., Rolfs, A., Taron, B., Isakoff, S. J., Brugge, J. S., and LaBaer, J. (2006). Functional proteomics approach to investigate the biological activities of cDNAs implicated in breast cancer. *J. Proteome Res.* **5,** 599–610.

CHAPTER FIVE

IMMORTALIZED MOUSE EPITHELIAL CELL MODELS TO STUDY THE ROLE OF APOPTOSIS IN CANCER

Robin Mathew,*,† Kurt Degenhardt,†,¶ Liti Haramaty,‡ Cristina M. Karp,†,§ *and* Eileen White*,†,§

Contents

* University of Medicine and Dentistry of New Jersey, Robert Wood Johnson Medical School, Piscataway, New Jersey
† Center for Advanced Biotechnology and Medicine, Rutgers University, Piscataway, New Jersey
‡ Institute of Marine and Coastal Sciences, Rutgers University, New Brunswick, New Jersey
§ Department of Molecular Biology and Biochemistry, Rutgers University, Piscataway, New Jersey
¶ Department of Pharmaceutical Sciences, St. John's University College of Pharmacy, Queens, New York

Methods in Enzymology, Volume 446

ISSN 0076-6879, DOI: 10.1016/S0076-6879(08)01605-4

Abstract

Human cancer cell lines are widely used to model cancer but also have serious limitations. As an alternate approach, we have developed immortalized mouse epithelial cell model systems that are applicable to different tissue types and involve generation of immortalized cell lines that are genetically defined. By applying these model systems to mutant mice, we have extended the powerful approach of mouse genetics to *in vitro* analysis. By use of this model we have generated immortal epithelial cells that are either competent or deficient for apoptosis by different gain- and loss-of-function mutations that have revealed important mechanisms of tumor progression and treatment resistance. Furthermore, we have derived immortalized, isogenic mouse kidney, mammary, prostate, and ovarian epithelial cell lines to address the issues of tissue specificity. One of the major advantages of these immortalized mouse epithelial cell lines is the ability to perform biochemical analysis, screening, and further genetic manipulations. Moreover, the ability to generate tumor allografts in mice allows the integration of *in vitro* and *in vivo* approaches to delineate the mechanistic aspects of tumorigenesis. These model systems can be used effectively to determine the molecular requirements of epithelial tumorigenesis and tumor-promoting functions. This approach provides an efficient way to study the role of apoptosis in cancer and also enables the interrogation and identification of potential chemotherapeutic targets involving this pathway. Applying this technology to other mouse models can provide insight into additional aspects of oncogenesis.

1. INTRODUCTION

Tumorigenesis is a multistep process characterized by step-wise and sequential mutational events that cooperate in a compounding manner resulting in resistance to apoptosis, uncontrolled proliferation, and invasiveness (Hanahan and Weinberg, 2000). Ideally, the best system to study the evolution of a tumor is the human body itself, where the physiology and the tumor–host interaction represent the most relevant situation. Understandably, this approach has limitations and raises ethical issues, as well as practical impediments. Moreover, there can be person-to-person variations that influence tumor physiology, therapeutic response, immunity, and other factors that may complicate assessment. Therefore, understanding the molecular mechanisms underlying tumorigenesis necessitates the use of appropriate model systems that can recapitulate this transition *in vitro* as well as *in vivo*. Animal models, where individual steps of tumorigenesis can be assessed in a more simplified setting, represent an excellent option. Alternatively, a large number of cell lines have been generated from surgically resected human tumor samples, which have been widely used for *in vitro* experiments, as well as in generating human tumor xenografts in immune deficient mice (Hahn and Weinberg, 2002; Masters, 2000).

Despite immense convenience because of easy availability, human cancer cell lines and xenograft models suffer from limitations. First, excised human tumors are difficult to establish in tissue culture because of altered demands for growth *in vitro*. Although cancer is characterized by unlimited growth potential, establishment of human cancer cell lines in culture is far from simple. Even among cancers that are relatively easy to grow *in vitro*, it is often the metastatic cancers that are most amenable to establishment as continuous cultures (Hsu, 1999). Tumors evolve to grow in conjunction with the neighboring matrix-associated stroma and vasculature, which may partially explain their poor growth *in vitro*. Moreover, the capacity of normal cells to grow beyond a small number of divisions is often limited by insufficient culture conditions and replicative senescence imposed by telomerase attrition, and cell lines that pass this stage are likely to have accompanying compensatory mutations (Hahn and Weinberg, 2002). All of these and other factors limit the availability of biologically pure, defined, and renewable sources of both normal and cancer cells for biochemical assays and functional analysis *in vitro*.

Although established human cancer cell lines have been greatly instrumental in improving our understanding of the tumorigenic process, they suffer from the lack of genetic definition, and human systems that represent progression from normal to metastatic stages are limited. The complex mutational history of established human cancer cell lines makes systematic analysis of the stepwise transition from normalcy to neoplasia and assessment of treatment response difficult. Furthermore, human tumor xenografts in the altered host environment of immune compromised mice suffer from the lack of physiologic relevance from the disease perspective.

It is advantageous for a cancer model to use isogenic cell lines with well-defined genetic identity and the flexibility of further manipulation. Recent developments in mutant mouse technology have resulted in the generation of numerous transgenic and knockout mice with loss- or gain-of-function mutations in genes involved in cancer. Despite several morphologic and genetic differences, most of the important pathways implicated in cancer are conserved between mice and humans. Mouse models have long played a key role in explaining the molecular characteristics associated with the development of tumors. The availability of several permutations and combinations of knockout and transgenic mice renders it possible to derive multiple, immortalized, and independent cell lines from different mouse tissues of defined genetic background (Degenhardt and White, 2006). This extends the usefulness of these mouse models where compounding different combinations of mutations by conventional genetic approaches has been increasingly difficult.

Mouse cell models offer the advantage of being manipulable, reproducible, and self-replenishing, thus offering a powerful strategy to not only tease out the molecular events that lead to cancer but also to identify novel therapeutic strategies to fight the disease (Degenhardt and White, 2006; Mathew et al., 2007a; Tan et al., 2005). Mouse embryonic fibroblasts (MEFs) are the most commonly used mouse-derived cell lines. However, fibroblasts are not the most suitable cell type to model the physiology of human cancers, because most human tumors are epithelial in origin. Therefore, among various mouse cell types to model cancer, immortalized mouse epithelial cells with a defined genetic background are a superior alternative to MEFs.

To bridge the gap between mouse models and human cancers, we have developed the technical means to isolate primary mouse epithelial cells from multiple tissue types, to introduce genetic mutations to immortalize them, and to establish them as stable cell lines. By use of these techniques, we have generated immortalized baby mouse kidney (iBMK) epithelial cells, mouse mammary epithelial cells (iMMECs), mouse prostate epithelial cells (iMPECs), and mouse ovarian surface epithelial cells (iMOSECs) that retain their epithelial characteristics (Tables 5.1 to 5.8) (Bray et al., 2008; Degenhardt and White, 2006; Degenhardt et al., 2006; Karantza-Wadsworth and White, 2008; Karantza-Wadsworth et al., 2007a; Karp and White, 2008; Mathew et al., 2007b; Nelson et al., 2004; Shimazu et al., 2007; Tan et al., 2005).

Major advantages of genetically defined, immortalized mouse epithelial cell lines are the ability to perform biochemical analysis, screening, and further genetic manipulations, thereby extending the utility of existing mutant mouse models. These cells are suitable for RNAi-mediated knockdown of specific gene expression using siRNA oligos targeting the gene of interest (Degen-hardt, et al., 2006). Stable gene knockdown using shRNA constructs or gain-of-function expression of transgenes are also possible in these cells using retrovirus-mediated gene transfer. These cells can be cultured and implanted back in mice after defined and controlled genetic manipulations for their potential use in tumorigenicity and other assays. Because these cells are nontumorigenic when immortalized, they are ideal for examining the molecular signatures involved in the oncogenic transformation into their tumorigenic counterparts. Furthermore, because these cells can be generated from any mutant mouse that survives to near birth, it is possible to combine multiple gene alterations to examine the compounding effect of multiple mutations and to test for genetic epistasis (Degenhardt et al., 2002b; Nelson et al., 2004). This advantage allows the evaluation of the means by which multiple genetic defects cooperate to promote epithelial tumorigenesis that is a fundamental aspect of oncogenesis (Hanahan and Weinberg, 2000). It also provides a renewable source of immortalized cells that can be used for comparison of tumorigenicity and chemotherapeutic response between different genetic backgrounds (Tan et al., 2005). More importantly, the possibility of introducing compound mutations in immortalized cell lines overcomes the limitations of early postnatal mortality. Tumor allografts of

Table 5.1   Immortalized mouse (C57BL/6, wild-type or mutant) epithelial cell lines, E1A or myc and p53DD or *p53*$^{-/-}$ derived

Cell lines	Tissue type	Genotype	Reporter/vector	Reference
iBMK: C57B/6 neonatal kidney epithelia, E1A, p53DD derived:				
WTB/6.1–3	Kidney	+/+		Unpublished
iBMK: C57B/6, c-myc, p53DD derived				
WTB/ 6-myc.1–3	Kidney	+/+	LTR.H-myc	Unpublished
WTB/ 6-myc.1, BCL-2	Kidney	*Bcl-2*	LTR.H-myc, pcDNA3hBcl-2	Unpublished
WTB/ 6-myc.1, H-ras	Kidney	*H-ras*	LTR.H-myc, pcDNA3HrasV12	Unpublished
p53$^{-/-}$ iBMK cell lines, E1A derived				
W3, W4 (E1A, p53DD controls)	Kidney	*p53*$^{+/+}$;		Degenhardt, 2002a
P53$^{-/-}$1, p53$^{-/-}$2	Kidney	*p53*$^{-/-}$		Degenhardt, 2002a
P53$^{-/-}$A	Kidney	*p53*$^{-/-}$		Tan, 2005

these cells in mice also allow the study of *in vivo* tumorigenic processes in a more physiologically relevant microenvironment. Tumors generated from immortalized epithelial cells can be excised and used to derive tumor-derived cell lines (TDCL). This allows the direct comparison of these cell lines with the unselected, nontumorigenic, parental cell lines to facilitate *in vitro* biochemical and functional analysis to identify *in vivo* tumor-promoting functions (Karp *et al.*, 2008). This chapter describes the main procedures and assays to study the role of apoptosis in cancer by use of the iBMK, iMPEC, and iMOSEC cell models. A detailed description of the iMMEC model can be found in the accompanying chapter (Karantza-Wadsworth and White, 2008).

2. IMMORTALIZATION OF BABY MOUSE KIDNEY, MAMMARY, PROSTATE, AND OVARIAN SURFACE EPITHELIAL CELLS

Immortalization of rat and mouse epithelial cells requires simultaneous inactivation of the retinoblastoma (Rb) and p53 pathways (White, 2001; 2006). The Rb pathway regulates the G_1/S transition, which is essential for

Table 5.2 Immortalized mouse (C57BL/6, wild-type or mutant) kidney (iBMK) epithelial cell lines, E1A, and p53DD derived. W2 and D3 cells were used to derive vector controls and the corresponding activated oncogene expressing cell lines (below)

Cell lines	Tissue type	Genotype	Reporter/ vector	Reference
iBMK: Bax/Bak deficient iBMK cell lines, E1A, p53DD derived				
W1, W2, W3	Kidney	$bak^{+/+}$, $bax^{+/-}$		Degenhardt *et al.*, 2002a,b
X1	Kidney	$bak^{+/+}$, $bax^{-/-}$		Degenhardt *et al.*, 2002a,b
X2, X3	Kidney	$bak^{+/-}$, $bax^{-/-}$		Degenhardt *et al.*, 2002a,b
K1, K2, K3	Kidney	$bak^{-/-}$, $bax^{+/-}$		Degenhardt *et al.*, 2002a,b
D1, D2, D3	Kidney	$bak^{-/-}$, $bax^{-/-}$		Degenhardt *et al.*, 2002a,b
W2.3.1-2, -5, -6 Control	Kidney	WT	pcDNA.3.1	Nelson; 2004; Tan, 2005
W2.Hras-2, -3, -7	Kidney	*H-ras V12*	pcDNA1. HrasV12	Degenhardt *et al.*, 2002a,b
W2.Raf-13, -15, -16	Kidney	*Raf-CAAX*	pcDNA3.Raf. CAAX	Degenhardt, 2006; Tan, 2005
W2-A1, -B1, -D2 Control	Kidney	WT	pcDNA3	
W2.AKT-C1. -D1, -E4	Kidney	*Myr-AKT*	pcDNA3. Myr-AKT	Degenhardt, 2006; Tan, 2005
W2.Bcl2-3, -14, -15	Kidney	*Bcl-2*	pcDNA3hBcl-2	Nelson, 2004
W2.19K-4, -7, -8	Kidney	*E1B 19K*	pcDNA3.1/V5/ His- TOPO19K	Nelson, 2004
D3.zeo-1, -2, -3 Control	Kidney	$bak^{-/--}$, $bax^{-/-}$	pcDNA3.1zeo	Nelson, 2004; Tan, 2005

Table 5.2 (*continued*)

Cell lines	Tissue type	Genotype	Reporter/ vector	Reference
D3.Hras-1, -3, -6	Kidney	$bak^{-/-}$, $bax^{-/-}$; H-ras V12	pcDNA3.1zeo	Degenhardt, 2006
D3-pcDNA3.1 Control	Kidney	$bak^{-/-}$, $bax^{-/-}$	pcDNA3.1	
D3.Raf-CAAX-1, -2, -3	Kidney	$bak^{-/-}$, $bax^{-/-}$; Raf-CAAX	pcDNA3.Raf-CAAX	
D3.-B4, -C4 Control	Kidney	$bak^{-/-}$, $bax^{-/-}$	pcDNA3zeo	
D3.Akt-D5, -F7, -G1	Kidney	$bak^{-/-}$, $bax^{-/-}$; myr-AKT	pcDNA3. Myr-AKT	Degenhardt, 2006
D3.Bcl2-3, -4, -6	Kidney	$bak^{-/-}$, $bax^{-/-}$; Bcl-2	pcDNA3hBcl-2 pcDNA3zeo	Nelson, 2004
D3.19K-4, -7, -8	Kidney	$bak^{-/-}$, $bax^{-/-}$; E1B 19K	pcDNA3.1/V5/ His-TOPO19K pcDNA3zeo	Nelson, 2004
iBMK cells expressing EGFP and the autophagy marker EGFP-LC3				
D3 EGFP	Kidney	$bak^{-/-}$, $bax^{-/-}$	GFP/pEGFP	Unpublished
D3 EGFP-LC3	Kidney	$bak^{-/-}$, $bax^{-/-}$	GFP/pEGFP-LC3	Degenhardt, 2006
iBMK cells expressing pDsRED-C1				
W2RED	Kidney	RFP	pDsRed-C1	Nelson, 2004; Tan, 2005
W2.3.1-5RED-4, -6,-11	Kidney	RFP	pDsRed-C1	Nelson et al., 2004
W2.Bcl2-3RED	Kidney	RFP, Bcl-2	pDsRed-C1	Nelson et al., 2004
W2.HRAS-3RED-3, -6,-8	Kidney	RFP, Bcl-2	pDsRed-C1	

(continued)

Table 5.2 (*continued*)

Cell lines	Tissue type	Genotype	Reporter/ vector	Reference
D3.zeo-2RED 1,2,4,10	Kidney	RFP, $bak^{-/-}$, $bax^{-/-}$	pDsRed-C1 pPUR	Nelson, 2004
D3.Hras-1RED-2, -7,-10	Kidney	RFP, $bak^{-/-}$, $bax^{-/-}$ H-ras,	pDsRed-C1	
D3.Bcl2-4RED-5, -9,-14	Kidney	RFP, $bak^{-/-}$, $bax^{-/-}$; Bcl-2	pDsRed-C1	

the cell proliferation, whereas p53 pathway activates the G_1–S checkpoint to prevent a premature entry into S phase because of pRb inactivation (White, 2001; 2006). The most common methods of inactivation of the Rb pathway are the direct inhibition of Rb by viral oncoproteins such as E1A or SV40 T antigen that promote epithelial proliferation (Berk, 2005; Helt and Galloway, 2003; White, 2001). In cells where p53 is functional, this Rb inactivation results in p53 mediated–growth arrest and apoptosis, which necessitates the simultaneous inactivation of both the pathways for epithelial cell immortalization (White, 2001; 2006). The p53 pathway can be inactivated by expression of a dominant negative p53 mutant (p53DD [Shaulian *et al.*, 1992]) (Degenhardt *et al.*, 2002b), or by use of epithelial cells from $p53^{-/-}$ mice (Degenhardt *et al.*, 2002a). We adapted this strategy to mutant mouse models, generating primary mouse epithelial cells where numerous apoptotic and tumor suppressor genes have been targeted for gene disruption (Tables 5.1 to 5.8) (Degenhardt and White, 2006; Degenhardt *et al.*, 2002a,b; Karantza-Wadsworth *et al.*, 2007a; Mathew *et al.*, 2007b; Shimazu *et al.*, 2007; Tan *et al.*, 2005). Coexpression of E1A or c-myc and p53DD successfully immortalizes epithelial cells from rat or mouse kidney, and mouse mammary, ovarian, and prostate tissues to form colonies (Fig. 5.1) from which immortalized cell lines are generated that retain their epithelial characteristics while remaining nontumorigenic (Degenhardt *et al.*, 2002b; Sakamuro *et al.*, 1995).

2.1. Protocol for the generation of iBMK cells

2.1.1. Solutions and reagents required

1. Collagenase/dispase: collagenase (125 mg) and dispase (1.25 g) (Sigma-Aldrich Co., St. Louis, MO) are dissolved in 500 ml sterile

Table 5.3 Control and BH-3-only (*bim$^{-/-}$, puma$^{-/-}$, noxa$^{-/-}$, nbk$^{-/-}$/bik$^{-/-}$*) deficient iBMK (C57BL/6) epithelial cell lines, E1A, and p53DD derived

Cell lines	Tissue type	Genotype	Reporter/vector	Reference
Bim–deficient iBMK cell lines, E1A, p53DD derived:				
BIM$^{+/+}$ iBMK cell lines				
BIM$^{+/+}$ A, B1,B2, C1, C2,C3	Kidney	+/+		Tan, 2005
BIM$^{+/+}$A derived vector controls				
BIM$^{+/+}$ A3.1A1, 1A2,1B1, 1D1,1D2, 1E,1F	Kidney	+/+	pcDNA3.1	
BIM$^{+/+}$A derived H-rasV12 expressing cell lines				
Bim$^{+/+}$A. Hras-B, -C, -D	Kidney	+/+; *H-ras V12*	pcDNA3. HrasV12	Tan, 2005
Bim$^{+/-}$ cells iBMK cell lines, E1A, and p53DD derived				
Bim$^{+/-}$A, B, C	Kidney	*bim$^{+/-}$*		Tan, 2005
BIM$^{-/-}$ iBMK cell lines, E1A, and p53DD derived				
Bim-/-A, B. C Control	Kidney	*bim$^{-/-}$*		Tan, 2005
Bim-/-A.3.1- C1,-C2	Kidney	*bim$^{-/-}$*	pcDNA3.1	Tan, 2005
M5-A1,-B2, -B3	Kidney	*bim$^{-/-}$*	pcDNA3.1	Tan, 2005
M2-A2,-B1, -D1	Kidney	*bim$^{-/-}$*	pcDNA3.1	Tan, 2005
M5-A1 (Bim$^{-/-}$) vector control				
M5A1C1-E4, -G6,-F5, M5A1C2- A1,-F2,-G3	Kidney	*bim$^{-/-}$*	pcDNA3.1	Tan, 2005
M5-A1 (BIM$^{-/-}$) derived H-ras expressing cell lines				
M5-A1Hras -A,-B,-C, -D,-F,	Kidney	*bim$^{-/-}$; H-ras V12*	pcDNA3HrasV12	Tan, 2005

(continued)

Table 5.3 (*continued*)

Cell lines	Tissue type	Genotype	Reporter/vector	Reference
M5-A1 (BIM$^{-/-}$A) derived Raf expressing cell lines				
M5A1.Raf-1H4,-C18,-2B5, -D	Kidney	$bim^{-/-}$; Raf-$CAAX$	pcDNA3.Raf-CAAX	Tan, 2005
Puma $^{+/+}$ and $^{-/-}$ iBMK cell lines, E1A, p53DD derived				
Puma$^{+/+}$A, B	Kidney	$puma^{+/+}$		Tan, 2005
Puma$^{-/-}$A, B	Kidney	$puma^{-/-}$		Tan, 2005
Noxa $^{+/+}$ and $^{-/-}$ iBMK cell lines, E1A, p53DD derived				
Noxa$^{+/+}$A, B, C	Kidney	$noxa^{+/+}$		Tan, 2005
Noxa$^{-/-}$A, B, C	Kidney	$noxa^{-/-}$		Tan, 2005
Nbk/Bik $^{+/+}$ and $^{-/-}$ iBMK cell lines				
Nbk$^{+/+}$A, B, C	Kidney	$nbk^{+/+}$		Shimazu, 2007
Nbk$^{-/-}$A, B. C	Kidney	$nbk^{-/-}$		Shimazu, 2007

phosphate-buffered saline (PBS) in a 2-L conical flask by gentle mixing. Filter sterilized collagenase/dispase solution is frozen as 50-ml aliquots at $-20\,^{\circ}C$.

2. Linearized E1A and p53DD plasmid DNA: adenoviral E1A (pCMVE1A [White *et al.*, 1991]) and dominant negative mouse p53 expression vectors (p53DD [Degenhardt *et al.*, 2002a; Shaulian *et al.*, 1992]) (10 μg each) are separately linearized by restriction digestion by use of Sca1 (1 μl; 10 U) in 50-μl reactions by incubating for 3 to 4 h at 37 $^{\circ}C$. The restriction digestion is stopped by either heat inactivation at 65 $^{\circ}C$ or by the addition of 2 μl of 0.2 M EDTA, pH 8.0.

2.1.2. Protocol for the isolation and immortalization of baby mouse kidney epithelial cells

1. Primary epithelial cells are isolated from wild-type baby C57B/6 mouse kidneys: Five-day-old mouse litter is sacrificed by asphyxiation in a CO_2 chamber followed by cervical dislocation according to an IACUC-approved protocol.

Table 5.4   Control and caspase-3/caspase-7 deficient iBMK (C57BL/6) cell lines, E1A, and p53DD derived

Cell lines	Tissue type	Genotype	Reporter/ vector	Reference
Caspase-3/caspase-7 mutant iBMK cell lines, E1A, p53DD derived				
Caspase-3$^{+/+}$/ Caspase-7$^{-/-}$ -1, -6, -8	Kidney	caspase-3$^{+/+}$ caspase-7$^{-/-}$		Karp, 2008
Caspase-3$^{-/-}$/ Caspase-7$^{-/-}$ -3	Kidney	caspase-3$^{-/-}$ caspase-7$^{-/-}$		Karp, 2008
Caspase-3$^{+/+}$/ Caspase-7$^{+/-}$ -7, -10	Kidney	caspase-3$^{+/+}$ caspase-7$^{+/-}$		Karp, 2008

2. The litter is then doused with ethanol, and the skin is carefully removed from neck down to the feet to expose the dorsal surface.

3. Both kidneys are extracted by carefully running a fine–point curved forceps down along both sides of the spinal column. When the genotypes of the mice are known, kidneys of the same genotypes are pooled and transferred into a 15-ml conical tube containing sterile PBS with 1% Pen Strep (Invitrogen, Carlsbad, CA) and placed on ice. Epithelial cells are extracted from the pooled kidneys and immortalized with E1A and p53DD as per the protocols described for kidneys from single mouse in sections 2.1.3 and 2.1.4, respectively. Volumes are scaled up, depending on the total number of pairs of kidneys used. If the genotypes of the mice are not known at this point, care is taken to process each pair of kidneys from a single mouse pup separately as described below in sections 2.1.3 and 2.1.4.

2.1.3. Protocol for the isolation of kidney epithelial cells

1. Both kidneys from the same baby mouse are transferred into a petri dish containing sterile PBS and washed twice with chilled sterile PBS containing antibiotics (1% Pen Strep) with gentle shaking.

2. The washed single pair of kidneys is then transferred into another petri dish containing 5 ml collagenase/dispase solution and mechanically minced into small pieces of approximately 1 × 1 mm size using two sets of sterile forceps (pieces should be small enough to pass through the mouth of a 10-ml pipette).

3. All of the minced tissue from a pair of kidneys from a single mouse along with the collagenase/dispase solution (5 ml) is then transferred into 20 ml

Table 5.5 Control and autophagy-deficient iBMK (C57BL/6) epithelial cell lines, E1A and p53DD derived

Cell lines	Tissue type	Genotype	Reporter/ vector	Reference
Beclin1 $^{+/+}$ and $^{-/-}$ iBMK cell lines, E1A, p53DD derived:				
Beclin1 wild-type iBMK cell lines				
W4-1, -2, -A1, -B1, -C1, -E1	Kidney	+/+	pCEP-4	Mathew, 2007b
Beclin 1 $^{+/-}$ iBMK cell lines				
BLN-4-1, -34-3, -A3, -B2, -C3, -D1	Kidney	*Beclin1*$^{+/-}$	pCEP-4	Mathew, 2007b
Beclin1 $^{+/+}$ iBMK cell lines expressing Bcl-2				
WB-3, -5, -10, -13, -A2, -B1, -D1, -D3	Kidney	+/+; *Bcl-2*	pCEP-4. Bcl-2	Mathew, 2007b
Beclin1$^{+/-}$ iBMK cell lines expressing Bcl-2				
BLNB1, -4, -12, -13, -A4, -B1, -C2, -D1	Kidney	*Beclin1*$^{+/-}$; *Bcl-2*	pCEP-4. Bcl-2	Mathew, 2007b
Beclin1 iBMK cell lines expressing EGFP-LC3				
WB-13- EGFP-LC3	Kidney	+/+, *Bcl-2*; *EGFP-LC3*	pCEP-4. Bcl-2, pEGFP-LC3	Mathew, 2007b
BLNB-13- EGFP-LC3	Kidney	*Beclin1*$^{+/+}$; *EGFP-LC3*	pCEP-4. Bcl-2 pEGFP-LC3	Mathew, 2007b
ATG5$^{+/+}$ iBMK cell lines, E1A, and p53DD derived				
6.1, 6.2, 6.3, 6.4	Kidney	+/+		Mathew, 2007b
ATG5$^{+/-}$ iBMK cell lines, E1A, and p53DD derived				
5.1, 5.2, 5.3, 5.4	Kidney	*atg5*$^{+/-}$		Mathew, 2007b
ATG5$^{-/-}$ iBMK cell lines, E1A, and p53DD derived				
7.1, 7.2, 7.3, 7.4	Kidney	*atg5*$^{-/-}$		Mathew, 2007b
ATG5$^{+/+}$ iBMK cell lines, E1A, and p53DD derived expressing Bcl-2				
6.1/pcDNA3 Cl.1-2, 6.2/ pcDNA3 Cl.1-4	Kidney	+/+	pcDNA3	Mathew, 2007b

Table 5.5 (*continued*)

Cell lines	Tissue type	Genotype	Reporter/ vector	Reference
6.1B2, –B3, –B5, 6.2B1, –B5, -B8, –B10	Kidney	$+/+; Bcl$-2	pcDNA3. Bcl-2	Mathew, 2007b
ATG5$^{+/-}$ iBMK cell lines, E1A, and p53 DD derived expressing Bcl-2				
5.1/pcDNA3 Cl.1-2 Control	Kidney	$atg5^{+/-}$	pcDNA3	
5.1B2	Kidney	$atg5^{+/-};$ Bcl-2	pcDNA3. Bcl-2	
ATG5$^{-/-}$ iBMK cell lines E1A and p53 DD derived, expressing Bcl-2				
7.1/pcDNA3 Cl.1-2, 7.2/ pcDNA3 Cl.1-4	Kidney	$atg5^{-/-}$	pcDNA3	Mathew, 2007b
7.1B1,–B2, –B4, –B5, 7.2B2, -B3, –B6, –B9	Kidney	$atg5^{-/-};$ Bcl-2		Mathew, 2007b

of collagenase/dispase solution in a sterile 100-ml bottle equipped with a sterile magnetic stirring bar (total volume is 25 ml). The mixture is then stirred vigorously in a warm room at 37 °C for 30 to 40 min or until cells form a homogenous suspension. The completeness of digestion is checked by aseptically removing small aliquots onto a glass slide and then mounted with a glass coverslip for low-power microscopic examination ($40\times$ magnification).

4. Tissue debris is allowed to settle, and the cell suspension containing single cells is transferred into a 50-ml centrifuge tube (polypropylene), and cells are pelleted for 5 min at 1000 rpm ($\sim$300 $\times$ g).

5. The cell pellet is then resuspended in 1 ml (per one pair of kidneys) of prewarmed tissue culture medium (DMEM/5%FBS) supplemented with 1% Pen Strep (0.5 ml per kidney) and subjected to electroporation as described in section 2.1.4.

2.1.4. Protocol for the immortalization of kidney epithelial cells

1. For immortalization with E1A and p53DD, 500 μl of the preceding cell suspension, after gentle mixing, is transferred into a sterile Eppendorf tube containing linearized DNA (10 μg E1A [pCMVE1A] and 10 μg p53DD [p53DD] linearized with Sca I as described in section 2.1.1). Total amount of DNA in each transfection is adjusted to 120 μg by

Table 5.6 Immortalized mouse (C57BL/6; wild-type or mutant) mammary epithelial cell lines (iMMECs), E1A, and p53DD derived

Cell lines	Tissue type	Genotype	Reporter/vector	Reference
iMMEC: C57B/6 mammary epithelia E1A, p53DD derived				
WT-A (21), -B, -C, -D, WT-3, -5	Mammary	+/+		Karantza–Wadsworth, 2007a
WTA.V, WT3.V Control	Mammary	+/+, *pcDNA3*	pcDNA3.1	Karantza–Wadsworth, 2007a
Beclin1⁺/⁻ iMMECs: C57B/6 mammary epithelia E1A, p53DD derived				
BLN2 (2.1), BLN4	Mammary	*Beclin1⁺/⁻*		Karantza–Wadsworth, 2007a
BLN2.V	Mammary	*Beclin1⁺/⁻; pcDNA3.1*	pcDNA3.1	Karantza–Wadsworth, 2007a
Beclin1⁺/⁻ iMMECs: C57B/6 mammary epithelia E1A, p53DD derived, expressing Bcl-2				
BLN2.B2, .B4, .B5, .B8	Mammary	*Beclin1⁺/⁻, Bcl-2*	pcDNA3.1.Bcl-2	Karantza–Wadsworth, 2007a
iMMEC: C57B/6 mammary epithelia E1A, p53DD derived expressing Bcl-2				
WTA.B1, .B4	Mammary	*+/+, Bcl-2*	pcDNA3.1.Bcl-2	Karantza–Wadsworth, 2007a
WT3.B1, .B2, .B3, .B8	Mammary	*+/+, Bcl-2*	pcDNA3.1.Bcl-2	Karantza–Wadsworth, 2007a
iMMEC: C57B/6 mammary epithelia E1A, p53DD derived expressing autophagy marker EGFP-LC3				
WTA-LC3, WT3-LC3	Mammary	*+/+; EGFP-LC3*	pEGFP-LC3	Karantza–Wadsworth, 2007a
WTA.B4-LC3 WT3.B3-LC3	Mammary	*+/+; Bcl-2; EGFP-LC3*	pcDNA3.1.Bcl-2, pEGFP-LC3	Karantza–Wadsworth, 2007a

BLN2-LC3	Mammary	*Beclin1*$^{+/-}$; *EGFP-LC3*	pEGFP-LC3	Karantza-Wadsworth, 2007a
BLN2.B4-LC3.5	Mammary	*Beclin1*$^{+/-}$, *Bcl-2 EGFP-LC3*	pcDNA3.1.Bcl-2, pEGFP-LC3	Karantza-Wadsworth, 2007a
iMMEC: C57B/6 mammary epithelia E1A, p53DD derived expressing HER2/neu				
WTA.H2, WTA.H3	Mammary	+/+, *HER2/neu*	pcDNA3.1Her2	Karantza-Wadsworth, 2007a
iMMEC: C57B/6 mammary epithelia E1A, p53DD derived expressing Myr-AKT				
WTA.A5, WTA.A7	Mammary	+/+, *AKT*	Myr-AKT	Karantza-Wadsworth, 2007a
iMMEC: C57B/6 mammary epithelia E1A, p53DD derived expressing H-Ras				
WTA.R3, WTA.R5	Mammary	+/+, *H-rasV12*	pcDNA3HrasV12	Karantza-Wadsworth, 2007a

Table 5.7 Immortalized mouse (C57BL/6) ovarian surface epithelial (iMOSECs) cell lines, E1A and p53DD derived

Cell lines	Tissue type	Genotype	Reporter	Reference
iMOSEC: C57B/6; ovarian surface epithelia E1A, p53DD derived				
MOSEC-1, -2, -3, -4, -5, -6, -7, -8	Ovary	+/+		Karp, 2008
iMOSEC: C57B/6; ovarian surface epithelia E1A, p53DD derived, expressing Bcl-2				
MOSEC-1, -3, -6 Control	Ovary	+/+	pcDNA3. Bcl-2	Karp, 2008
MOSEC-1, -3, -6/ BCL-2	Ovary	Bcl-2	pcDNA3. Bcl-2	Karp, 2008
iMOSEC: C57B/6; ovarian surface epithelia E1A, p53DD derived, expressing H–Ras				
MOSEC-1/H–Ras, -6/ H–Ras	Ovary	H–rasV12	pcDNA3. H–ras	Karp, 2008

Table 5.8 Immortalized mouse (C57BL/6) prostate epithelial (iMPECs) cell lines, E1A and p53DD derived

Cell lines	Tissue type	Genotype	Reporter	Reference
iMPEC: C57B/6; prostate epithelia E1A, p53DD derived				
MPEC-1, -2, -3, -4, 5, - 6, -7, -8	Prostate	+/+		Bray, 2008

adding 100 μg salmon sperm DNA to the mixture, and the cells are transfected by electroporation as described in the following.

2. For electroporation, 250 μl of the preceding suspension is transferred into a 0.4 cm electroporation cuvette (BIO-RAD, Hercules, CA) and pulsed at 0.22 V and 950 μF by use of Gene Pulser II (BIO-RAD, Hercules, CA). The electroporated mixture is allowed to sit for 10 min and is resuspended in 40 ml of culture medium (DMEM/5%FBS/1% Pen Strep) and 5 ml each is plated in eight 6-cm plates. A duplicate electroporation is also performed using the remaining 250 μl cell suspension from step 1 (section 2.1.4) above, and plated in parallel to ensure sufficient independent immortalized clones that can be cloned and expanded.

3. The remaining 500 μl cell suspension from section 2.1.3, step 5 is divided into two equal halves of 250 ls and each is transfected separately with E1A and p53DD respectively, as negative controls for immortalization.

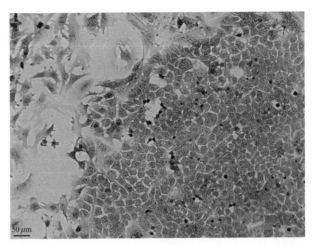

Figure 5.1   Morphology of an emerging colony of an E1A and p53DD transfection of primary baby mouse kidney cells after approximately 3 weeks of growth *in vitro*. These colonies can be recognized by their distinctive epithelial morphology, typified by smooth, round edges and a characteristic cobblestone appearance. Colonies can be cloned and expanded to generate stable iBMK cell lines (magnification 100×).

4. No drug selections are necessary, because this is a functional selection for immortalization. This provides the additional advantage of being able to use any selectable marker for introduction of genes at a later point. Doubly transfected epithelial cells overcome p53 mediated growth arrest and grow into colonies giving rise to approximately 5 to 50 colonies per pair of kidneys per plate, which are visible at approximately 7 to 10 days (Fig. 5.1). Colonies are composed of tightly packed, cuboidal cells with typical epithelial morphology. They can be distinguished from surrounding fibroblasts by their compact appearance and sharply defined boundaries (Fig. 5.1).

5. Multiple, independent colonies are recovered from independent plates and expanded after 3 to 4 weeks. Typically, up to 20 colonies are picked and expanded to obtain approximately 5 to 10 viable colonies. Multiple early passage stocks are cryopreserved in the vapor-phase of a liquid nitrogen freezer as described in section 2.1.5.

2.1.5. Protocol for cloning, expansion, and preservation of iBMK cells

1. When each colony is approximate 0.6 to 0.8 cm in diameter, tissue culture plates are marked for ring cloning. Multiple plates for the same genotype are selected to ensure that truly independent colonies are derived.

2. Plates are washed in PBS, and well-separated colonies are marked on the bottom of the plates. Sterile cloning rings (6 × 8 mm to 10 × 10 mm;

Belco Biotechnology, Vineland, NJ) are carefully placed around the colonies with the support of autoclaved vacuum grease (VWR), to hold them in place.

3. Cells are then carefully trypsinized by use of a small amount of 0.05% trypsin-EDTA (Invitrogen, Carlsbad, CA) from the cloning rings and transferred into a 96-well plate to expand the colonies that are sequentially expanded into 24- and 12-well plates and then to 6-cm and finally into a 10-cm plate. Protein lysates are made from each clone to confirm that the clones express E1A, and p53DD by Western blotting by use of anti–E1A (Oncogene/Calbiochem Immunochemicals, San Diego, CA) and anti-p53 (Ab-1; Oncogene/Calbiochem Immunochemicals, San Diego, CA) antibodies, respectively. iBMK cells are also evaluated by genotyping and for the expression of the transgene or the loss of expression of a particular protein when the cells are derived from mutant mice.

4. For long-term storage, cells are trypsinized from five 10-cm plates at 90% confluency (approximately 10×10^6 cell/10-cm plate), resuspended in 2 ml freezing medium (90% FBS and 10% sterile tissue culture grade dimethyl sulfoxide [DMSO]) per each 10-cm plate, and placed in $-70\ °C$ freezer wrapped in paper towels overnight and then cryopreserved in vapor phase nitrogen.

3. DEVELOPMENT OF iBMK CELL LINES FOR IDENTIFYING THE ROLE OF APOPTOSIS IN CANCER

Apoptosis is an effective tumor suppressor mechanism that is inactivated in many human cancers (Adams and Cory, 2007; Hanahan and Weinberg, 2000; Karantza-Wadsworth, 2007b). Identifying gain- or loss-of-function mutations capable of targeting this pathway for inactivation is extremely important to gain insight into the molecular mechanisms of tumor growth and to identify novel therapeutic targets for cancer. A common obstacle to achieving this goal is the lack of availability of genetically defined epithelial cell lines that have an intact apoptotic pathway and are nontumorigenic, but rendered tumorigenic by defects in apoptosis.

The Bcl-2 family members Bax and Bak are two functionally redundant, universal downstream regulators of apoptosis, the simultaneous deletion of both of which is one of the most efficient ways to inactivate apoptosis (Degenhardt *et al.*, 2002b; Gelinas and White, 2005; Wei *et al.*, 2001). The proapoptotic BH3-only Bcl-2 family members (e.g., Bim, Puma, Noxa, Nbk/Bik) promote apoptosis by activating Bax and Bak or antagonizing the survival activity of Bcl-2 (Gelinas and White, 2005). However, more than 95% of the $bax^{-/-}/bak^{-/-}$ mice die shortly after birth (Wei *et al.*, 2001), making it difficult to study a cancer phenotype associated with the double deletion of *bax* and *bak*. Our method offers an alternative by allowing the generation of iBMK cells from *bax/bak* double knockout mutant mice, BH3-only mutant mice, and mice with deletions of the executioner caspases (−9 and −3) down stream of Bax and Bak (Table 5.4), providing an apoptosis-deficient epithelial cell model system to study the role of apoptosis in tumorigenesis. To this end, we have generated $Bax^{-/-}/Bak^{-/-}$, BH3-only mutants $Bim^{-/-}$, $Puma^{-/-}$, $Noxa^{-/-}$, and $Nbk^{-/-}/Bik^{-/-}$ (Shimazu et al., 2007; Tan et al., 2005), as well as $caspase\text{-}3^{+/+}/caspase\text{-}7^{-/-}$, $caspase\text{-}3^{-/-}/caspase\text{-}7^{+/-}$, and $caspase\text{-}3^{+/+}/caspase\text{-}7^{+/-}$ (Karp and White, 2008) iBMK cell lines (Tables 5.3 and 5.4).

3.1. Protocol for the immortalization of baby mouse kidney epithelial cells from $bax^{-/-}$, $bak^{-/-}$, and $bax^{-/-}/bak^{-/-}$ Mice

Breeding pairs are set up, crossing various *bax* and *bak* mutant mice ($bax^{+/-}/bak^{-/-}$ with $bax^{+/-}/bak^{-/-}$; $bax^{+/-}/bak^{+/-}$ with $bax^{+/-}/bak^{-/-}$; and $bax^{+/-}/bak^{+/+}$ with $bax^{+/-}/bak^{+/+}$). Newly born litters of pups are collected and numbered. Both kidneys from each pup are removed separately under sterile conditions, washed, and processed by use of the general immortalization protocol described in section 2.1 to generate iBMK cell lines that are either $bax^{+/+}/bak^{+/-}$, $bax^{+/-}/bak^{-/-}$, or $bax^{-/-}/bak^{-/-}$. Special care is taken to process each pair of kidneys from each animal separately, because the genotypes of the pups are not known at this stage. Colonies are expanded and cryopreserved as described in section 2.1.5. Tail snips from each pup are collected and processed for DNA isolation and PCR genotyping as described elsewhere (Lindsten et al., 2000; Shindler et al., 1997). Genotyping is verified on the DNA isolated from the iBMK cell lines themselves (Degenhardt et al., 2002a).

4. APOSCREEN: AN IBMK CELL-BASED SCREEN FOR THE IDENTIFICATION OF APOPTOSIS-INDUCING COMPOUNDS AS POTENTIAL ANTI-CANCER AGENTS

In addition to being an effective tumor-suppressor mechanism, apoptosis is also an important determinant of treatment response. Compounds that specifically and irreversibly activate apoptotic pathway have the potential for use as anti-cancer agents (Fesik, 2005). Therefore, iBMK cell lines with defined and differential capacities for apoptosis, such as the apoptosis competent wild-type (W2 cells) and $bax^{-/-}/bak^{-/-}$ apoptosis-deficient (D3 cells) are extremely useful to identify novel and specific apoptosis inducers as potential anti-cancer agents (Andrianasolo *et al.*, 2007; Degenhardt *et al.*, 2002b).

Secondary metabolites produced by marine invertebrates are a rich source of yet unknown natural products with potential proapoptotic activity. We adapted the iBMK cell system for a bioassay-guided screen (ApoScreen) for the identification, isolation, and purification of novel compounds with potential anticancer activity (Andrianasolo *et al.*, 2007). Because W2 cells are competent, and D3 cells are deficient for apoptosis, compounds that selectively kill W2 but not D3 cells are specific inducers of apoptosis and, therefore, potential anti-cancer agents. Those compounds that kill both W2 and D3 cells can be excluded as nonspecifically toxic (Andrianasolo *et al.*, 2007). As a simple and effective screening method useful for crude extracts, this assay can identify potential fractions isolated from natural sources before compound purification. Fractions showing potent bioactivity can be subjected to further chemical purification, structural elucidation, and functional validation in the ApoScreen bioassay (Andrianasolo *et al.*, 2007).

By use of this ApoScreen-guided purification we have screened and identified four novel diterpenes from the soft coral *Xenia elongata* with specific proapoptotic and, therefore, potential anti-cancer activities (Andrianasolo *et al.*, 2007). All four compounds are capable of differentially inducing cell death in the apoptosis competent W2 cells and not in the apoptosis deficient D3 cells (Andrianasolo *et al.*, 2007). Elucidation of the mechanism of apoptosis induction by these novel diterpenes, as well as further screening for proapoptotic natural products, is currently underway.

4.1. Protocol for the iBMK cell-based aposcreen bioassay

1. Apoptosis-competent W2 cells (7500 cells in 100 μl per well) and apoptosis-deficient *bax* and *bak* double knockout (D3 cells) (5000 cells in 100 μl per well) are plated in 96-well plates to achieve 50% confluency in 24 h.
2. Compounds are dissolved in DMSO and diluted in growth medium (DMEM) to make 2× suspensions of serial concentrations. 100 μl each of the preceding suspensions are added to cells at various concentrations to achieve final concentrations in the range of 100 μg/ml to 10 μg/ml, keeping the DMSO concentration uniform at 0.5% in all the wells. In separate wells, Staurosporine (0.1 μM), a strong apoptosis inducer, is used as the positive control, and DMSO (0.5%) is used as the vehicle control.
3. Cells are incubated at 37 °C for 24, 48, or 72 h at which time cell viability is assayed by use of a modification of a 3-(4,5-dimethylthiazol-2-yl)-2, 5-diphenyltetrazolium bromide (MTT) assay (Mosmann, 1983). MTT solution (50 μl of 2.5 mg/ml) (Sigma-Aldrich Co., St. Louis, MO) is added to each well to a final concentration of 0.5 mg/ml, and final volume of 250 μl. The plate is then incubated at 37 °C for 3 h.

4. After 3 h of incubation in MTT, the supernatant is aspirated, and 150 μl of 100% DMSO is added to dissolve the formazan crystals, the reduction product of MTT in mitochondria. After an additional 30-min incubation at 37 °C on a shaker, absorbance is measured at 570 nm by use of a spectrophotometer. Absorbances from five wells are averaged for each concentration point as a measure of cell viability, along with standard deviation from the mean. Relative cell viability for each concentration of compound is calculated as a difference in absorbance values between time 0 (addition of the compound) and 48 h. Apoptosis induction is defined as at least 20% cell death of W2 (apoptosis competent) cells and a 10% or higher growth in D3 (apoptosis deficient) cells.

5. ADAPTATION OF THE iBMK CELL MODEL SYSTEM FOR THE EVALUATION OF TISSUE-SPECIFIC TUMOR-PROMOTING FUNCTIONS

One of the strengths of the iBMK model system is its adaptability to multiple tissue types and both transgenic and knockout mouse models. By deriving immortalized epithelial cells from different tissues from an appropriate mouse model it is possible to recapitulate the tissue-specific characteristics *in vitro* with several advantages. For example, adaptation of iBMK model to mouse mammary epithelia enabled us to derive immortalized mouse mammary epithelial cells (iMMEC) that express surface markers and demonstrate functional properties that are characteristic of mammary epithelia (Karantza-Wadsworth *et al.*, 2007a). Importantly, iMMEC undergo three-dimensional ductal morphogenesis when cultured in the presence of extracellular matrix (Karantza-Wadsworth *et al.*, 2007a). iMMEC are nontumorigenic, but become tumorigenic by the activation of HER-2/*neu*, consistent with the findings in human breast cancer (Karantza-Wadsworth *et al.*, 2007a; Karantza-Wadsworth and White, 2008). For a detailed description of the iMMEC model, please see the accompanying chapter (Karantza-Wadsworth and White, 2008).

We have established similar isogenic and immortalized but nontumorigenic epithelial models from mouse ovarian surface (iMOSEC) and prostate (iMPEC) epithelia (Tables 5.7 and 5.8). Establishment of immortalized mouse hepatocyte cell lines (iHEP) is currently in progress. All these models mimic their own corresponding tissue-specific phenotype, including marked tropism in tumorigenic assays to specific anatomic locations resembling corresponding tumors in humans, thus allowing the evaluation of tissue-specific requirements for cancer formation.

5.1. Immortalization and establishment of iMMEC, iMOSEC, and iMPEC models

Generation of iMMEC cell lines (Table 5.6) are described in detail in the accompanying chapter (Karantza-Wadsworth and White, 2008). Mouse ovarian surface and prostate epithelial cells are immortalized as described in the following.

5.1.1. Protocol for the generation of iMOSEC cell lines: Isolation of primary mouse ovarian surface epithelial cells

1. Twenty adult female C57B/6 mice (11 weeks old) are sacrificed by asphyxiation in a CO_2 chamber and cervical dislocation according to an IACUC-approved protocol.
2. Both ovaries are extracted from each mouse by use of antiseptic techniques and placed in a dish with ice-cold Hank's balanced salt solution (HBSS; Invitrogen, Carlsbad, CA) (Roby *et al.*, 2000).
3. After rinsing with HBSS, the ovaries are placed in a 15-ml conical tube containing 10 ml of 0.2% trypsin in HBSS, incubated at 37 °C under humidified atmosphere (5% CO_2) for 30 min. The tube containing the ovaries is placed in a horizontal position with the ovaries distributed over the length of the tube, taking care not to agitate the ovaries.
4. After 30 min, medium containing epithelial cells is transferred to a fresh tube with 5 ml DMEM supplemented with 4% FBS, 1% Pen Strep, 5 μg/ml insulin (Sigma-Aldrich Co., St. Louis, MO), 5 μg/ml transferrin (Sigma-Aldrich Co., St. Louis, MO), and 5 ng/ml sodium selenite (Sigma-Aldrich Co., St. Louis, MO), and centrifuged at 1000 rpm ($\sim$300 g) for 10 min at room temperature and resuspended in 2 ml complete medium.
5. The cells are then plated in special CellBIND 6-well plate (Corning CellBIND Surface; Corning Incorporated Life Sciences, Lowell, MA) for facilitating cell attachment. Cells reach 95% confluency in approximately 4 days and typically provide approximately 1×10^5 cells per 10 ovaries.

5.1.2. Protocol for the generation of iMOSEC cell lines: Immortalization of primary mouse ovarian surface epithelial cells

Primary mouse ovarian epithelial cells obtained from step 5 (section 5.1.1) are transfected with 2.5 μg each of E1A and p53DD expression vectors (linearized with Sca1 as described in section 2.1.1) by use of the Amaxa nucleofection protocol (Amaxa Inc., Gaithersburg, MD) as described in the following.

1. Primary mouse ovarian surface epithelial cells growing in log-phase are harvested by trypsinization after three passages and resuspended in PBS at a density of 3×10^6 cells/ml. For transfection, 1×10^6 cells are transferred into a 15-ml conical tube and centrifuged at $\sim$1000 rpm ($\sim$300 g).

2. The cell pellet is resuspended in 100 μl of prewarmed nucleofector solution, mixed with Sca1 linearized, E1A, and p53DD expression plasmids ($2.5 \mu g/3 \times 10^6$ cells/transfection), transferred into Amaxa cuvettes, and transfected by use of a preset program optimized for iBMK cells (G-16). The nucleofection program and the appropriate transfection reagent need to be optimized for each cell type.

3. Transfected cells are plated immediately in prewarmed DMEM supplemented with 10% FBS, 1% Pen Strep, 5 μg/ml insulin, 5 μg/ml transferrin, and 5 ng/ml sodium selenite and incubated at 37 °C.

4. Transfected cells are plated into special CellBIND petri dishes and allowed to grow into colonies. Round colonies with smooth edges, composed of cells with cobblestone morphology characteristic to epithelial cells, become visible in 7 to 10 days and are very similar in appearance to colonies formed by iBMK cells (Fig. 5.1).

5. Well-separated colonies are cloned and expanded separately in special 12-well CellBIND tissue culture plates and eventually transferred into uncoated 6-cm dishes by use of the general protocol described in section 2.1.5.

6. Cells are allowed to grow into monolayer for a week, after which cells are cultured in DMEM containing 10% FBS and 1% Pen Strep.

7. iMOSECs cell lines (Table 5.7) are trypsinized from three 10-cm plates at 90% confluency (approximately 10×10^6 cell/15-cm plate), harvested, and cryopreserved (one 10-cm plate per vial) as described in section 2.1.5. Protein lysates are made from each clone and analyzed for expression of E1A, p53DD, and epithelial markers by use of Western blot analysis (section 2.1.5).

5.1.3. Protocol for the generation of iMPEC cell lines: Isolation of primary mouse prostate epithelial cells

1. Ten to fifteen adult male C57B/6 mice (6 weeks old) are sacrificed by asphyxiation in a CO_2 chamber and cervical dislocation according to an IACUC-approved protocol.

2. Mouse prostate glands (Abate-Shen and Shen, 2002) are carefully isolated under an inverted dissection microscope, pooled, and transferred into petri dishes containing sterile PBS and washed twice with chilled sterile PBS with gentle shaking.

3. Washed prostate glands are transferred to 25 ml of collagenase/dispase solution into sterile 100-ml bottles equipped with magnetic stirrer and

agitated on an orbital shaker at 125 rpm for 10 min 37 °C until cells have formed a homogenous suspension and examined under the microscope.
4. The suspension is allowed to settle, the supernatant is discarded, and the residue is resuspended in 25 ml of fresh collagenase/dispase solution by pipetting several times.
5. The suspension is incubated on an orbital shaker at 37 °C at 150 rpm with agitation for 40 min. After digestion, the cells are mixed by pipetting and centrifuged at 1000 rpm ($\sim$300 g) for 6 min at room temperature. A total of approximately 3 to 5×10^6 cells are recovered by this procedure from 10 to 15 mice.

5.1.4. Protocol for the generation of iMPEC cell lines: Immortalization of primary mouse prostate epithelial cells

1. The cell pellet obtained is resuspended in 500 μl of DMEM with 5% FBS. Primary mouse prostate epithelial cells are transfected with 10 μg each of E1A and p53DD expression vectors (linearized with Sca1; section 2.1.1) by electroporation as described in section 2.1.4.
2. After transfection, the cells are plated at a 1:3 ratio into 10-cm dishes in DMEM containing 5% FBS and 1% Pen Strep and allowed to grow into colonies for 2 weeks. Colonies appear round, with smooth edges composed of cells with a cobblestone morphology similar to iBMK cell colonies (Fig. 5.1).
3. Well-separated colonies are cloned and expanded separately in 96–well tissue culture plates and eventually transferred into 10-cm dishes by use of the general protocol described in section 2.1.5.
4. Multiple vials of iMPEC cell lines (Table 5.8) are harvested and cryopreserved as described in section 2.1.5. Protein lysates are made from each clone and analyzed for E1A and p53DD expression (section 2.1.1) and epithelial markers such as E-cadherin and cytokeratin by Western blot analysis (Bray *et al.*, 2008).

6. TUMORIGENICITY ASSAY

Immortalized primary iBMK cell, iMMEC, iMOSEC, and iMPEC lines from wild-type (C57B/6) or from adult mouse are nontumorigenic (Bray *et al.*, 2008; Degenhardt *et al.*, 2002a; Karantza-Wadsworth *et al.*, 2007a). However, they are rendered tumorigenic by cooperating functions such as blockade of apoptosis or oncogene activation (HER-2/*neu*, myr-Akt, RAF-CAAX, H-ras, Bcl-2) (Tables 5.1 to 5.8) (Bray *et al.*, 2008; Degenhardt *et al.*, 2002a; 2006; Karantza-Wadsworth *et al.*, 2007a; Nelson *et al.*, 2004; Tan *et al.*, 2005). Thus, tumorigenicity resulting from specific genetic manipulations can be directly assessed to isogenic and nontumorigenic cell lines for comparison.

6.1. Protocol for tumor formation in mice

1. Isogenic iBMK cells of the genotype to be studied, along with their controls, growing in log-phase in normal culture condition are trypsinized, harvested.
2. Cells are grown to 90% confluency in approximately five 15-cm tissue culture plates (roughly one 15-cm plate per animal to be injected). Cells are pooled, washed twice with PBS, and viable cell number is determined by trypan blue exclusion.
3. Cells are then resuspended to a cell density to 1×10^8 cells/ml in sterile PBS.
4. 10^6 to 10^7 cells are injected into the subcutaneous space in the abdominal flank of five athymic nude mice per cell line (5 weeks old, male, NCR Nu/Nu; Taconic, German Town, NY) as per an IACUC-approved protocol.
5. Tumor growth is monitored twice a week by measurement with a 6-inch dial caliper (General Tools mfg. Co., New York, NY), and tumor growth rates are compared by calculating tumor volumes with the following formula (Streit et al., 1999):

$$Volume \ (mm^3 = [(4\pi/3(0.5 \times smaller \ diameter)^2)$$
$$\times \ (0.5 \times larger \ diameter)]$$

6. We use cell lines stably expressing fluorescence or chemiluminescence markers to assess the role of apoptosis regulators in tumorigenicity and treatment response (Tan et al., 2005). In addition, tumors from cells stably expressing fluorescence reporter genes such as RFP may be monitored for tumorigenicity and treatment response noninvasively by use of Illumitool imaging and camera system (see section 7) (Lightool Research, Encinitas, CA) equipped with an RFP filter (Nelson et al., 2004; Tan et al., 2005).
7. When tumors reach a volume of 1000 mm^3, animals are sacrificed and tumors are excised under sterile conditions, divided into three sections, and preserved for immunohistochemistry, Western blotting, and DNA/RNA isolation, respectively.

6.2. Protocol for tail vein injections for tumor formation in mice

1. Isogenic iBMK cells, growing in log-phase in tissue culture plates are trypsinized, harvested, and resuspended in PBS at a density of 2.5×10^6 cells/ml.

2. The mice are weighed and transferred to a mouse restrainer (PLASLABS, Lansing, MI), which restrains the mouse while allowing access to the tail vein.

3. The lateral tail vein is identified on either side of the tail nerve. Start the injection at the tip of the tail and move closer to the body if you need to inject the mouse more then once. The injection site is disinfected with 70% ethanol and a needle (30G) containing the cell suspension is inserted into the vein at a slight angle. The cell suspension (0.5×10^6 cells in $200 \ \mu l$ volume) is injected slowly and carefully into the vein starting at the tip of the tail keeping the needle as flat and parallel to the vein as possible. If incorrect positioning occurs (which results in bulging in the tail), the process is repeated proximal to the previous site. Cells systemically migrate to various organs and anatomic sites and colonize as tumors that are visible in approximately 5 to 6 weeks. Animals are monitored for apparent tumor formation and are sacrificed after approximately 5 to 6 weeks for the assessment of tumor formations at various anatomical sites. Cell lines stably expressing fluorescence or chemiluminescence markers (Tables 5.2, 5.5, and 5.6) (also see section 7) can be tracked and monitor tumor growth noninvasively *in vivo*.

7. NONINVASIVE *IN VIVO* MONITORING OF TUMOR GROWTH BY EXPRESSION OF FLUORESCENCE AND CHEMILUMINESCENCE MARKERS

Because these immortal mouse epithelial cells are genetically manipulable, they can be easily engineered to express fluorescence or luminescence reporters such as green or red fluorescent protein (GFP or RFP respectively) to noninvasively monitor tumor growth and chemotherapeutic response *in vivo* (Table 5.2) (Nelson *et al.*, 2004; Tan *et al.*, 2005) or GFP or RFP fusions of proteins of interest (Table 5.2) (Degenhardt *et al.*, 2006; Karantza-Wadsworth *et al.*, 2007a; Mathew *et al.*, 2007b). Immortalized epithelial cell lines can also be engineered to stably express luciferase reporter genes under the control of a specific promoter element for *in vivo* imaging and examination of activation of transcription factors such as NF-κB in tumors (Degenhardt *et al.*, 2006).

We have successfully used immortalized epithelial cell lines expressing the fluorescence protein RFP to demonstrate that loss of Bax and Bak or Bim or gain of Bcl-2 blocks cell death and promotes tumor growth in nude mice (Nelson *et al.*, 2004; Tan *et al.*, 2005). iBMK cells expressing RFP that are either wild-type (W2.3.1-5), double deficient in Bax and Bak (D3.zeo-2), or expressing Bcl-2 (W2.Bcl-2-3) injected subcutaneously into nude mice. Each mouse is ear-tagged and monitored individually for tumor growth. For monitoring tumor growth *in vivo* noninvasively, animals are individually imaged by use of an Illumitool imaging and camera system

(Lightools Research, Encinitas, CA) equipped with an RFP and GFP filters (Nelson *et al.*, 2004). Only those cells with defects in apoptosis formed tumors within 20 to 60 days.

8. TUMOR-DERIVED CELL LINES (TDCLs) AS A MODEL SYSTEM FOR EPITHELIAL TUMOR-PROMOTING FUNCTIONS

Another powerful application of our immortalized mouse epithelial cell models is the ability to select and screen for *in vivo* genetic and epigenetic events that promote tumor growth by rederiving cell lines from tumors. As described earlier, immortalized epithelial cell lines (iBMKs, iMMECs, iMO-SECs, and iMPECs) are nontumorigenic. However, tumors emerge with long latency (>3 months) because of clonal outgrowth caused by tumor-promoting mutations (Karp *et al.*, 2008; Nelson *et al.*, 2004). Because these cells are initially adapted for growth *in vitro*, they are capable of returning efficiently to growth *in vitro* when removed from the tumor after *in vivo* selection. These tumor cells have acquired stable genetic and epigenetic alterations that render them highly tumorigenic in subsequent tumor formation assays (Karp *et al.*, 2008). Because these tumors are generated from immortalized, nontumorigenic epithelial cells, when cells are isolated from these tumors and established as cultures, these tumor-derived cell lines (TDCLs) show remarkable phenotypic alterations consistent with their enhanced tumorigenicity acquired *in vivo* (Karp *et al.*, 2008). Clonal selection for tumor-promoting functions provides a powerful tool to analyze the molecular events regulating epithelial tumor progression.

iBMK cells are selected *in vivo* for tumorigenicity and tumors are extracted to generate TDCLs. By use of this TDCL-assisted *in vivo* selection we have identified a mammalian ortholog of the *Drosophila* tumor suppressor and polarity regulator, *crumbs* as a gene whose loss of expression disrupted tight junction formation, apicobasal polarity, and contact inhibition and promoted tumor progression in mice. Restoration of *crumbs* expression restored junctions, polarity, and contact inhibition and suppressed migration and metastasis (Karp *et al.*, 2008). Additional and different genetic events regulate tumor growth in distinct organs and anatomic sites after metastasis, which remains to be investigated.

8.1. Protocol for the generation of TDCLs

1. Isogenic iBMK cells of the genotype to be studied, along with their controls, growing in log-phase in normal culture condition, are trypsinized and harvested.

2. Cells are pooled, washed twice with PBS, and cells are then resuspended to a cell density to 1×10^7 cells/ml in PBS and allowed to form tumors in appropriate animal models and locations to generate tumor allografts as described in section 8.1.

3. Once the tumor reaches a volume of 1000 mm³, the tumor is excised and collected in a petri dish in cold sterile PBS.

4. The excised tumor is divided into three equal pieces. One piece is immediately fixed in 10% buffered formalin solution (Buffered Formalde-Fresh; Fisher Scientific, Fair Lawn, NJ) at 4 °C over night, paraffin embedded, and sliced into 5-μm sections for immunohistochemical (IHC) analysis. The second tumor piece is weighed and snap-frozen in liquid nitrogen for DNA/RNA isolation and analysis.

5. The third tumor piece is chopped into several small pieces of approximately 1×1-mm sizes and transferred into a 50-ml conical tube with sterile PBS.

6. Minced tumor pieces from step 5 are washed several times in sterile PBS and broken into smaller pieces in 10 ml tissue culture medium by pipetting up and down by use of a 10-ml pipette.

7. The preceding suspension is diluted 1:4 in tissue culture medium and is transferred into four 10-cm tissue culture dishes in appropriate growth medium containing antibiotics (DMEM/10%FBS/1% Pen Strep).

8. Dissociated tissue pieces are incubated in at 37 °C (5% CO_2) for a week without any disturbance until proliferative cells emerge from small tumor fragments.

9. Cells derived from tumors are allowed to develop into monolayer, at which point cells are harvested from each plate by trypsinization, expanded, and frozen down in multiple aliquots in the vapor phase liquid nitrogen as described in section 2.1.5.

10. TDCLs are then subjected to further phenotypic characterization. Gene expression profiling is also performed by comparing the RNA isolated from the TDCLs, with that from parental cells, for the elucidation of any gain of oncogenic activations or loss of tumor suppressor functions by use of microarray analysis (Karp *et al.*, 2008).

9. Concluding Remarks and Future Perspectives

Derivation of these immortalized mouse epithelial cell line model systems described here combines the power of mouse genetics and convenience of genetically defined nontumorigenic cell systems to address the molecular requirements in epithelial tumorigenesis. The availability of mutant mice with specific gene manipulations makes it possible to derive immortalized mouse epithelial cancer cell lines from many tissues and genotypes of interest with the inherent advantage of having comparable isogenically-derived wild-type cells. The use of these cell lines have been instrumental in the demonstration of the role of a large number of key apoptosis regulators and have provided necessary insight into how this pathway is regulated in cancer (Tables 5.1 to 5.8) (Karp *et al.*, 2008; Mathew *et al.*, 2007a; Nelson *et al.*, 2004; Shimazu *et al.*, 2007; Tan *et al.*, 2005). We are in the process of expanding our cell line panels to include mutants of an increasing number of essential genes implicated in novel pathways such as autophagy (Tables 5.5 and 5.6) (Karantza-Wadsworth *et al.*, 2007a; Mathew *et al.*, 2007a,b). This will provide the necessary impetus in establishing the role of these pathways in cancer progression.

ACKNOWLEDGMENTS

We thank Dr. Zhenyu Yue (Mount Sinai School of Medicine, New York, NY) and Dr. Shengkan Jin (UMDNJ-Robert Wood Johnson Medical School, Piscataway, NJ) for providing the $beclin1^{+/+}$ and $beclin1^{+/-}$ mice, Dr. Noboru Mizushima (Tokyo Metropolitan Institute of Medical Sciences, Tokyo, Japan) for providing $atg5^{+/+}$, $atg5^{+/-}$ and $atg5^{-/-}$ mice, Drs. Tullia Lindsten and Craig B. Thompson (University of Pennsylvania, Philadelphia, PA) for providing $bax^{+/-}/bak^{-/-}$, $bax^{+/-}/bak^{+/-}$ and $bax^{+/-}/bak^{+/+}$ mice, Dr. Jerry Adams (The Walter and Eliza Hall Institute of Medical Research, Melbourne, Australia) for $bim^{+/+}$, $bim^{+/-}$, $puma^{+/+}$, $puma^{+/-}$, $noxa^{+/+}$ and $noxa^{+/-}$ mutant mice, Dr. Andreas Strasser (The Walter and Eliza Hall Institute of Medical Research, Melbourne, Australia) for providing $nbk^{+/+}/bik^{+/+}$ and $nbk^{-/-}/bik^{-/-}$ mutant mice, and Dr. Rick Flavell (Yale University, New Haven, CT) for providing $caspase$-$3^{+/-}/caspase$-$7^{+/-}$ and $caspase$-$3^{+/-}/caspase$-$7^{-/-}$ mice.

REFERENCES

Abate-Shen, C., and Shen, M. M. (2002). Mouse models of prostate carcinogenesis. *Trends. Genet.* **18**, S1–S5.

Adams, J. M., and Cory, S. (2007). Bcl-2-regulated apoptosis: Mechanism and therapeutic potential. *Curr. Opin. Immunol.* **19**, 488–496.

Andrianasolo, E. H., *et al.* (2007). Induction of apoptosis by diterpenes from the soft coral *Xenia elongata. J. Nat. Prod.* **70**, 1551–1557.

Berk, A. J. (2005). Recent lessons in gene expression, cell cycle control, and cell biology from adenovirus. *Oncogene* **24**, 7673–7685.

Bray, K., *et al.* (2008). Targeting Bcl-2 in prostate cancer therapy. In Preparation.

Degenhardt, K., *et al.* (2002a). BAX and BAK mediate p53-independent suppression of tumorigenesis. *Cancer Cell* **2**, 193–203.

Degenhardt, K., *et al.* (2006). Autophagy promotes tumor cell survival and restricts necrosis, inflammation, and tumorigenesis. *Cancer Cell* **10**, 51–64.

Degenhardt, K., *et al.* (2002b). Bax and Bak independently promote cytochrome C release from mitochondria. *J. Biol. Chem.* **277**, 14127–14134.

Degenhardt, K., and White, E. (2006). A mouse model system to genetically dissect the molecular mechanisms regulating tumorigenesis. *Clin. Cancer Res.* **12**, 5298–5304.

Fesik, S. W. (2005). Promoting apoptosis as a strategy for cancer drug discovery. *Nat. Rev. Cancer* **5**, 876–885.

Gelinas, C., and White, E. (2005). BH3-only proteins in control: Specificity regulates MCL-1 and BAK-mediated apoptosis. *Genes Dev.* **19**, 1263–1268.

Hahn, W. C., and Weinberg, R. A. (2002). Rules for making human tumor cells. *N. Engl. J. Med.* **347,** 1593–1603.

Hanahan, D., and Weinberg, R. A. (2000). The hallmarks of cancer. *Cell* **100,** 57–70.

Helt, A. M., and Galloway, D. A. (2003). Mechanisms by which DNA tumor virus oncoproteins target the Rb family of pocket proteins. *Carcinogenesis* **24,** 159–169.

Hsu, M.-Y., Elder, D. A., and Herlyn, M. (1999). *In* "Human Cell Cultures Volume 1, Cancer Cell Lines Part 1 Vol. 1." (B Masters, Ed.), pp. 259–274. Kluwer Academic, Dordrecht.

Karantza-Wadsworth, V., *et al.* (2007a). Autophagy mitigates metabolic stress and genome damage in mammary tumorigenesis. *Genes Dev.* **21,** 1621–1635.

Karantza-Wadsworth, V., and White, E. (2007b). Programmed Cell Death. *In* "Cancer: Principles and Practice of Oncology." (L. T. DeVita and S. A. Rosenberg, eds.), Lippincott, Williams & Wilkins, Baltimore.

Karantza-Wadsworth, V., and White, E. (2008). A mouse epithelial cell model for investigation of mammary tumor-promoting functions. *Methods Enzymol.* This issue.

Karp, C. M., *et al.* (2008). Role of the Polarity Determinant Crumbs in Suppressing Mammalian Epithelial Tumor Progression Cancer Research (In Press).

Karp, C. M., and White, E. (2008). Unpublished.

Lindsten, T., *et al.* (2000). The combined functions of proapoptotic Bcl-2 family members bak and bax are essential for normal development of multiple tissues. *Mol. Cell.* **6,** 1389–1399.

Masters, J. R. (2000). Human cancer cell lines: Fact and fantasy. *Nat. Rev. Mol. Cell Biol.* **1,** 233–236.

Mathew, R., *et al.* (2007a). Role of autophagy in cancer. *Nat. Rev. Cancer* **7,** 961–967.

Mathew, R., *et al.* (2007b). Autophagy suppresses tumor progression by limiting chromosomal instability. *Genes Dev.* **21,** 1367–1381.

Mosmann, T. (1983). Rapid colorimetric assay for cellular growth and survival: Application to proliferation and cytotoxicity assays. *J. Immunol. Methods* **65,** 55–63.

Nelson, D. A., *et al.* (2004). Hypoxia and defective apoptosis drive genomic instability and tumorigenesis. *Genes Dev.* **18,** 2095–2107.

Roby, K. F., *et al.* (2000). Development of a syngeneic mouse model for events related to ovarian cancer. *Carcinogenesis* **21,** 585–591.

Sakamuro, D., *et al.* (1995). c-Myc induces apoptosis in epithelial cells by both p53-dependent and p53-independent mechanisms. *Oncogene* **11,** 2411–2418.

Shaulian, E., *et al.* (1992). Identification of a minimal transforming domain of p53: Negative dominance through abrogation of sequence-specific DNA binding. *Mol. Cell Biol.* **12,** 5581–5592.

Shimazu, T., *et al.* (2007). NBK/BIK antagonizes MCL-1 and BCL-XL and activates BAK-mediated apoptosis in response to protein synthesis inhibition. *Genes Dev.* **21,** 929–941.

Shindler, K. S., *et al.* (1997). Bax deficiency prevents the increased cell death of immature neurons in bcl-x–deficient mice. *J. Neurosci.* **17,** 3112–3119.

Streit, M., *et al.* (1999). Overexpression of thrombospondin-1 decreases angiogenesis and inhibits the growth of human cutaneous squamous cell carcinomas. *Am. J. Pathol.* **155,** 441–452.

Tan, T. T., *et al.* (2005). Key roles of BIM-driven apoptosis in epithelial tumors and rational chemotherapy. *Cancer Cell.* **7,** 227–238.

Wei, M. C., *et al.* (2001). Proapoptotic BAX and BAK: A requisite gateway to mitochondrial dysfunction and death. *Science* **292,** 727–730.

White, E. (2001). Regulation of the cell cycle and apoptosis by the oncogenes of adenovirus. *Oncogene* **20,** 7836–7846.

White, E. (2006). Mechanisms of apoptosis regulation by viral oncogenes in infection and tumorigenesis. *Cell Death Differ.* **13,** 1371–1377.

White, E., *et al.* (1991). Adenovirus E1B 19-kilodalton protein overcomes the cytotoxicity of E1A proteins. *J. Virol.* **65,** 2968–2978.

CHAPTER SIX

DNA DAMAGE RESPONSE AND APOPTOSIS

Dragos Plesca,*,† Suparna Mazumder,* *and* Alexandru Almasan*,‡

Contents

Abstract

A number of methods have been developed to examine the morphologic, biochemical, and molecular changes that happen during the DNA damage response that may ultimately lead to death of cells through various mechanisms that include apoptosis. When cells are exposed to ionizing radiation or chemical DNA-damaging agents, double-stranded DNA breaks (DSB) are generated that rapidly result in the phosphorylation of histone variant H2AX. Because phosphorylation of H2AX at Ser 139 correlates well with each DSB, phospho-H2AX is a sensitive marker to used to examine the DNA damage and its repair. Apoptotic cells are characterized on the basis of their reduced DNA content and morphologic

* Department of Cancer Biology, The Lerner Research Institute, Cleveland Clinic, Cleveland, Ohio
† School of Biomedical Sciences, Kent State University, Kent, Ohio
‡ Department of Radiation Oncology, Cleveland Clinic, Cleveland, Ohio

Methods in Enzymology, Volume 446
ISSN 0076-6879, DOI: 10.1016/S0076-6879(08)01606-6

changes, including nuclear condensation, which can be detected by flow cytometry (sub-G1 DNA content), trypan blue, or Hoechst staining. The appearance of phosphatidylserine on the plasma membrane with annexin V–fluorochrome conjugates indicates the changes in plasma membrane composition and function. By combining it with propidium iodide staining, this method can also be used to distinguish early versus late apoptotic or necrotic events. The activation of caspases is another well-known biochemical marker of apoptosis. Finally, the Bcl-2 family of proteins and the mitochondria that play a critical role in DNA damage-induced apoptosis can be examined by translocation of Bax and cytochrome *c* in and out of mitochondria. In this chapter, we discuss the most commonly used techniques used in our laboratory for determining the DNA damage response leading to apoptosis.

1. INTRODUCTION

The cellular response to DNA damage has been an important component of many cytotoxic therapeutics commonly used in cancer therapy, as well as for physiologic processes that occur during DNA replication, DNA repair, and recombination. Among various DNA-damaging agents, ionizing radiation is a prototypical DNA-damaging agent that has been used in many laboratories to define our mechanistic understanding of the DNA damage response. DNA damage activates checkpoint mechanisms in mammalian cells that arrest cell cycle to allow time to repair the DNA damage, or if that is too severe to induce cell death, most commonly by apoptosis.

Formation of foci of different nuclear proteins is a widely used method for assessing DNA double-strand breaks (DSB) formation and their repair. Among these are 53BP1, Nbs1, Rad51, and BRCA1 (Paull *et al.*, 2000), with γ-H2AX being the most frequently used (Fig. 6.1). Histone H2AX is phosphorylated in response to DNA damage on serine 139 (Rogakou *et al.*, 1998) forming a nuclear focus. γ-H2AX foci serve as flags on the chromatin for further DNA repair. It has been shown that each γ-H2AX focus represents an individual DSB (Rothkamm and Lobrich, 2003). Quantification of γ-H2AX foci has become a standard in the detection and documentation of DNA damage and its repair.

Apoptosis is a universal genetic program of cell death in higher eukaryotes that represents a basic process involved in cellular development and differentiation (Danial and Korsmeyer, 2004; vol. 322). Alternate models of programmed cell death (PCD) have been proposed, including necrosis, autophagy, paraptosis, mitotic catastrophe, and the descriptive model of apoptosis-like and necrosis-like PCD. Cell morphology still remains an important criterion for distinguishing these various forms of cell death from classical apoptosis, because, often, similar molecular mechanisms are involved in their execution.

The Bcl-2 family of proteins has been highly conserved during evolution; its members (24 to date) are critical regulators of apoptosis (Adams and

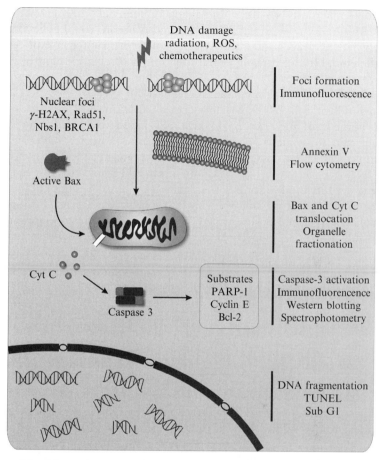

Figure 6.1 Schematic representation of the DNA damage and apoptotic responses. The most commonly used techniques for examining them are represented on the right side.

Cory, 2007). Pro-apoptotic members of this family, such as Bax, after their activation, promote apoptosis by causing the release of cytochrome *c* from mitochondria into the cytosol. Cytochrome *c* acts as a cofactor to stimulate the complexing of Apaf-1 with caspase-9, which then initiates activation of the caspase cascade (Chen *et al.*, 2000; Liu *et al.*, 1996; Wang, 2001). Genotoxic stressors, such as ionizing radiation or topoisomerase inhibitors, induce Bax expression and activation, as well as its mitochondrial localization (Gong *et al.*, 1999; Ray and Almasan, 2003).

Caspases are synthesized as inactive precursors, which are activated by proteolytic cleavage to generate active enzymes. The activation of caspases is a common and critical regulator of the execution phase of apoptosis, triggered by many factors, including treatment with radiotherapeutic and chemotherapeutic DNA-damaging agents (Chen *et al.*, 2000; Gong *et al.*, 1999; Mazumder *et al.*, 2002; 2007; Ray and Almasan, 2003). They further proteolytically cleave proteins that are essential for maintenance of cellular cytoskeleton, DNA repair, signal transduction, and cell cycle control. More than 300 *in vivo* caspase substrates exist (Fischer *et al.*, 2003); among them are poly (ADP-ribose) polymerase (PARP-1), ICAD/DFF45, Bcl-2, and cyclin E (Chen *et al.*, 2000; Mazumder *et al.*, 2002).

A number of methods have been developed to identify the DNA damage and subsequent apoptosis by their morphological, biochemical, and molecular alterations. γ-H2AX is the most recognized assay to measure the DNA damage, with two other methods, MTS and Hoechst staining, offering an initial indication for the occurrence of cell death. These initial observations need to be followed up by more specific assays. Changes in plasma membrane composition and function are detected by the appearance of phosphatidylserine, which reacts with annexin V–fluorochrome conjugates on the external plasma membrane surface. Apoptotic cells are recognized either on the basis of their reduced DNA-associated fluorescence as cells with diminished DNA content (sub-G1) or morphologic changes. Activation of caspases can be examined through a variety of methods: colorimetric, immunoblot, or immunohistochemical. Activated caspases cleave many cellular proteins, and the resulting fragments may also serve as useful biomarkers. This chapter provides a few standard protocols that we have successfully used in our laboratory for a number of experimental systems, including cells grown in culture (Chen *et al.*, 2000; Mazumder *et al.*, 2002; Ray and Almasan, 2003), xenografts (Ray and Almasan, 2003), or *ex vivo* grown specimens (blood or bone marrow) derived from patients with hematologic (Chen *et al.*, 2001) or other malignancies (Masri *et al.*, 2003). MTS and clonogenic assays provide a more general readout for the cumulative effect of various forms of cell death, as well as the ability of cells to proliferate (Brown and Attardi, 2005).

 ## 2. Methods

2.1. Treatment with DNA-damaging agents

The sources of ionizing radiation most commonly used to induce DNA damage are as follows:

1. A radioactive ^{137}Cs γ-ray source, typically with a fixed dose rate of 2 to 3 Gy/min. We use a Shepherd Mark II Cesium-137 irradiator (Gong *et al.*, 1999).
2. As X-ray source, we use a Pantak HF320 Cabinet X-ray irradiator (320kVP, 20 A, half-value layer 2 mm Cu, East Haven, CT) (Crosby *et al.*, 2007).

3. As low-dose-rate irradiator (LDRI), an iridium-192 irradiator (Ir-192 has a half-life of $\sim$74 days). Loss from radioactive decay is appreciable, amounting to approximately 1% per day.

Ionizing radiation mimetic drugs, such as bleomycin or its derivatives, peliomycin, talisomycin, and peplomycin, can also be used. If the DNA-damaging agent is a chemotherapeutic agent, such as an ionizing radiation mimetic drug, topoisomerase inhibitors (e.g., etoposide [VP16], camptothecin and its derivatives (e.g., CPT-11), or another therapeutic (e.g., fludarabine), this is added to the freshly replaced culture medium. If ionizing radiation is used to induce the DNA damage (for doses used see NOTE 1), the culture dishes may need to be transported to and/or from the irradiator and irradiated on ice to avoid phosphorylation of H2AX and other biochemical processes that do not take place at 4 °C.

2.2. DNA damage detection by use of γ-H2AX

We will only briefly describe one method used for assessing the DNA damage, because checkpoint responses and DNA repair have been recently covered in Volumes 408 and 409. Another method used for examining DNA damage and repair in single cells, the comet assay, has been also described (Piperakis et al., 1999). The detection of γ-H2AX foci can be performed by immunofluorescence (the most common method), Western blotting, or flow cytometry (for more detailed description, see Nakamura et al. [2006]).

1. Depending on the cell type (cell size and doubling time), cells are plated 24 h before the treatment on coverslips in 6-well plates or 60-mm dishes at a density that would reach 50 to 75% confluency at the time of treatment. Cells are maintained in a tissue culture incubator at 37 °C, 5% CO_2 overnight.
2. At the end of the incubation period cells are washed three times with PBS (phosphate-buffered saline) at room temperature. TBS (Tris-buffered saline) may also be used instead of PBS. Fix the cells at room temperature with 4% paraformaldehyde in PBS solution for 10 to 30 min depending on the cell line. Wash again three times with PBS, then permeabilize with ice-cold 0.5% Triton X-100 in PBS for 5 min. After permeabilization, wash three times for 5 min with PBS. Block by incubating the cells in a solution of 3% BSA (bovine serum albumin) in PBS for at least 30 min.
3. Primary anti-γ-H2AX antibody is prepared in 1% BSA in PBS. One of the most commonly used antibodies for detection of γ-H2AX is the mouse monoclonal antiphospho-histone γ-H2AX (Ser139) raised against the synthetic peptide CKATQA[pS]QEY available from Upstate (Temecula, CA, cat.#05-636). The antibody is specific for human origin material but has also been used successfully in other species. Typically,

1:100 dilution of anti-γ-H2AX should be sufficient, but this may be variable depending on the cell type or temperature of incubation. Place the coverslips on a paper towel and add 200 μl of the diluted primary antibody. Coverslips are covered with the lids of the 60-mm dishes and incubated overnight at 4 °C. Covering them with the lids makes sure that the solution will not evaporate, thus leaving the coverslips dry. After the incubation, place the coverslips back in the 60-mm culture dishes and wash three times for 5 min with PBS.

4. A solution of secondary antibody is prepared by diluting 1:500 to 1:1000 anti-mouse Alexa fluor 488 (Molecular Probes; cat.#A11029) or anti-mouse Alexa fluor 594 (cat.#A11032) in 1% BSA in PBS. Add 200 μl of the secondary antibody solution to the coverslips, as in the preceding, and incubate for 1 h at room temperature. At the end of the incubation time wash the cells three times for 5 min with PBS.

5. Coverslips are counterstained with 4′,6′-diamidino-2-phenylindole hydrochloride (DAPI)–containing mounting medium Vectashield (Vector Laboratories, Burlingame, CA, cat.#H-1200) and placed upside down on slides. The edges may be sealed with nail polish to make the storage easier. The slides can now be stored at −20 °C until visualization by microscopy.

2.3. MTS assay as an easy, first-hand method for assessing viability and cell proliferation

MTS [3-(4,5-dimethylthiazol-2-yl)-5-(3-carboxymethoxyphenyl)-2-(4-sulfophenyl)-2H-tetrazolium] assay is a colorimetric method used to determine the number of viable cells in proliferation or chemosensitivity assays. MTS (or as an alternative MTT) is a tetrazolium compound that is bioreduced by the cells into formazan, and the quantity of formazan produced is directly proportional to the number of living cells in culture. The absorbance of formazan at 490 nm can be measured directly in 96–well plates. Although very easy and amenable to larger screens, the MTS data obtained alone are not a proof for apoptosis, because they measures changes in metabolic activity that could be also caused by a change in cell proliferation. To assess long-term outcome, clonogenic assays have been commonly used (see NOTE 2).

1. A 96-well plate is used that contains cells cultured in a 100 μl volume of media.
2. Seed cells at a concentration of 2 to 3 $\times$ 10^3 to 10^4 cells/ml, depending on cell type and growth characteristics, in 100 μl medium.
3. Add 20 μl MTS solution (CellTiter 96® Aqueous One Solution Reagent; Promega) to each well. MTT (Promega) can be used as an alternative.
4. Incubate the plate for 1 to 4 h in a humidified 5% CO_2 atmosphere.
5. Record the absorbance at 490-nm on an ELISA plate reader.

2.4. Hoechst staining

Illuminating DNA-bound Hoechst results in blue-light fluorescence. Apoptotic cells appear first as brightly stained with condensed chromatin that are later partitioned into blue beads in fragmented nuclei. This assay allows examination of chromatin condensation of the apoptotic nuclei, as a quick and easy way to differentiate between normal and apoptotic cells on the basis of their fluorescence (UV-excitation < 350 nm).

1. All cells (untreated and treated) are pooled by centrifugation and washed one time with PBS.
2. The PBS is decanted, and the cells are mixed and incubated with the Hoechst 33258 dye (1 μg/ml; Sigma) for 5 min.
3. The fluorescence of the apoptotic cells is determined with a UV-equipped fluorescence microscope.
4. The number of apoptotic cells is scored in several fields (minimum 200 cells/field should be scored).

2.5. Plasma membrane changes detected by annexin-V staining

During the early phases of apoptosis, phosphatidylserine (PS), a protein usually located on the inner leaflet of the plasma membrane in healthy cells, translocates to the outer layer, where it is exposed on the external surface of the cells. Annexin-V has a high affinity for PS, and fluorochrome-tagged annexin-V staining is used as an indicator of apoptosis. The staining is done in combination with PI, a nucleic acid–specific stain that is excluded from live and early apoptotic or necrotic cells but stains DNA and RNA once the plasma membrane is disrupted in these cells. Therefore, it is possible to distinguish live, healthy cells (negative for both annexin-V and PI) from early apoptotic cells (annexin-V positive/PI negative) and late apoptotic or necrotic cells (positive for both annexin-V and PI) by flow cytometry.

1. Remove medium and rinse cells with PBS.
2. Trypsinize attached cells by use of trypsin/EDTA. Only mild trypsinization should be used to get a single cell suspension, because trypsin may damage the PS on the plasma membranes. Alternative methods should be used (e.g., EDTA w/o trypsin) whenever these are effective in dissociating attached cells.
3. Wash cells twice with cold PBS and resuspend 1×10^6 cells in 1 ml of $1\times$ binding buffer (10 mM HEPES/NaOH, pH 7.4, 140 mM NaCl, 2.5 mM CaCl$_2$).
4. Transfer 100 μl of the solution (1×10^5 cells) to a 5-ml polystyrene tube.

5. Add 5 μl of annexin-V (fluorescein or other fluorophore-conjugated annexin-V) and 10 μl of PI (propidium iodide; Sigma): 50 μg/ml stock solution of PI in PBS after which gently vortex the cells.
6. Incubate at room temperature for 15 min in the dark.
7. Add 300 μl of 1× binding buffer to each tube.
8. Analyze by flow cytometry within 1 h.

2.6. Flow cytometry–based assays for cellular morphology and DNA fragmentation

The sub-G1 method for detecting cell death relies on the principle that after DNA endonucleolytic cleavage, the fragmented low-molecular-weight DNA is released from cells during prolonged fixation. That will yield a population of cells that binds a quantitative DNA stain, PI, to a lesser extent than what is characteristic for G1 cells; G1 represents the longest phase of the cell cycle and, therefore, the largest fraction of cells are typically in G1. As a result, there will be a population of cells that appears to the left of the G1 peak (see NOTE 3).

Flow cytometry has the advantage of rapidly examining a large cell population. An analysis of the forward and side light scatter signals of the cells provides an additional method for identification of apoptotic cells on the basis of their physical properties. Thus, for example, as cells shrink during the early stages of apoptosis, the intensity of light that is scattered by these cells in a forward direction along the laser beam will also decrease. As chromatin condenses and apoptotic bodies are formed during the later stages of apoptosis, these cells reflect and refract more light, which can be determined by the increase in the intensity of light scattered at a 90 °C angle (side scatter).

1. Collect 1×10^6 cells by centrifugation at 500g.
2. Prepare a second set of 1-ml centrifuge tubes containing 900 μl of 100% absolute methanol. Methanol needs to be chilled to −20 °C.
3. Resuspend the cells in 100 μl of PBS.
4. Add the cells drop wise by use of a Pasteur pipette to the tubes containing the 100% methanol, while gently vortexing to ensure that the cells are in a single-cell suspension.
5. Place the tubes in a −20 °C freezer and allow fixation to proceed for at least 12 h.
6. Centrifuge the cells at 1000g and aspirate the methanol.
7. Wash the fixed cells two times with 1 ml of cold PBS, again centrifuging at 1000g and being careful to not aspirate cells.
8. Treat the cells with RNase A (100 μg/ml, Gentra Systems) for 20 min at 37 °C.
9. Turn off the lights and stain cells with 50 μg/ml per sample of PI for at least 30 min at room temperature or 15 min at 37 °C.

10. Keep the samples in the dark and run the samples on the flow cytometer, measuring the emission wavelength (617 nm) with a 600- or 610-nm filter.

2.7. Immunocytochemistry to detect active bax/bak, caspase-3, and cytochrome *c*

Whenever Bax and Bak get activated, there is a conformation change that takes place when an internal epitope is exposed, which can be detected by specific antibodies. Although active Bax can be also detected by its mitochondrial translocation, Bak is present mostly in the mitochondria so this is the only way to examine its activation. After their activation, Bax and Bak permeabilize the mitochondrial outer membrane thus facilitating the release of cytochrome *c* into the cytosol.

1. For plating the cells we use glass coverslips (sterilized by dipping in ethanol and passing through flame) that are placed into 6-well plates. Cells (1×10^5 cells/well) are seeded and grown overnight.
2. Remove media and rinse cells with PBS warmed up to 37 °C.
3. To fix the cells, add 1 to 2 ml of 4% formaldehyde to each well. Leave cells to fix for 20 min at room temperature.
4. Wash with PBS for 5 min, three times, each well.
5. Incubate cells in blocking buffer for 5 to 10 min at room temperature.
6. Dilute the primary antibody (two primary antibodies could be added at the same time, but they need to be of different origin (e.g., one rabbit and the other mouse) in 100 to 200 μl blocking buffer, according to the recommended dilution (Primary antibodies: active-Bax, cytochrome *c* [BD Pharmingen], active-Bak [Calbiochem], active Caspase-3 [Cell Signalling]).
7. Use a different 6-well dish to incubate the antibodies. Soak filter paper (3-cm diameter circles) in PBS and place them into the wells; this is necessary to maintain the humidity in the chamber. Place the coverslips on top of the soaked filter paper. Add the primary antibody carefully to cover the entire coverslip. Incubate at room temperature for 1 to 2 h.
8. Wash with PBS for 5 min, three times, each well.
9. Add the secondary antibody (fluorochrome conjugated) in blocking buffer and incubate for 30 to 45 min at room temperature in the dark. For dual staining, the fluorophores need to have different emissions spectra for each individual antibody (e.g., FITC at 525 nM and Phycoerythrin at 578 nM). A set of very sensitive and stable Alexa dyes are available (Molecular Probes, now part of Invitrogen); consult *The Handbook—A Guide to Fluorescent Probes and Labeling Technologies* for a comprehensive resource for fluorescence technology and its applications (http://probes.invitrogen.com/handbook/).

10. Wash with PBS for 5 min, three times, each well.
11. Pick up coverslips with a forceps and drain away excess PBS.
12. For mounting, add a drop of Vectashield to a clean microscope slide and gently lay the coverslip on top.
13. Remove excess Vectashield (mounting medium for fluorescence [Vector Laboratories, Inc.[), with or without DAPI by blotting with Kim wipe and seal with nail polish.
14. After adding the secondary antibody, it is important to keep slides in the dark at all times. A similar protocol can be used for tissue sections (see *NOTE 4*).
15. Store slides in a −20 °C freezer.

2.8. Caspase-3 activity determination: Colorimetric assay

A simple colorimetric assay can measure the release of the chromogenic group from the synthetic substrate, most commonly *p*-nitroanilide (pNA) by activated caspases. Ac-DEVD-pNA is most frequently used, with the cleaved pNA being monitored colorimetrically through its absorbance at 405 to 410 nm. Although DEVD-based substrates are called caspase-3-specific, they are in fact cleaved by most caspases, with caspase-3 being the most efficient. *In vitro* titration experiments and/or use of specific inhibitors may be required to distinguish the activity of various caspases (Gong *et al.*, 1999; vol. 322).

1. Wash cells (1×10^6) with cold PBS and resuspend them in 50 μl of cold lysis HEPES (pH 7.5), 4 mM EDTA. Just before use, add the following protease inhibitors: aprotinin (10 μg/ml), leupeptin (10 μg/ml), pepstatin (10 μg/ml), and PMSF (1 mM), vortex, and keep on ice for 30 min.
2. Centrifuge the cell lysates at 12,000g for 10 min at 4 °C, collect the supernatant in fresh tubes, and assay the protein concentration for each sample. Keep on ice.
3. To a 96-well plate add reaction buffer (100 mM HEPES [pH 7.5], 20% v/v glycerol, 5 mM dithiothreitol (DTT), 0.5 mM EDTA), caspase substrate (100 μM final concentration; Ac-DEVD-pNA, Calbiochem), 20 mM stock in DMSO, and 20 to 50 μg cell lysates for a final 200-μl reaction volume.
4. Incubate samples at 37 °C for 1 to 2 h and monitor the enzyme-catalyzed release of pNA at 405 nm with a microtiter plate reader.

2.9. Immunoblot detection of active bax and caspase-3

In case of Bax-mediated apoptosis, active Bax can be also detected by immunoblot by use of either active Bax-specific antibody or by applying crosslinking agents. Moreover, in most cases, pro-caspase (inactive caspase)-3 is processed into active caspase-3 that can be similarly detected by Western blot analyses. One of its many cellular substrates is PARP-1. In apoptotic cells, PARP-1 (116-kDa) is cleaved into two fragments, most frequently the 86-kDa fragment being detected by the available commercial antibodies.

2.10. Immunoblot analyses

1. Collect the treated and untreated cells (1×10^6) by centrifugation ($500g$ for 5 min). Decant the medium and resuspend the cell pellet in cold PBS very gently and spin it down ($500g$ for 5 min). Decant the supernatant and repeat the process one more time. Remove the PBS carefully without disturbing the cell pellet.

2. Lyse the cells in a buffer (20 mM HEPES, pH 7.5, 1 mM EDTA, 150 mM NaCl, 1% NP-40, 1 mM DTT with protease inhibitors (1 mM PMSF, 10 μg/ml leupeptin), and incubate for 30 min on ice with occasional vortexing.

3. Centrifuge the cells for 15 min at 15,000g and collect the supernatants.

4. The protein estimation of these samples will be performed by use of a spectrophotometric method with the Bio Rad Protein Assay reagent (working solution, 1:10 dilution) at 595 nm (1 to 2 μl of the sample will be mixed with 1 ml of diluted Bio-Rad protein assay reagent), and the concentration of unknown samples will be measured from the BSA standard curve (the curve can be drawn from the spectrophotometric readings of known concentrations of BSA).

5. Load 50 to 100 μg protein with SDS-sample buffer (finally 1×) containing beta-mercaptoethanol (as well as the protein standard marker on a 8 to 15% SDS-PAGE to separate the proteins under denaturing conditions.

6. Transfer the proteins to a nitrocellulose membrane either by the wet or semidry transfer method.

7. Block the membrane with 5% milk for 1 h at room temperature or overnight at 4 °C.

8. Incubate the membrane with primary antibodies (PARP-1, active Bax, or active caspase-3) for 2 h at room temperature or overnight at 4 °C (following the company's recommended dilution; PARP-1, active caspase-3 (Cell Signalling), or active Bax (6A7; Pharmingen), antibodies; see NOTE 5).

9. Wash the blot three times with PBST (PBS with 0.1% Tween 20) at room temperature at 10-min intervals.

10. Add the appropriate secondary antibody (anti-mouse or anti-rabbit depending on the primary antibody) with a 1:2000 dilution to the blot and incubate for 1 to 1.5 h at room temperature.

11. Wash the blot five times with PBST at room temperature at 10 min intervals.

12. Wash the blot with double-distilled water for very short time to get rid of Tween 20 and develop it by use of chemiluminescent reagents, such as Lumiglo or ECL, following the company's suggested protocol.

2.11. Cell fractionation to detect protein translocation to and from mitochondria

During apoptosis, Bax is activated and translocates to the mitochondria. At the same time, cytochrome c is released from the mitochondria into the cytosol. The detection of these and other apoptosis-relevant proteins (Danial and Korsmeyer, 2004; vol. 322) in various subcellular compartments can be used as an important indication of apoptosis.

1. 5×10^7 cells (untreated and treated) are washed first with cold medium (without FBS), then, with ice-cold mitochondria isolation buffer (20 mM Hepes-KOH, pH 7.2, 10 mM KCl, 1.5 mM MgCl$_2$, 1 mM EGTA, 1 mM EDTA, 250 mM sucrose, 1 mM EDTA, 1 mM EGTA, with protease inhibitors (1 mM PMSF, 10 μg/ml aprotinin, 10 μg/ml, leupeptin, and 10 μg/ml pepstatin).

2. Five volumes of the preceding buffer are added to the cell pellet and incubated on ice for 30 min. The cell suspension is then homogenized gently with a Dounce homogenizer, with the extent of lysis being determined by trypan blue exclusion (80 to 90% of cells should be lysed; see NOTE 6). By this procedure, no or minimal cytochrome c should be present in the cytosolic extracts of healthy, untreated cells.

3. The homogenate is centrifuged at 750g for 10 min to remove unbroken cells, large debris, and nuclei.

4. The supernatant is again centrifuged at 10,000g for 15 min.

5. The pellet containing mitochondria, designated as P10, is used for further experiments.

6. The supernatant is subjected to ultracentrifugation at 100,000g for 30 min.

7. The resulting pellet, designated as P100, represents the cellular membranes.

8. The supernatant consists of the cytosolic fraction, designated as S100.

9. The proteins in the cell lysates are separated by SDS-PAGE and transferred to a nitrocellulose membrane.

10. Western blot analysis is performed with primary anti-cytochrome c (Pharmingen) and Bax (Santa Cruz), antibodies. As controls, cytochrome c oxidase I or VDAC (Molecular Probes) are used as mitochondrial markers, with β-actin (Sigma) as a marker for the cytosol to indicate any possible mitochondrial or cytoplasmic contamination in the cellular fractions.

NOTES

1. For hematopoietic cells, 1 to 4 Gy induces DNA damage that leads to apoptosis, whereas for epithelial cells, 5 to 10 Gy is the usual dose used. Ionizing radiation is administered in the clinic not only as external beam radiation (γ- and X-rays) but also as brachytherapy, which can be mimicked by use of low-dose rate irradiators (up to 10 to 50 cGy per h). The dose of ionizing radiation required to induce DNA damage in model organisms (e.g., *C. elegans*, yeast) is quite high compared with mammalian cells.

2. Clonogenic assays provide a more general readout for the cumulative effect of various forms of cell death, as well as the ability of cells to proliferate (Gupta *et al.*, 2008; Morrison *et al.*, 2002; Ray *et al.*, 2007). This assay has been extremely useful for epithelial cells and fibroblasts for which it could well predict the clinical tumor response (Brown and Attardi, 2005). There has been one notable exception in lymphoma cells (Schmitt *et al.*, 2000). Less suitability of hematopoietic cells for this assay could be because their ability to proliferate or even to survive very much depends on cell density; a low density may result in low efficiency of plating (below the 10%, a limit under which this assay is not recommended) (Brown and Attardi, 2005). A lack of a difference in clonogenic survival could mean that compensatory mechanisms have shifted the cellular response from apoptosis to necrosis, autophagy, or another form of cell death. Although the outcome may seem at a first glance to be the same by this assay, how the cells die is important, particularly *in vivo*, because diverting cells from apoptotic to necrotic cell death profoundly alters tumor microenvironment, producing inflammatory cell infiltration and cytokine signaling (Degenhardt *et al.*, 2006). This inflammatory response could extend well beyond the tumor that is targeted by radiation therapy.

3. If cells enter apoptosis at phases other than G1, or if aneuploidy is present, there may not be a sub-G1 peak. In addition, microscopic examination should discern among debris, intact single cells, or doublets. Unless otherwise mentioned, keep the samples on ice throughout staining for flow cytometric studies. In addition, sometimes a sub-G2 DNA content can be detected that could be misinterpreted as S-phase content by PI staining, but which becomes clear once it is shown that cells are negative for BrdU incorporation (Crosby *et al.*, 2007).

4. Formalin-fixed and paraffin-embedded mouse (Ray and Almasan, 2003) or patient-derived human (Masri *et al.*, 2003) tissue sections can be also examined. The slides are deparaffinized with xylene and graded alcohol and treated with citrate buffer (pH 6) for 20 min for antigen retrieval before incubation with primary antibodies. Sections are counterstained with hematoxylin before being examined under the microscope.

Immunohistochemistry for caspase-3 can be combined with the *in situ* detection of apoptotic cells by terminal deoxynucleotide transferase-mediated dUTP nick-end labeling (TUNEL) (Masri *et al.*, 2003). TUNEL or Comet (Piperakis *et al.*, 1999) assays are methods that detect DNA strand breaks that are associated with apoptosis. For TUNEL, the samples are first immunostained for caspase-3 and, after washing in PBS, with a horseradish peroxidase (HRP)-linked secondary antibody. Immunoreactivity is visualized by a 10-min incubation with the HRP substrate diaminobenzidine. After staining for caspase-3, the same slides are then processed for *in situ* detection and localization of apoptosis at the level of single cells. Sections are then stained with anti-fluorescein antibodies linked with alkaline phosphatase, developed with Fast Red substrate and counterstained with hematoxylin.

5. The molecular weight of native PARP-1 is 116 kDa and that of cleaved PARP-1 is of 86 kDa. The molecular weight of pro caspase-3 is 32-kDa, whereas active caspase-3 migrates at 17, as well as 12 kDa. Some antibodies recognize only the pro-form of caspase-3, some recognize only the active form, and some can recognize both. The primary antibodies can be reused a couple of times if they are stored at 4 °C in the presence of sodium azide (0.01%, w/v).

6. For cell fractionation protocols, kits from different companies (e.g., Qiagen, Pierce) are available that are practical and work well for separation of proteins from the various subcellular compartments. Optimal conditions for cell homogenization will depend on the cell type and Dounce homogenizer used. Therefore, the number of strokes required to detect cytochrome *c* in the cytosol of apoptotic but not in control cells will need to be determined.

ACKNOWLEDGMENTS

We thank previous members of our laboratory, particularly Drs. Meredith Crosby and Marcela Oancea, for development of some of the methods presented. The work described in this article was supported by a research grant from the US National Institutes of Health (CA81504).

REFERENCES

Adams, J. M., and Cory, S. (2007). The Bcl-2 apoptotic switch in cancer development and therapy. *Oncogene* **26**, 1324–1337.

Brown, J. M., and Attardi, L. D. (2005). The role of apoptosis in cancer development and treatment response. *Nat. Rev. Cancer* **5**, 231–237.

Chen, Q., Gong, B., and Almasan, A. (2000). Distinct stages of cytochrome *c* release from mitochondria: Evidence for a feedback amplification loop linking caspase activation to mitochondrial dysfunction in genotoxic stress induced apoptosis. *Cell Death Differ.* **7,** 227–233.

Chen, Q., Gong, B., Mahmoud-Ahmed, A., Zhou, A., Hsi, E. D., Hussein, M., and Almasan, A. (2001). Apo2L/TRAIL and Bcl-2-related proteins regulate type I interferon- induced apoptosis in multiple myeloma. *Blood* **98,** 2183–2192.

Crosby, M. E., Jacobberger, J., Gupta, D., Macklis, R. M., and Almasan, A. (2007). E2F4 regulates a stable G(2) arrest response to genotoxic stress in prostate carcinoma. *Oncogene* **26,** 1897–1909.

Danial, N. N., and Korsmeyer, S. J. (2004). Cell death. Critical control points. *Cell* **116,** 205–219.

Degenhardt, K., Mathew, R., Beaudoin, B., Bray, K., Anderson, D., Chen, G., Mukherjee, C., Shi, Y., Gelinas, C., Fan, Y., Nelson, D. A., and Jin, S. *et al.* (2006). Autophagy promotes tumor cell survival and restricts necrosis, inflammation, and tumorigenesis. *Cancer Cell* **10,** 51–64.

Fischer, U., Janicke, R. U., and Schulze-Osthoff, K. (2003). Many cuts to ruin: A comprehensive update of caspase substrates. *Cell Death Differ.* **10,** 76–100.

Gong, B., Chen, Q., Endlich, B., Mazumder, S., and Almasan, A. (1999). Ionizing radiation-induced, Bax-mediated cell death is dependent on activation of serine and cysteine proteases. *Cell Growth Diff.* **10,** 491–502.

Gupta, D., Crosby, M. E., Almasan, A., and Macklis, R. M. (2008). Regulation of CD20 expression by radiation-induced changes in intracellular redox milieau. *Free Radic. Biol. Med.* **44,** 614–623.

Liu, X., Kim, C. N., Yang, J., Jemmerson, R., and Wang, X. (1996). Induction of apoptotic program in cell-free extracts: Requirement for dATP and cytochrome c. *Cell* **86,** 147–157.

Masri, S. C., Yamani, M. H., Russell, M. A., Ratliff, N. B., Yang, J., Almasan, A., Apperson-Hansen, C., Li, J., Starling, R. C., McCarthy, P., Young, J. B., and Bond, M. (2003). Sustained apoptosis in human cardiac allografts despite histologic resolution of rejection. *Transplantation* **76,** 859–864.

Mazumder, S., Chen, Q., Gong, B., Drazba, J. A., Buchsbaum, J. C., and Almasan, A. (2002). Proteolytic cleavage of cyclin E leads to inactivation of associated kinase activity and amplification of apoptosis in hematopoietic cells. *Mol. Cell. Biol.* **22,** 2398–2409.

Mazumder, S., Plesca, D., Kintner, M., and Almasan, A. (2007). Interaction of a Cyclin E fragment with Ku70 regulates Bax-mediated apoptosis in hematopoietic cells. *Mol. Cell Biol.* **27,** 3511–3520.

Morrison, B. H., Bauer, J. A., Hu, J., Grane, R. W., Ozdemir, A. M., Chawla-Sarkar, M., Gong, B., Almasan, A., Kalvakolanu, D. V., and Lindner, D. J. (2002). Inositol hexakisphosphate kinase 2 sensitizes ovarian carcinoma cells to multiple cancer therapeutics. *Oncogene* **21,** 1882–1889.

Nakamura, A., Sedelnikova, O. A., Redon, C., Pilch, D. R., Sinogeeva, N. I., Shroff, R., Lichten, M., and Bonner, W. M. (2006). Techniques for gamma-H2AX detection. *Methods Enzymol.* **409,** 236–250.

Paull, T. T., Rogakou, E. P., Yamazaki, V., Kirchgessner, C. U., Gellert, M., and Bonner, W. M. (2000). A critical role for histone H2AX in recruitment of repair factors to nuclear foci after DNA damage. *Curr. Biol.* **10,** 886–895.

Piperakis, S. M., Visvardis, E. E., and Tassiou, A. M. (1999). Comet assay for nuclear DNA damage. *Methods Enzymol.* **300,** 184–194.

Ray, S., and Almasan, A. (2003). Apoptosis induction in prostate cancer cells and xenografts by combined treatment with Apo2 ligand/tumor necrosis factor-related apoptosis-inducing ligand and CPT-11. *Cancer Res.* **63,** 4713–4723.

Ray, S., Shyam, S., Fraizer, G. C., and Almasan, A. (2007). S-phase checkpoints regulate Apo2 ligand/TRAIL and CPT-11-induced apoptosis of prostate cancer cells. *Mol. Cancer Ther.* **6,** 1368–1378.

Rogakou, E. P., Pilch, D. R., Orr, A. H., Ivanova, V. S., and Bonner, W. M. (1998). DNA double-stranded breaks induce histone H2AX phosphorylation on serine 139. *J. Biol. Chem.* **273,** 5858–5868.

Rothkamm, K., and Lobrich, M. (2003). Evidence for a lack of DNA double-strand break repair in human cells exposed to very low x-ray doses. *Proc. Natl. Acad. Sci. USA* **100,** 5057–5062.

Schmitt, C. A., Rosenthal, C. T., and Lowe, S. W. (2000). Genetic analysis of chemoresistance in primary murine lymphomas. *Nat. Med.* **6,** 1029–1035.

Wang, X. (2001). The expanding role of mitochondria in apoptosis. *Genes Dev.* **15,** 2922–2933.

CHAPTER SEVEN

Phorbol Ester–Induced Apoptosis and Senescence in Cancer Cell Models

Liqing Xiao,[*,1] M. Cecilia Caino,[*,1] Vivian A. von Burstin,[*,1] Jose L. Oliva,[†] and Marcelo G. Kazanietz[*]

Contents

Abstract

Protein kinase C (PKC) isozymes catalyze the phosphorylation of substrates that play key roles in the control in proliferation, differentiation, and survival. Treatment of cells with phorbol esters, activators of classical and novel PKC isozymes, leads to a plethora of responses in a strict cell-type–dependent specific manner. Interestingly, a few cell models undergo apoptosis in response to phorbol ester

* Department of Pharmacology and Institute for Translational Medicine and Therapeutics (ITMAT), University of Pennsylvania School of Medicine, Philadelphia, Pennsylvania
† Unidad de Biología Celular, Centro Nacional de Microbiología, Instituto de Salud Carlos III, Carretera Majadahonda-Pozuelo, Madrid, Spain
1 These three authors contributed equally

Methods in Enzymology, Volume 446
ISSN 0076-6879, DOI: 10.1016/S0076-6879(08)01607-8

stimulation, including androgen-dependent prostate cancer cells. This effect involves the autocrine secretion of death factors and activation of the extrinsic apoptotic cascade. We have recently found that in other models, such as lung cancer cells, phorbol esters lead to irreversible growth arrest and senescence. This chapter describes the methods we use to assess these phorbol ester responses in cancer cell models, focusing on apoptosis and senescence.

1. INTRODUCTION

Despite early studies suggesting a role for protein kinase C (PKC) serine/threonine kinases in mitogenesis and transformation, it became clear that individual PKC isozymes can also have important roles either as modulators of pro-apoptotic or growth inhibitory responses (Black, 2000; Detjen *et al.*, 2000, Frey *et al.*, 1997, Gavrielides *et al.*, 2004; Oster *et al.*, 2006). Studies with phorbol ester tumor promoters have unambiguously implicated PKC as a family of tumor-promoting kinases, because PKC activation promotes the clonal expansion of an initiated cell. Many years have passed since these initial observations, and the field has gone a long way to establish that the paradigm is not as simple as initially thought, primarily because individual members of the PKC family play distinct (or sometimes overlapping) roles in mitogenesis, differentiation, survival, and cell death (Dempsey *et al.*, 2000; Jaken *et al.*, 2000; Mischak *et al.*, 1993; Ron *et al.*, 1999; Schechtman *et al.*, 2001; Weinstein *et al.*, 1997). Because phorbol esters target classical (calcium-dependent) PKCs α, βI, βII, and γ, as well as novel (calcium-independent) PKCs δ, ε, η, and θ, responses in most cases will depend on the distinctive expression pattern of PKC isozymes in a particular cell type, as well as to the differential expression of PKC targets and downstream effectors, which unfortunately have not yet been fully defined. PKCs differ not only in their biochemical properties but also in tissue distribution and intracellular localization, and they are subject to translocation to membrane compartments upon stimulation with either phorbol esters or ligands that activate plasma membrane receptors that generate diacylglycerol (DAG). Responses to phorbol esters can also be mediated by proteins unrelated to PKC isozymes, including protein kinase D (PKD) isozymes, chimaerins, and Ras GRPs (Colon-Gonzalez *et al.*, 2006).

2. PKCδ AS A PRO-APOPTOTIC KINASE

Studies in hematopoietic cells have recognized that activation of PKC can signal toward apoptosis. In 1995, Emoto *et al.* demonstrated that in U-937 cells, ionizing radiation proteolytically cleaves the novel PKCδ isozyme to generate a ~40 kDa constitutively active fragment (Emoto *et al.*, 1995). Cleavage of PKCδ was found to occur adjacent to an Asp residue at a site with homology to that involved in proteolytic activation of interleukin-1-beta-converting enzyme (ICE). Subsequent studies demonstrated that overexpression of the catalytic fragment of PKCδ in cells is associated with an apoptotic response, as revealed by characteristic chromatin condensation, nuclear fragmentation, and accumulation of cells in sub G0/G1 phase (Bharti *et al.*, 1998; Denning *et al.*, 1998; 2002, Ghayur *et al.*, 1996, Sitailo *et al.*, 2004). A similar link between the generation of a PKCδ catalytic fragment and apoptosis was later demonstrated in several other cell types, including keratinocytes, neurons, and salivary gland cells. PKCδ cleavage occurs in response to various stimuli, such as DNA damage, UV radiation, and death factors (tumor necrosis factor-alpha or TNFα) (Basu *et al.*, 2001; Denning *et al.*, 1998; 2002; DeVries *et al.*, 2002). Cleavage was found to be dependent on caspase-3, as caspase inhibitors prevented PKCδ catalytic fragment formation as blocked the apoptotic response (Reyland *et al.*, 1999; Sitailo *et al.*, 2004). Studies from Reyland and coworkers determined that nuclear translocation of the PKCδ catalytic fragment is highly relevant for apoptosis, and a nuclear localization signal in the C-terminal region of PKCδ has been identified as responsible for nuclear shuttling in response to etoposide. PKCδ mutants lacking a functional nuclear localization signal are, indeed, unable to trigger an apoptotic response (DeVries *et al.*, 2002). In some models, mitochondrial localization of PKCδ is required for its apoptotic activity (Denning *et al.*, 2002).

It also became evident that tyrosine phosphorylation of specific residues in PKCδ plays a modulatory role in apoptosis induced by DNA damaging agents, as initially described by Brodie, Blumberg and coworkers (Blass *et al.*, 2002). The PKCδ caspase cleavage site located in the hinge region that connects the regulatory and kinase domains is flanked by tyrosines 311 and 332. A recent study in glioma cells has determined that phosphorylation of tyrosine 311 in PKCδ by c-Abl, but not Src, Lyn, or Yes, contributed to the apoptotic effect of H_2O_2, and a phosphomimetic PKCδ mutation in that position (Y→E) induced glioma cell apoptosis through the p38 MAPK pathway. Tyrosine phosphorylation in position 332 is enhanced by TRAIL and cisplatin and modulates the cleavage of the catalytic domain, as well as apoptotic function, suggesting a key regulatory role for tyrosine phosphorylation in the hinge region (Brodie *et al.*, 2003).

3. PKCδ AS A MEDIATOR OF APOPTOTIC RESPONSES IN PROSTATE CANCER CELLS: THE EMERGENCE OF AN AUTOCRINE PARADIGM

One of the first models in which phorbol esters were found to trigger an apoptotic response was the androgen-dependent LNCaP prostate cancer cell line. Early studies by Powell and coworkers, as well as several others, have established that treatment of LNCaP cells with phorbol 12-myristate (PMA or TPA) leads to apoptotic cell death (Powell *et al.*, 1996). A typical experiment is shown in Fig. 7.1(left panel). LNCaP apoptosis is preceded by $p21^{cip1}$ up-regulation and Rb dephosphorylation (Blagosklonny *et al.*, 1997). Moreover, $p21^{cip1}$ depletion with siRNA markedly impaired the apoptotic response of PMA in these cells (Xiao and Kazanietz, unpublished observations). Surprisingly, only androgen-dependent cells undergo apoptosis in response to phorbol esters.

Our laboratory has extensively studied the molecular basis of the apoptotic effect of phorbol esters in prostate cancer cells. Adenoviral delivery of PKCδ markedly potentiates the apoptotic response of PMA in LNCaP cells. On the other hand, expression of a dominant-negative (kinase-deficient) PKCδ mutant in LNCaP cells reduced the phorbol ester response (Fujii *et al.*, 2000). Similarly, transient PKCδ depletion with siRNA or stable expression of an shRNA against this novel PKC impaired PMA-induced apoptosis (Gonzalez-Guerrico *et al.*, 2005). Notably, we found that induction of LNCaP cell death by PMA does not depend on PKCδ proteolytic cleavage by caspase-3. PMA causes a marked translocation of PKCδ to the plasma and nuclear membranes, although the site of action of this PKC in LNCaP cells is still a matter of debate. Most probably, kinase activation depends on an allosteric mechanism rather than proteolytic activation (Fujii *et al.*, 2000).

Recent studies from our laboratory have determined a key role for the extrinsic apoptotic cascade in phorbol ester–induced apoptosis in LNCaP cells. Conditioned medium (CM) collected from PMA-treated cells (CM-PMA) has apoptogenic activity when added to an untreated LNCaP culture Fig. 7.1(right panel). CM collected from androgen-independent cell lines DU-145 and PC3 cells treated with PMA also triggered cell death when

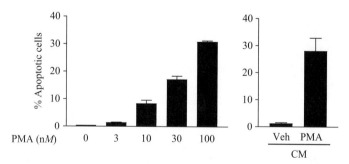

Figure 7.1 Apoptosis induction by PMA in LNCaP prostate cancer cells involves the secretion of autocrine factors. *Left panel*, LNCaP cells were treated with different concentrations of PMA (0–100 nM) for 1 h, and apoptosis was assessed 24 h later by DAPI staining. *Right panel*, LNCaP cells were incubated for 24 h with conditioned medium (CM) collected from LNCaP cultures treated either with vehicle (*Veh*) or PMA (100 nM, 1 h). Apoptotic cells were scored by DAPI staining.

added to LNCaP cells. On the other hand, CM-PMA from NIH-3T3 cells lacks apoptogenic activity. Repetitive washings after addition of the phorbol ester, which prevents the accumulation of apoptotic factors, abolishes the apoptotic response, suggesting that factors released to the medium are essential in conferring cell death. It seems that PKCδ plays a significant role in death factor release, because PKCδ inhibition or depletion by RNAi impairs the apoptotic activity of the CM-PMA (Gonzalez-Guerrico et al., 2005).

An extensive analysis that used cytokine-neutralizing antibodies, death-receptor blocking antibodies, and RNAi for death receptors revealed that TNFα and TRAIL play a significant role in this response. However, the limited ability of these two cytokines to induce apoptosis in LNCaP cells is suggestive of additional factors being released to the CM in response to phorbol esters that sensitize cells to the death factors. The involvement of TNFα was further demonstrated by the ability of TAPI-2, an inhibitor of the metalloproteinase TACE/ADAM17 (TNFα-converting enzyme), and TAPI-2 RNAi, to reduce the apoptotic activity of the CM from PMA-treated LNCaP cells. PMA also causes a significant up-regulation of TNFα mRNA levels and promotes the secretion of this cytokine to the medium. TNFα secretion is markedly reduced on PKCδ RNAi (Gonzalez-Guerrico et al., 2005).

Further experimental support for the involvement of the extrinsic apoptotic cascade in PMA-induced apoptosis comes from signaling studies. It is well established that stimulation of death receptors by TNFα and TRAIL activates a number of signaling cascades, including the JNK, p38 MAPK, and NF-κB pathways (Deng et al., 2003; Lin et al., 2000; 2004; Luschen et al., 2004; Takada et al., 2004; Weldon et al., 2004). Emerging evidence argues for the involvement of p38 MAPK and JNK in prostate cancer cell apoptosis (Park et al., 2003; Shimada et al., 2003; Tanaka et al., 2003). Our studies found that CM-PMA causes a significant activation of JNK and p38 MAPK in LNCaP cells, and in addition it promotes caspase-8 cleavage. Pharmacologic inhibition of p38 MAPK and JNK significantly blocked the apoptotic effect of the CM-PMA but did not affect the release of apoptotic factors by PMA, suggesting that these pathways are primarily involved as effectors of the released death factors. Furthermore, RNAi depletion of caspase-8 or the adaptor FADD impairs the ability of the CM-PMA to promote apoptosis (Gonzalez-Guerrico et al., 2005). Altogether, these studies argue for the contribution of the extrinsic apoptotic cascade in the autocrine effect triggered by PKC activation in LNCaP cells.

4. ROLES OF PKCα AND PKCε IN LNCaP CELL DEATH AND SURVIVAL

LNCaP cells do not only express PKCδ but also the classical PKCα, the novel PKCε, and the atypical PKCζ (Fujii et al., 2000). Early studies by Powell and coworkers have established a correlation between membrane-associated

PKCα and apoptosis (Powell *et al.*, 1996). The diacylglycerol analog HK654, which competes with phorbol esters for binding to PKCs, triggers an apoptotic response in LNCaP cells by specifically activating PKCα. This analog causes a differential relocalization of PKCs to intracellular compartments, and its ability to cause apoptosis correlates with its ability to target PKCα to the plasma membrane. Consistently, PKCα overexpression potentiates PMA-induced apoptosis and, conversely, a dominant-negative (kinase-deficient) PKCα mutant inhibits PMA-induced apoptosis (Garcia-Bermejo *et al.*, 2002). Signaling studies revealed that this classical PKC is responsible for causing Akt dephosphorylation in LNCaP cells, probably by sensitizing cells to death factors. Because cell death can be rescued by an active Akt1 mutant (Myr-Akt1), it is believed that Akt dephosphorylation is an obligatory event for the apoptotic effect of phorbol esters (Tanaka *et al.*, 2003).

The role of PKCε in LNCaP cells is less clear, but this PKC probably mediates survival signaling, as shown in several other cellular models (Ding *et al.*, 2002; Gillespie *et al.*, 2005; Lu *et al.*, 2006; Okhrimenko *et al.*, 2005; Pardo *et al.*, 2006). Terrian's laboratory demonstrated that PKCε overexpression enhances mitogenesis through ERK and favors the transition of LNCaP cells to an androgen-independent state (Wu *et al.*, 2002a; 2002b). Unpublished studies from our laboratory also suggest that PKCε has a pro-survival role in LNCaP cells, because PKCε siRNA significantly potentiates the apoptotic effect of PMA in these cells (Meshki, J., M. C. C., and M. G. K., unpublished observations).

5. REGULATION OF CELL CYCLE AND SENESCENCE BY PKC

Cellular senescence describes the permanent withdrawal from the cell cycle in response to diverse stresses, such as dysfunctional telomeres, DNA damage, strong mitogenic signals, or disrupted chromatin (Campisi, 2005). Increasing evidence indicates that cellular senescence is a critical effector program in response to DNA-damaging chemotherapeutic agents. Despite promising results of early-life survival in cancer patients, which justifies the therapeutic application of premature senescence as a tumor suppressor mechanism, senescent cells can contribute to aging and age-related diseases in renewable tissues can contribute to aging and age-related diseases (Campisi, 2005; Schmitt, 2007). Dimri and colleagues introduced a specific senescence-associated-β-galactosidase (SA-β-Gal) marker (Dimri *et al.*, 1995), which is currently used as a biomarker of senescence. There is consensus that this method can be used to distinguish between senescent cells and quiescent cells in heterogeneous cell populations and aging tissues, such as skin biopsies from older individuals (Itahana *et al.*, 2007).

PKC activation by phorbol esters has been associated with cell cycle arrest, either in G1 or G2, and it can cause a marked change in the expression of cyclins and Cdk inhibitors such as p21^{cip1}, as well as in the activity of various Cdks (Table 7.1). Recently, two independent laboratories reported the involvement of PKC in senescence of human fibroblasts and melanoma cells (Cozzi *et al.*, 2006; Takahashi *et al.*, 2006). In our laboratory, we have found that phorbol ester treatment induces a senescence-like phenotype in lung cancer and colon cancer cell lines, which is characterized by morphologic changes (cell enlargement, flattening of cells, increased vacuolization) (Fig. 7.2A), irreversible cell cycle arrest, inhibition of DNA synthesis, and the expression of senescence-specific markers (Campisi, 2005). Non-small cell lung carcinoma (NSCLC) cells treated with PMA display senescence-like morphologic changes as soon as 1 day after treatment. SA-β-Gal activity is increased approximately the third day after phorbol ester stimulation (Fig. 7.2B). By use of RNAi and pharmacological approaches we established that the phorbol ester effect is mediated by the classical PKCα, and it involves the up-regulation of p21^{cip1} at the transcriptional level (Oliva *et al.*, 2008).

6. DETERMINATION OF PHORBOL ESTER–INDUCED APOPTOSIS IN PROSTATE CANCER CELLS

For these assays we use LNCaP human prostate cancer cells and media from the American Type Culture Collection (ATCC; Manassas, VA). Cells are cultured in RPMI 1640 medium supplemented with 10% fetal bovine serum (FBS; HyClone, Logan, UT) and penicillin (100 U/ml)-streptomycin (100 μg/ml) (Gibco, Grand Island, NY) at 37deg;C in a humidified 5% CO_2 atmosphere. We normally use low passage cells (<8 passages), as growth properties change with subsequent passages and cells start becoming androgen-independent and resistant to apoptosis. We have noticed marked differences in LNCaP growth properties depending on the serum. Therefore,

Table 7.1. Cell cycle distributions 24 h after treatment with PMA (does not include sub-G0/G1)

Cell line	G0/G1 (%)	S (%)	G2/M (%)
H358 (lung cancer)	Veh: 49	Veh: 18	Veh: 21
	PMA: 22	PMA: 12	PMA: 60
HT29 (colon cancer)	Veh: 43	Veh: 22	Veh: 32
	PMA: 63	PMA: 11	PMA: 23

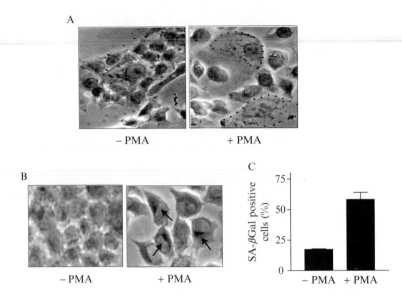

Figure 7.2 Lung cancer cells undergo senescence in response to PMA. Panel A, morphology of H358 cells treated for 30 min with either vehicle (−PMA) or 100 n*M* PMA (+PMA), assessed 3 days later by light field microscopy. The borders of some cells are highlighted by dotted lines. Panels B and C, H460 cells were treated for 30 min with either vehicle (−PMA) or 100 n*M* PMA (+PMA). Three days later, cells were fixed and stained for the senescence associated-β-galactosidase (SA-β-Gal) marker. Positive (senescent) SA-β-Gal cells in Panel B show typical perinuclear staining (arrows). In Panel C the percentage of senescent cells 3 days after PMA treatment is shown. Results are expressed as mean ± SEM of three independent experiments.

we strongly recommend testing different brands and lots of serum. The following protocol is used:

1. Plate LNCaP cells at density of 2×10^5 cells per well in a 6-well plate (~70% confluence) in RPMI 1640 supplemented with 10% FBS.
2. Allow cells to attach. Leave cells for 48 h in complete RPMI medium.
3. Prepare a 1 m*M* PMA (LC Laboratories, Woburn, MA) stock solution in either dimethyl sulfoxide (DMSO) or 100% ethanol. Aliquot the solution and keep it at −80 °C. Even after freezing, we have noticed that PMA activity is reduced with time. We recommend preparing a fresh stock every 3 to 4 weeks.
4. Treat cells with PMA for 1 h. Maximum apoptosis is normally observed with 100 n*M* PMA.
5. If inhibitors are used, such as the PKC inhibitor GF109203X (bisindo-lylmaleimide I), they are normally added 30 min before PMA and left in the medium during the incubation with the phorbol ester.

6. After 1 h incubation with PMA, wash cells twice with prewarmed (37°C) complete medium to remove the PMA. Wash gently to minimize cell detachment.

7. Fresh medium is added and cells are grown for 24 h.

8. Twenty four hours after PMA treatment, the supernatant containing floating cells is collected in a 15-ml conical tube. The plate is washed with 2 ml of PBS and also transferred to the same conical tube.

9. The attached cells are trypsinized and resuspended in 2 ml of PBS, and the cell suspension mixed with that containing the floating cells.

10. Cells are then pelleted at $800\,g$ for 5 min at $4\,°C$. Remove the supernatant and resuspend the cell pellet in 200 μl of PBS.

11. Add 100 μl of the cell suspension on a glass slide (Micro Slides, Single Frosted, Corning; Corning, NY), spread it, and then air-dry. It is possible to let them air-dry for up to 48 h.

12. Fix cells in precooled ($-20\,°C$) 70% ethanol for 20 min.

13. Stain cells with 1 $\mu g/ml$ 4′,6–diamidino-2-phenylindole (DAPI; Sigma, Saint Louis, MO) in PBS for 20 min at $4\,°C$ protected from light. Let the samples air-dry in the dark. The DAPI stock solution is stored at $4\,°C$ in the dark.

14. Finally, mounting medium Fluoromount-G (SouthernBiotech, Birmingham, AL) is added, and the cells are examined by fluorescence microscopy.

15. We normally count 300 cells per preparation and determine the percentage of cells with condensed or fragmented chromatin. PMA 100 nM will normally cause ~30% of cells to undergo apoptosis. The percentage of apoptosis normally matches that observed by flow cytometry (sub G0/G1 population).

7. ADENOVIRAL EXPRESSION OF PKC ISOZYMES

To express wild-type or dominant-negative PKCs in prostate cancer cells we normally use an adenoviral approach. Although adenoviral delivery of PKCδ is insufficient to trigger an apoptotic response, overexpression of this PKC greatly enhances the PMA effect. Generation of PKC AdVs is described elsewhere (Berkner, 1992; Li *et al.*, 1999). Following is a protocol that we normally use to infect prostate cancer cells, but it is similar for other cancer models that we use in the laboratory, such as lung cancer cells.

1. Plate LNCaP cells at a density of 2×10^5 cell/well in a 6-well plate (~70% confluence) in RPMI 1640 supplemented with 10% FBS (see earlier).

2. After 48 h, infect cells with AdV encoding for a PKC isozyme (Berkner, 1992; Fujii, 2000; Kuroki, 1999; Li, 1999; Ohba 1998) or control LacZ AdV for 14 h at different multiplicities of infection (MOIs) in 2%

FBS-supplemented RPMI 1640 medium. We normally use a MOI of 1 to 300 pfu/cell. For the calculation of the volume of adenoviral stock to be added use the following formula:

$$Volume(\mu l) = \frac{MOI(pfu/cell) \times Number\ of\ cells}{Viral\ titer(pfu/\mu 1)}$$

3. After removal of the virus, cells are incubated in RPMI 1640 medium supplemented with 10% FBS for 24 h.
4. Check expression of the corresponding PKC by Western blot. For PKCα we use an antibody from Upstate (Lake Placid, NY) at a 1:1000 dilution. For PKCδ we use an antibody from Cell Signaling (Denvers, MA) at a 1:1000 dilution. Normally, expression of PKCs can be detected 24 h after infection, remaining stable for several days.

8. RNA INTERFERENCE OF PKC ISOZYMES

To interfere with PKC expression we have used 21 bp dsRNAs from Dharmacon, Inc. (Gonzalez-Guerrico *et al.* 2005). The following target sequences were used: CCATCCGCTCCACACTAA (PKCα), CCATGA GTTTATCGCCACC (PKCδ), GTGGAGACCTCATGTTTCA (PKCε). As a control we used CATCGCTGTAGCATCGTCT. For transfection of siRNAs into LNCaP cells we use either Oligofectamine (Invitrogen) or the A maxa nucleofector at a final concentration of 100 nM, although we recommend doing a concentration–response analysis for optimization in each case.

8.1. Transfection with oligofectamine

1. Seed 2×10^5 LNCaP cells/well in a 6-well plate.
2. Change medium to serum-free RPMI 1640 medium for 4 h.
3. Prepare solution #1: For each well, add 200 pmol of the dsRNA (10 μl of a 20 μM solution) to 340μl of Optimen medium for each well.
4. Prepare solution #2 by mixing 12 μl Oligofectamine with 38 μl Opti-mem for each well.
5. Combine solutions #1 and #2. Mix gently and incubate 15 min at room temperature.
6. While complexes form remove media, and add 1400 μl media/well.
7. Add the mix (400 μl/well) to each well.
8. After 4 h, add 200 μl FBS to each well, the final volume should be 2 ml/well.
9. Check expression of PKCs by Western blot. We have observed maximum depletion at 48 h, lasting up to 96 h in some cases.

8.2. Transfection with the amaxa nucleofector

1. LNCaP cells are seeded 2 to 4 days before transfection, to reach ~60 to 80% confluence at the time of transfection.
2. For transfection of LNCaP cells we use Amaxa cell line nucleofector Kit R, according to the standard protocol provided by the manufacturer. After trypsinization, resuspend cells in 100 μl/reaction of nucleofector Solution R. For 2 plates (6-well each), mix 2×10^6 cells with 240 pmol of the dsRNA (6 μl of a 40 μM solution).
3. Add the cell/siRNA mix to the cuvette. Select and start the T-009 Nucleofector program, according to the manual.
4. Transfer cells from cuvette to complete prewarmed RPMI 1640 medium immediately. Plate cells into the 6-well plates that have been coated with poly-L-lysine from Sigma (St. Louse, MO, see standard protocol).
5. Forty-eight hours later, check expression of PKCs by Western blot.

9. COLLECTION AND STORAGE OF CM FROM LNCaP CELLS

CM from PMA-treated prostate cancer cells was shown to have pro-apoptotic activity when added to LNCaP cells (Gonzalez-Guerrico, 2005). Although the nature of the apoptotic factors is not yet fully characterized, it became clear that they act mainly through the activation of the extrinsic apoptotic cascade. For the collection of CM we use the following protocol:

1. Plate LNCaP cells in complete medium in 10-cm dishes, as described previously.
2. Two or 3 days later, when the culture reaches ~70% confluence, treat cells with either 100 nM PMA or vehicle for 1 h.
3. Gently wash the cells twice with prewarmed (37 °C) medium to remove the phorbol ester, trying to minimize cell detachment. Add fresh complete medium.
4. At 24 h, collect CM and pass it through a 13-mm syringe filter (0.45-μm pore size, Fisher Scientific). CM can be dialyzed by use of 12 to 14 kDa cutoff membrane.
5. To assess the apoptogenic activity of the CM, 2 ml/well of the CM are applied to LNCaP cells in 6-well plates, and cells are collected 24 h later for the assessment of apoptosis by DAPI staining.
6. Although we found that CM retained its apoptotic activity up to a week at 4 °C, we strongly recommend the use of fresh CM for each experiment. Freezing and thawing may cause significant loss of apoptotic activity.

10. DETERMINATION OF SENESCENCE IN RESPONSE TO PHORBOL ESTERS

We found that a short–term treatment with PMA leads to a senescent phenotype in H358 lung carcinoma cells (Oliva *et al.*, 2008). We normally use a combination of flow cytometry, cell counting, [³H] thymidine incorporation, and SA-β-Gal staining for routine assessment of senescence. For this analysis it is important that cells do not reach confluency to avoid inhibition by contact.

10.1. H358 cell culture and PMA treatment

1. Plate H358 lung cancer cells in RPMI 1640 medium supplemented with 10% FBS, penicillin (100 U/ml)-streptomycin (100 μg/ml). Cell number for each type of readout are indicated in Table 7.2.
2. The next day treat cells with PMA (100 n*M*, 30 min).
3. Wash three times with PBS to remove PMA.
4. Add complete medium.

10.2. Flow cytometry

Cells are seeded in duplicates (see Table 7.2) and treated with PMA as described previously. At the desired times (e.g., 24, 48, and 72 h after treatment), process the cells for flow cytometry, as follows:

1. Trypsinize the cells and resuspend them in complete medium.
2. Transfer the cells from each plate to 15-ml conical tubes.
3. Pellet cells by centrifugation at 800g (5 min, 4 °C).
4. Discard the supernatant and resuspend the pellet in 500 μl PBS.
5. Add 4.5 ml of 70% ethanol. Allow the cells to fix at 4 °C for at least 30 min before continuing the staining.
6. Collect the cells by centrifugation.
7. Resuspend pellet in 300 to 500 μl of staining solution (0.1 mg/ml propidium iodide, 1 mg/ml RNase A in 0.1% Triton-X 100).

Table 7.2. Cell number for H358 lung cancer cells

Assay	Plate	Cell number	Volume of medium
Flow cytometry	60-mm	400,000 cells/plate	4 ml
Proliferation	12-well plate	100,000 cells/well	1 ml
SA-β-Gal	6-well plate	200,000 cells/well	2 ml

8. Perform the FACS analysis. Samples from different time points should be stored and processed at the same time. Typical results for different cell lines are indicated in Table 7.2.

10.3. [³H]Thymidine incorporation

Seed cells in triplicate wells (see Table 7.2). Treat cells with PMA as described previously. We recommend thymidine incorporation to be determined at various times after PMA treatment (1 to 3 days). Measurements are carried out as follows:

1. Add 3 μCi/ml of [methyl-^{3}H] thymidine (Amersham). Incubate 4 h at 37 °C.
2. Wash cells twice with PBS (2 ml/well).
3. Fix the cells in cold ethanol for 5 min (2 ml/well).
4. Wash twice with PBS.
5. Precipitate the DNA by adding 1 ml/well of ice-cold 20% trichloroacetic acid (TCA, 20 min, 4 °C).
6. Wash with PBS.
7. Add 0.5 N NaOH (600 μl/well).
8. Store the plates at -20 °C until all time point samples are ready for measurement.
9. Measure radioactivity in each sample in a scintillation counter (300 μl of sample in 2 ml scintillation fluid/ vial). A 70% inhibition of DNA synthesis is normally observed in H358 cells 24 h after PMA treatment.

10.4. SA-β-Gal staining

Cells are seeded as described in Table 7.2. Treat cells with PMA as described previously. Staining for SA-β-Gal is conducted 3 days later, as follows:

1. Wash cells once with PBS.
2. Fix cells in 2% formaldehyde/0.2% glutaraldehyde in PBS (5 min, room temperature).
3. Wash cell twice with PBS.
4. Incubate cells overnight with 1 ml/well of staining solution: 1 mg/ml X-Gal (5-bromo-4-chloro-3-indolyl-β-D galactopyranoside), 5 mM potassium ferrocyanide, 5 mM potassium ferricyanide, 150 mM NaCl, and 2 mM MgCl$_2$ in 40 mM citric acid/sodium phosphate buffer, pH 6.0. The pH conditions are extremely important for development of the reaction. Therefore, a fresh staining solution should be prepared each time, and incubations should be performed at 37 °C in an incubator without CO$_2$ injector. Stocks solutions of potassium ferrocyanide/ferricyanide (500 mM in H$_2$O) should be aliquoted and stored at -20 °C.
5. Wash cells twice with PBS.

6. Determine the percentage of blue cells by phase contrast microscopy. We normally count ∼300 cells/well. Fig. 7.2B shows the typical perinuclear staining (arrows) in positive cells.
7. Plates can be stored in 70% glycerol at 4 °C.

REFERENCES

Basu, A., Mohanty, S., and Sun, B. (2001). Differential sensitivity of breast cancer cells to tumor necrosis factor-alpha: Involvement of protein kinase C. *Biochem. Biophys. Res. Commun.* **280**, 883–891.

Berkner, K. L. (1002). Expression of heterologous sequences in adenoviral vectors. *Curr. Top. Microbiol. Immunol.* **158**, 39–66.

Bharti, A., Kraeft, S. K., Gounder, M., Pandey, P., Jin, S., Yuan, Z. M., Lees-Miller, S. P., Weichselbaum, R., Weaver, D., Kufe, L. B., Chen, D., and Kharbanda, S. (1998). Inactivation of DNA-dependent protein kinase by protein kinase Cdelta: Implications for apoptosis. *Mol. Cell. Biol.* **18**, 6719–6728.

Black, J. D. (2000). Protein kinase C-mediated regulation of the cell cycle. *Front. Biosci.* **5**, D406–423.

Blagosklonny, M. V., Prabbu, N.S, and El-Eeiry, W.S (1997). Defects in p21[cip1]WAF1/CIP1, Rb, and c-myc signaling in phorbol ester-resistant cancer cells. *Cancer Res.* **57**, 320–325.

Blass, M., Kronfeld, I., Kazimirsky, G., Blumberg, P. M., and Brodie, C. (2002). Tyrosine phosphorylation of protein kinase Cdelta is essential for its apoptotic effect in response to etoposide. *Mol. Cell Biol.* **22**, 182–195.

Brodie, C., and Blumberg, P.M. (203) Regulation of cell apoptosis by protein kinase C delta Apoptosis. 8 19-27.

Campisi, J. (2005). Senescent cells, tumor suppression, and organismal aging: Good citizens, bad neighbors. *Cell* **120**, 513–522.

Colon-Gonzalez, F., and Kazanietz, M. G. (2006). C1 domains exposed: from diacylglycerol binding to protein-protein interactions,. *Biochim. Biophys. Acta.* **1761**, 827–837.

Cozzi, S. J., Parsons, P. G., Ogbourne, S. M., Pedley, J., and Boyle, G. M. (2006). Induction of senescence in diterpene ester-treated melanoma cells via protein kinase C-dependent hyperactivation of the mitogen-activated protein kinase pathway, *Cancer Res.* **66**, 10083–10091.

Dempsey, E. C., Newton, A. C., Mochly-Rosen, D., Fields, A. P., Reyland, M. E., Insel, P. A., and Messing, R. O. (2000). Protein kinase C isozymes and the regulation of diverse cell responses. *Am. J. Physiol. Lung Cell Mol. Physiol.* **279**, L429–438.

Deng, Y., Ren, X., Yang, L., Lin, Y., and Wu, X. (2003). A JNK-dependent pathway is required for TNFalpha-induced apoptosis. *Cell* **115**, 61–70.

Denning, M. F., Wang, Y., Tibudan, S., Alkan, S., S., Nickoloff.S., and Qin, J. Z. (2002). Caspase activation and disruption of mitochondrial membrane potential during UV radiation-induced apoptosis of human keratinocytes requires activation of protein kinase C. *Cell Death Differ.* **9**, 40–52.

Denning, M. F., Wang, Y., Nickoloff, B. J., and Wrone-Smith, T. (198). Protein kinase Cdelta is activated by caspase-dependent proteolysis during ultraviolet radiation-induced apoptosis of human keratinocytes. *J. Biol. Chem.* **273**, 29995–30002.

Detjen, K. M., Brembeck, F. H., Welzel, M., Kaiser, A., Haller, H., Wiedenmann, B., and Rosewicz, S. (2000). Activation of protein kinase Calpha inhibits growth of pancreatic cancer cells via p21[cip1](cip)-mediated G(1) arrest. *J. Cell Sci.* **Pt 17**, 3025–3035.

DeVries, T. A., Neville, M. C., and Reyland, M. E. (200). Nuclear import of PKCdelta is required for apoptosis: Identification of a novel nuclear import sequence. *EMBO J.* **21,** 6050–6060.

Dimri, G. P., Lee, X., Basile, G., Acosta, M., Scott, G., Roskelley, C., Medrano, E. E., Linskensi, M., Rubelj, I., Pereira-Smith, O., Peacocke, M., and Campisi, J. A biomarker that identifies senescent human cells in culture and in aging skin *in vivo. Proc. Natl. Acad. Sci. USA* **92,** 9363–9367.

Ding, L., Wang, H. M., Lang, W. H., and Xiao, L. (2002). Protein kinase Cepsilon promotes survival of lung cancer cells by suppressing apoptosis through dysregulation of the mitochondrial caspase pathway. *J. Biol. Chem.* **277,** 35305–35313.

Emoto, Y., Manome, Y., Meinhardt, G., Kisaki, H., Kharbanda, S., Robertson, M., Ghayur, T., Wong, W. W., Kamen, R., Weichselbaum, R., *et al.* (1995). Proteolytic activation of protein kinase C delta by an ICE-like protease in apoptotic cells. *EMBO J.* **24,** 6148–6156.

Frey, M. R., Saxon, M. L., Zhao, X., Rollins, A., Evans, S. S., and Black, J. D. (1997). Protein kinase C isozyme-mediated cell cycle arrest involves induction of p21[cip1](waf1/cip1) and p27(kip1) and hypophosphorylation of the retinoblastoma protein in intestinal epithelial cells. *J. Biol. Chem.* **272,** 9424–9435.

Fujii, T., Garcia-Bermejo, M. L., Bernabo, J. L., Caamano, J., Ohba, M., Kuroki, T., Li, L., Yuspa, S. H., and Kazanietz, M. G. (2000). Involvement of protein kinase C delta (PKCdelta) in phorbol ester-induced apoptosis in LNCaP prostate cancer cells. Lack of proteolytic cleavage of PKCdelta. *J. Biol. Chem.* **275,** 7574–7582.

Garcia-Bermejo, M. L., Leskow, F. C., Fujii, T., Wang, Q., Blumberg, P. M., Ohba, M., Kuroki, T., Han, K. C., Lee, J., Marquez, V. E., and Kazanietz, M. G. (2002). Diacyl-glycerol (DAG)-lactones, a new class of protein kinase C (PKC) agonists, induce apoptosis in LNCaP prostate cancer cells by selective activation of PKCalpha. *J. Biol. Chem.* **277,** 645–655. Erratum in: *J. Biol. Chem.* (2004). **279,** 23846.

Gavrielides, M. V., Frijhoff, A. F., Conti., C. J., and Kazanietz, M. G. (2004). Protein kinase C and prostate carcinogenesis: Targeting the cell cycle and apoptotic mechanisms. *Curr. Drug Targets* **5,** 431–443.

Ghayur, T., Hugunin, M., Talanian, R. V., Ratnofsky, S., Quinlan, C., Emoto, Y., Pandey, P., Datta, R., Huang, Y., Kharbanda, S., Allen, H., Kamen, R., Wong., W., and Kufe, D. (1996). Proteolytic activation of protein kinase C delta by an ICE/CED 3-like protease induces characteristics of apoptosis. *J. Exp. Med.* **184,** 2399–2404.

Gillespie, S., Zhang, X. D., and Hersey, P. (2005). Variable expression of protein kinase C epsilon in human melanoma cells regulates sensitivity to TRAIL-induced apoptosis. *Mol. Cancer Ther.* **4,** 668–676.

Gonzalez-Guerrico, A. M., and Kazanietz, M. G. (2005). Phorbol ester-induced apoptosis in prostate cancer cells via autocrine activation of the extrinsic apoptotic cascade: A key role for protein kinase C delta. *J. Biol. Chem.* **280,** 38982–38991.

Itahana, K., Campisi, J., and Dimri, G. P. (2007). Methods to detect biomarkers of cellular senescence: The senescence-associated beta-galactosidase assay. *Methods Mol. Biol.* **371,** 21–31.

Jaken, S., and Parker, P. J. (2000). Protein kinase C binding partners. *Bioessays.* **22,** 45–54

Kuroki, T., Kashiwagi, M., Ishino, K., Huh, N., and Ohba, M. (1999). Adenovirus-mediated gene transfer to keratinocytes—A review. *J. Invest. Dermatol. Symp. Proc.* **4,** 153–157.

Li, L., Lorenzo, P. S., Bogi, K., Blemberg, P. M., and Yuspa, S. H. (1999). Protein kinase Cdelta targets mitochondria, alters mitochondrial membrane potential, and induces apoptosis in normal and neoplastic keratinocytes when overexpressed by an adenoviral vector. *Mol. Cell Biol.* **19,** 8547–8558.

Lin, Y., Choksi, S., Shen, H. M., Yang, Q. F., Hur, G. M., Kim, Y. S., Tran, J. H., Nedospasov, Z. G., and Liu, S. (2004). Tumor necrosis factor-induced nonapoptotic cell death requires receptor-interacting protein-mediated cellular reactive oxygen species accumulation. *J. Biol. Chem.* **279,** 10822–10828.

Lin, Y., Devin, A., Cook, A., Keane, M. M., Kelliher, M., Lipkowitz, S., and Liu, Z. G. (2000). The death domain kinase RIP is essential for TRAIL (Apo2L)-induced activation of IkappaB kinase and c-Jun N-terminal kinase. *Mol. Cell Biol.* **20,** 6638–6645.

Lu, D. M., Huang, J., and Basu, A. (2006). Protein kinase Cepsilon activates protein kinase B/Akt via DNA-PK to protect against tumor necrosis factor-alpha-induced cell death. *J. Biol. Chem.* **281,** 22799–22807.

Luschen, S., Scherer, G., Ussat, S., Ungefroren, H., and Adam-Klages, S. (2004). Inhibition of p38 mitogen-activated protein kinase reduces TNF-induced activation of NF-kappaB, elicits caspase activity, and enhances cytotoxicity. *Exp. Cell Res.* **293,** 196–206.

Mischak, H., Goodnight, J. A., Kolch, W., Martiny-Baron, G., Schaechtle, C., Kazanietz, M. G., Blumberg, P. M., Pierce, J. H., and Mushinski, J. F. (1993). Overexpression of protein kinase C-delta and -epsilon in NIH 3T3 cells induces opposite effects on growth, morphology, anchorage dependence, and tumorigenicity. *J. Biol. Chem.* **268,** 6090–6096.

Ohba, M., Ishino, K., Kashiwagi, M., Kawabe, S., Chida, K., Huh, N. H., and Kuroki, T. (1998). Induction of differentiation in normal human keratinocytes by adenovirus-mediated introduction of the eta and delta isoforms of protein kinase C. *Mol. Cell Biol.* **18,** 5199–5207.

Okhrimenko, H., Lu, H, W., Xiang, C., Hamburger, N., Kazimirsky, G., and Brodie, C. (2005). Protein kinase C-epsilon regulates the apoptosis and survival of glioma cells. *Cancer Res.* **65,** 7301–7309.

Oliva, J. L., Caino, M. C., Senderowicz, A. M., and Kazanietz, M. G. (2008). S-phase-specific activation of PKCalpha induces senescence in non-small cell lung cancer cells. *J. Biol. Chem.* **283,** 5466–5476.

Oster, H., and Leitges, M. (2006). Protein kinase C alpha but not PKC zeta suppresses intestinal tumor formation in ApcMin/+ mice. *Cancer Res.* **66,** 6955–6963.

Pardo, O. E., Wellbrock, C., Khanzada, U. K., Aubert, M., Arozarena, I., Davidson, S., Bowen, F., Parker, P. J., Filonenko, V. V., Gout, I. T., Sebire, N., Marais, R., Downward, J., and Seckl, M. J. FGF–2 protects small cell lung cancer cells from apoptosis through a complex involving PKCepsilon, B-Raf and S6K2. *EMBO J.* **25,** 3078–3088.

Park, J. I., Lee, M. G., Cho, K., Park, B. J., Chae, K. S., Byun, D. S., Ryu, B. K., Park, Y. K., and Chi, S. G. (2003). Transforming growth factor-beta1 activates interleukin-6 expression in prostate cancer cells through the synergistic collaboration of the Smad2, p38-NF-kappaB, JNK, and Ras signaling pathways. *Oncogene* **22,** 4314–4332.

Powell, C. T., Brittis, N. J., Stec, D., Hug, H., Heston, W. D., and Fair, W. R. (1996). Persistent membrane translocation of protein kinase C alpha during 12–0-tetradecanoyl-phorbol-13-acetate-induced apoptosis of LNCaP human prostate cancer cells. *Cell Growth Differ.* **7,** 419–428.

Reyland, M. E., Anderson, S. M., Matassa, A. A., Barzen, K. A., and Quissell, D. O. (1999). Protein kinase C delta is essential for etoposide-induced apoptosis in salivary gland acinar cells. *J. Biol. Chem.* **274,** 19115–19123.

Ron, D., and Kazanietz, M. G. (1999). New insights into the regulation of protein kinase C and novel phorbol ester receptors. *FASEB J.* **13,** 1658–1676.

Schechtman, D., and Mochly-Rosen, D. (2001). Adaptor proteins in protein kinase C-mediated signal transduction. *Oncogene.* **20,** 6339–6347.

Schmitt, C. A. (2007). Cellular senescence and cancer treatment. *Biochim. Biophys. Acta.* **1775,** 5–20.

Shimada, K., Nakamura, M., Ishida, E., Kishi, M., and Konishi, N. (2003). Roles of p38- and c-jun NH2-terminal kinase-mediated pathways in 2-methoxyestradiol-induced p53 induction and apoptosis. *Carcinogenesis.* **24,** 1067–1075.

Sitailo, L. A., Tibudan, S. S., and Denning, M. F. (2004). Bax activation and induction of apoptosis in human keratinocytes by the protein kinase C delta catalytic domain. *J. Invest. Dermatol.* **123,** 434–443.

Takada, Y., and Aggarwal, B. B. (2004). TNF activates Syk protein tyrosine kinase leading to TNF-induced MAPK activation, NF-kappaB activation, and apoptosis. *J. Immunol.* **173,** 1066–1077.

Takahashi, A., Ohtani, N., Yamakoshi, K., Iida, S., Tahara, H., Nakayama, K., Nakayama, K. I., Ide, T., Saya, H., and Hara, E. (2006). Mitogenic signalling and the p16INK4a-Rb pathway cooperate to enforce irreversible cellular senescence. *Nat. Cell Biol.* **8,** 1291–1297.

Tanaka, Y., Gavrielides, M. V., Mitsuuchi, Y., Fujii, T., and Kazanietz, M. G. (2003). Protein kinase C promotes apoptosis in LNCaP prostate cancer cells through activation of p38 MAPK and inhibition of the Akt survival pathway. *J. Biol. Chem.* **278,** 33753–33762.

Weinstein, I. B., Kahn, S. M., O'Driscoll, K., Borner, C., Bang, D., Jiang, W., Blackwood, A., and Nomoto, K. (1997). The role of protein kinase C in signal transduction, growth control and lipid metabolism. *Adv. Exp. Med. Biol.* **400A,** 13–21.

Weldon, C. B., Parker, A. P., Patten, D., Elliott, S., Tang, Y., Frigo, D. E., Dugan, C. M., Coakley, E. L., Butler, N. N., Clayton, J. L., Alam, J., Curiel, T. J., Beckman, B. S., Jaffe, B. M., and Burow, M. E. (2004). Sensitization of apoptotically-resistant breast carcinoma cells to TNF and TRAIL by inhibition of p38 mitogen-activated protein kinase signaling. *Int. J. Oncol.* **24,** 1473–1480.

Wu, D., Foreman, T. L., Gregory, C. W., McJilton, M. A., Wescott, G. G., Ford, O. H., Alvey, R. F., Mohler, J. L., and Terrian, D. M. (2002). Protein kinase Cepsilon has the potential to advance the recurrence of human prostate cancer. *Cancer Res.* **62,** 2423–2439.

Wu, D., and Terrian, D. M. (2002). Regulation of caveolin-1 expression and secretion by a protein kinase Cepsilon signaling pathway in human prostate cancer cells. *J. Biol. Chem.* **277,** 40449–40455.

CHAPTER EIGHT

Manipulation of PKC Isozymes by RNA Interference and Inducible Expression of PKC Constructs

Alakananda Basu, Shalini D. Persaud, *and* Usha Sivaprasad

Contents

Abstract

Protein kinase C (PKC), a family of serine/threonine kinases, plays an important role in apoptosis. Several members of the PKC family act as substrates for caspases. In addition, PKCs can also regulate caspase activation and cell death by apoptosis. The cleavage of PKCs separates the regulatory domain from the catalytic domain. The full-length, the catalytic domain, and the regulatory domain of PKC family members may have distinct function in apoptosis. Delineating the role of protein kinase C (PKC) isozymes in apoptosis has been challenging because of the lack of selective inhibitors of PKC isozymes and difficulty in generating stable cell lines expressing pro-apoptotic PKC isozymes. In this chapter, we describe the use of RNA interference (siRNA) technology and tetracycline-inducible expression of PKC isozymes to study their function in apoptosis.

1. Introduction

Protein kinase C (PKC) is a family of phospholipid–dependent serine/threonine protein kinases that can be categorized into three groups on the basis of their structure and regulation: conventional (α, βI, βII, and γ), novel

Department of Molecular Biology and Immunology, University of North Texas Health Science Center, Fort Worth, Texas

Methods in Enzymology, Volume 446
ISSN 0076-6879, DOI: 10.1016/S0076-6879(08)01608-X

(δ, ε, η, θ), and atypical (ζ and ι/λ) (Basu, 1993; Newton, 2003; Parker and Parkinson, 2001). PKC isozymes have been shown to play important roles during apoptosis. Although PKCδ and -θ are considered pro-apoptotic, PKCα, -β, -ε, -η, -ζ, and -ι are considered anti-apoptotic (Basu, 2003; Basu and Sivaprasad, 2007; Brodie, 2003; Griner and Kazanietz, 2007). However, the function of PKC isozymes during apoptosis varies, depending on the cell type and apoptotic stimuli. Several members of the PKC family have been identified as substrates for caspases (Basu, 2003; Basu and Sivaprasad, 2007; Brodie, 2003).

PKC isozymes contain an N-terminal regulatory domain and a C-terminal catalytic domain joined by a hinge region. PKC remains in an inactive state because of interaction of the regulatory domain with the catalytic domain through pseudosubstrate sequences (House and Kemp, 1987; Makowske and Rosen, 1989). Binding of phospholipids and diacylglycerol/tumor-promoting phorbol esters at the regulatory domain changes the conformation of PKCs, allowing access of substrates at the active site so that catalysis can take place. Several PKCs, including PKCδ, -θ, -ε, and -ζ contain caspase cleavage sites at the hinge region (Basu *et al.*, 2002; Datta, 1997; Emoto, 1995; Mizuno, 1997; Smith, 2000). Thus, cleavage of PKCs by caspases separates the autoinhibitory regulatory domain from the catalytic domain and results in cofactor-independent activation of PKCs. The proteolytic cleavage of PKCs has several consequences. Depending on the PKC isozyme, the active catalytic fragment can function as pro- or anti-apoptotic protein. For example, catalytic fragments of PKCδ and -θ were shown to induce apoptosis, whereas catalytic fragment of PKCζ is believed to be anti-apoptotic. However, the regulatory domain of PKCθ could also enhance apoptosis (Schultz *et al.*, 2003). Thus, the catalytic and regulatory domains of PKC isozymes may have distinct functions compared with the full-length protein.

Three major approaches could be used to delineate the function of individual PKC isozymes during apoptosis: (1) the use of pharmacologic inhibitors, (2) knockdown of PKC isozymes by RNA interference (siRNA), and (3) ectopic expression of PKC constructs. Most of the pharmacologic inhibitors affect multiple PKC isozymes. Although rottlerin is considered a specific inhibitor of PKCδ, it clearly has additional targets (Soltoff, 2001). Thus, one has to rely on molecular approaches to manipulate PKC isozymes to study their functions. The technique of siRNA silencing has advanced the field significantly, because it is possible to specifically knockdown an individual PKC isozyme. However, it is difficult to discern the function of individual domains of PKCs by use of siRNA knockdown. Another approach is to manipulate the levels of PKC isozymes by ectopic expression of different PKC constructs in cells. It is possible to introduce wild-type, constitutively active or dominant–negative PKC isozymes, as well as individual domains of PKCs and chimera between two different PKCs in cells. In most cell types, however, transient transfection results in introduction of PKCs to only a small fraction of cells. If a particular PKC isozyme induces apoptosis, then it may not be possible to generate stable cell lines expressing the PKC isozyme. Several ways are available to circumvent this problem. First, the use of viral vectors could enhance transfection efficiency. Second, PKC isozymes could be tagged with enhanced-green fluorescence protein (EGFP), and cell death could be monitored in EGFP expressing cells. Finally, stable cell lines could be generated by use of an inducible system, where the expression of a particular PKC isozyme could be turned on or off. Because each method has advantages and disadvantages, it is important to use multiple complementary approaches. In the following section, we have described siRNA knockdown and tetracycline-inducible expression of PKC isozymes to study their function during apoptosis.

 ## 2. METHODS

2.1. Knockdown of PKC isozymes with short interfering RNA

The use of short interfering RNA (siRNA) is a particularly effective "loss of function" approach to determine the function of a gene. The RNA interference pathway was originally identified in *Clostridium elegans* nearly a decade ago (Fire *et al.*, 1998). The authors found that exposure of cells to double-stranded RNA leads to suppression of gene expression in a sequence-specific manner. The mechanism and applications of RNA interference have been reviewed extensively (Hannon and Rossi, 2004; Kim, 2003; Meister and Tuschl, 2004) as have the design of siRNA and validation of the technique in mammalian cells (Huppi *et al.*, 2005; Peek and Behlke, 2007). Numerous transfection reagents are available, and each company provides specific instructions for each reagent. We have illustrated this section by use of one such reagent, Lipofectamine 2000 (Invitrogen, Carlsbad, CA) for the knockdown of PKCε in a 12-well plate. The transfection reagent, siRNA, growth media, and cell number should be scaled up or down on the basis of the size of the tissue culture plate. The general transfection procedure is followed by a description of important considerations for siRNA-mediated knockdown of target genes.

2.1.1. PKCε siRNA transfection procedure with Lipofectamine 2000

1. Plate adherent cells in antibiotic-free media in a 12-well plate 24 h before transfection such that they are 60 to 70% confluent at the time of transfection. The number of cells will vary with cell type.
2. On the day of transfection, examine cells and ensure that they are healthy and at the appropriate confluency.
3. In separate tubes, dilute PKCε siRNA and Lipofectamine 2000 (Invitrogen; Catalog #11668-019) in 50 μl prewarmed serum-free medium

(Opti–MEM; Invitrogen, Catalog #31985-070). We could knockdown PKCε expression in MCF-7 cells by use of 50 to 80 nM PKCε and 1 to 2 μl Lipofectamine 2000 (Fig. 8.1).

4. Add diluted Lipofectamine 2000 drop wise to diluted siRNA. Incubate at room temperature for 20 min.
5. Aspirate media from wells. Add 400 μl Opti-MEM to each well. Alternately, you can add the same volume of complete media containing fetal bovine serum (FBS) to each well.
6. Add the siRNA-Lipofectamine 2000 complex (100 μl) from step 3 into each well (drop wise). Rock the plate back and forth after each drop to ensure that the complexes are dispersed evenly.
7. If you use serum-free media, add 500 μl serum-containing media to each well after 6 h.
8. Process cells 24 to 72 h after transfection.
9. Knockdown of the target gene at the mRNA level can be detected by reverse transcription polymerase chain reaction (RT-PCR) and at the protein level by Western blotting.
10. PKCε knockdown takes 24 h after transfection at the mRNA level, but efficient knockdown of PKCε at the protein level may require 48 to 72 h after transfection.

2.1.2. Resources for siRNA

siRNAs can be custom designed or ordered from an existing catalog from many companies, including Ambion (Austin, TX), Dharmacon, Inc. (Lafayette, CO), IDT (Coralville, IA), Santa Cruz Biotechnology (Santa Cruz, CA), Sigma (St. Louis, MO), Upstate Biotechnology (Lake Placid, NY), and others. Table 8.1 shows the accession number of different PKC isozymes and the published siRNA sequences (or target sequences against which siRNAs were designed) used by various laboratories. siRNAs can be ordered from Dharmacon as purified duplexes with >97% purity. We use siRNA SMARTpools®, which are a set of four duplexes designed against various sites within the mRNA sequence for optimal targeting. The

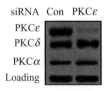

Figure 8.1 Knockdown of PKCε with siRNA. MCF-7 cells were plated in a 12-well plate at a density of 1 × 10^5 cells/well. 24 h after plating, cells were transfected with control or PKCε SMARTpool® siRNA with Lipofectamine 2000. Cells were processed after 48 h and analyzed by western blotting with the indicated antibodies.

Table 8.1 Accession numbers and published siRNA sequences or target sequences for PKC isozymes

PKC isozyme	Accession no.	siRNA sequence or target sequence	References[a]
PKCα	NM 002737	5'-CGACTGGGAAAAACTGGAGAAGCTTGTCCAGTTTTCCCAGTCG-3';	(Chang et al., 2006)
		5'-GGATCGACAGTCGAAAAACTGGACAAGCTTCTCCAGTTTTTCCCAGTCG-3';	
		5'-AAA GGC UTT-3'; 5'-AAA GGC UGA GGU UGC UGA UTT-3';	(Irie et al., 2002)
		5'-AUC AGC AAC CUC AGC CUU UTT-3';	
PKCβI,	NM 002738,	5'-GAAGATGAACTCTTCCAAGAAGCTTGTTGGAAGAGTTCATCTTC-3'	(Chang et al., 2006)
βII	NM 212535	5'-GGATGAAGATGAACTCTTCCAACAAGCTTCTTGGAAGAGTTCATCTTC-3';	
		5'-AAGCGCTGGCGTCATGAATGTT-3';	(Irie et al., 2002)
PKCγ	NM 002739	5'-TCTTTCCCCAGAGGCTCCAAGCTTGGGAGCCTCTGGGGAAAGA-3';	(Chang et al., 2006)
		5'-GGATTCTTTCCCCAGAGGCTCCAAGCTTCGGAGCCTCTGGGGAAAGA-3';	
PKCδ	NM 006254,	5'-GTGGTCCTGATCGACGACGAAGCTTGGTCGTCGATCAGGACCAC-3';	(Chang et al., 2006)
	NM 212539	5'-GGATGTGGTCCTGATCGACGACGAACCTTCGTCGTCGATCAGGACCAC-3';	
		5'-CGA CAA GAU CAU CGG CAG ATT-3'; 5'-UCU GCC GAU GAU CUU GUC GTT-3'	(Irie et al., 2002)
PKCε	NM 005400	5'-GATGACGTGGACTGCACAGAAGCTTGTGTGCAGTCCACGTCATC-3';	(Chang et al. 2006)
		5'-GGATGATGACGTGGACTGCACACAGAAGCTTCTGCAGTCCACGTCATC-3';	
		5'-GCC CCU AAA GAC AAU GAA GTT-3'; 5'-CUU CAU UGU CUU UAG GGG CTT-3'	(Irie et al., 2002)
PKCη	NM 006255	5'-GGAACTTTCAGATATCAAGAAGCTTGTTGTGATATCTGAAAGTTCC-3';	(Chang et al., 2006)
		5'-GGATGGAACTTTCAGATATCAACAAGCTTCTTGATATCTGAAAGTTCC-3'	
PKCθ	NM 006257	5'-gatcccGAGTATGTCGAATCAGAGAttcaagagaTCT	(Manicassamy et al., 2006)
		TGATTCGACATACTCttttggaaagct-3'	
		Seq1: 5'-AAACCACCGTGGAGCTCTACT-3'	(Srivastava et al., 2004)
		Seq2: 5'-AAGAGCCGACCTTCTGTGAA-3'	
PKCλ/ι	NM 002740	5'-AACTTCCTGAAGAACATGCCA-3'	(Jin et al., 2005)
		5'-GAAGAAGCCUUUAGACUUUTA-3'	(Kanayasu–Toyoda et al., 2007)
PKCζ	NM 001033581,	5'-TACACTCCTGCTTCCAGAGAAGCTTGTCTGGAAGCAGGAGTGTA-3'	(Chang et al., 2006)
	NM 002744,	5'-GGATTACACTCCTGCTTCCAGACAAGCTTCTCTGGAAGCAGGAGTGTA-3'	
	NM 001033582	Seq1: 5'-CGC CGC CAC GAC CUU CGA GTT-3'; 5'-CUC GAA GGU CGU GGC GGC GTT-3'	(Irie et al., 2002))
		Seq2: 5'-GAA CGA GGA CGC CGA CCU UTT-3'; 5'-AAG GUC GGC GUC CUC GUU CTT-3'	

[a] References

Chang, J. T., Lu, Y. C., Chen, Y. J., Tseng, C. P., Chen, Y. L., Fang, C. W., Cheng, A. J., (2006). hTERT phosphorylation by PKC is essential for telomerase holoprotein integrity and enzyme activity in head neck cancer cells. *Br. J. Cancer* **94**, 870–878.

Irie, N., Sakai, N., Ueyama, T., Kajimoto, T., Shirai, Y., Saito, N., (2002). Subtype- and species-specific knockdown of PKC using short interfering RNA. *Biochem. Biophys. Res. Commun.* **298**, 738–743.

Jin, Z., Xin, M., Deng, X., (2005). Survival function of protein kinase C[iota] as a novel nitrosamine 4-(methylnitrosamino)-1-(3-pyridyl)-1-butanone-activated bad kinase. *J. Biol. Chem.* **280**, 16045–52.

Kanayasu–Toyoda, T., Suzuki, T., Oshizawa, T., Uchida, E., Hayakawa, T., Yamaguchi, T., (2007). Granulocyte colony–stimulating factor promotes the translocation of protein kinase Ciota in neutrophilic differentiation cells. *J. Cell Physiol.* **211**, 189–96

Manicassamy, S., Sadim, M., Ye, R. D., Sun, Z., (2006). Differential roles of PKC–theta in the regulation of intracellular calcium concentration in primary T cells. *J. Mol. Biol.* **355**, 347–359.

Srivastava, K. K., Batra, S., Sassano, A., Li, Y., Majchrzak, B., Kiyokawa, H., Altman, A., Fish, E. N., Platanias, L. C., (2004). Engagement of protein kinase C–theta in interferon signaling in T-cells. *J. Biol. Chem.* **279**, 29911–20.

mixture of four different siRNAs targeted at distinct sites minimizes non-specific effects that may be associated with targeting of high concentrations of single siRNA to a specific site (Reynolds *et al.*, 2004). siRNA duplexes can be dissolved in $1 \times$ siRNA buffer (dilute $5 \times$ siRNA buffer available from Dharmacon with RNAse-free water). Allow siRNA to stand for 15 min in the $1 \times$ buffer before gently vortexing to resuspend the siRNA. Store siRNA at $-20\,^{\circ}C$ in small aliquots to avoid repeated freeze-thaw cycles.

2.1.3. Transfection reagent

The choice of transfection reagent will depend on the cell line and on empirical assessment of the appropriateness of the reagent by the investigator. Each manufacturer has tested their reagent in a panel of cell lines and provides a protocol for use of their product. The investigator must use this to optimize a procedure for the PKC isozyme and cell line of choice. We have used Lipofectamine 2000 (Invitrogen), SiIMPORTER (Upstate), and Dharmafect (Dharmacon) and found that Lipofectamine 2000 is effective in knocking down PKC isozymes in most cells. However, one limitation with Lipofectamine 2000 is that it could be toxic to cells (see later). In our hands, Dharmafect was also toxic to MCF-7 cells. Although siIMPORTER was not toxic, it was not very effective in knocking down PKC isozymes in MCF-7 cells, although it worked well in HeLa cells.

2.1.4. Toxicity/health of cells

Tied in to the choice of transfection reagent is the fact that many reagents are toxic to cells. Prolonged exposure or use of transfection reagents in excessive amounts can result in cytotoxicity, which can obscure the effect of gene knockdown. The ratio of transfection reagent/siRNA should be optimized for each gene and cell line under consideration. We have found that the siRNA–Lipofectamine complexes diluted in serum-free Opti-MEM can be added to media containing FBS. This diminishes the toxic effects of Lipofectamine 2000 to a large extent.

The maintenance and condition of cells also affects transfection efficiency. If cells that were formerly showing efficient knockdown begin to transfect poorly, we recommend the use of a fresh culture of cells of a lower passage number (Brazas and Hagstrom, 2005).

2.1.5. Media for preparing liposomes

Commonly, Opti-MEM is used for transfection and medium with FBS (complete medium) is added 6 h after transfection. Most serum-free media that do not contain polyanions (heparin or dextran sulfate) can be used. Polyanions inhibit transfection by interfering with liposome formation.

2.1.6. Confluency of cells at time of transfection

Cell confluency can affect the extent of toxicity of the reagent. It is recommended that cells be 30 to 50% confluent at the time of transfection with Lipofectamine 2000 to allow longer interval between transfection and processing of cells and to avoid loss of cell viability because of cell overgrowth. We have found that sparsely plated MCF-7 cells are particularly sensitive to transfection reagents. Therefore, we plate cells at a higher densitiy (60 to 70% confluent). When dealing with a protein displaying a long half-life, however, this can pose a problem, because the cells can get confluent and unhealthy within 48 h. In this case, cells can be replated 24 to 36 h after the first transfection and subsequently retransfected.

2.1.7. Ratio of reagent to siRNA

More is *not* always better when it comes to effective siRNA transfection. We have shown that a ratio of 300 ng Lipofectamine 2000:133 ng siRNA can effectively knockdown expression of PKC isozymes in HeLa cells (Mohanty *et al.*, 2005) and MCF-7 human breast cancer cells (Sivaprasad *et al.*, 2007). Concentrations of siRNA greater than 100 n*M* have greater potential of inducing off-target, nonspecific effects (Straka and Boese, 2005). Therefore, it is important to optimize the ratio of transfection reagent to siRNA at low concentrations of siRNA.

2.1.8. Controls

Chosing the appropriate controls for siRNA transfection is of utmost importance, especially when the readout of the knockdown will be apoptosis. Controls should include a non-transfected well, a well with transfection reagent alone, and one with non-targeting or scrambled siRNA. Although scrambled siRNA has been used extensively, it can bind elsewhere in the genome. Therefore, some researchers prefer the use of a non-targeting control, which is available commercially from Dharmacon. Detailed analysis has confirmed that this control does not cause changes in gene expression in transfected cells (Brazas and Hagstrom, 2005). Non-targeting siRNA has been specifically designed to distinguish general cellular responses to delivery methods and potential off-target effects from siRNA-mediated-sequence-specific silencing.

2.1.9. Specificity of knockdown

When the target belongs to a structurally similar family like the PKCs, it is important to show specificity of the knockdown by confirming that at least one other isoform is not affected (Fig. 8.1). Furthermore, it is important to use more than one siRNA to confirm that the observed effects are specific to knockdown of the target gene and not because of non-specific effects. If a single siRNA rather than SMARTpool® is used to knockdown a specific

PKC isozyme, it is important to examine the effect of at least two siRNAs that target distinct sites.

2.1.10. Stable knockdown with plasmids/viral delivery

Introduction of synthetic siRNA is expected to have a less non-specific effect than delivery of siRNA in a vector. A disadvantage of the use of synthetic siRNA is the short-term nature of the effect. In some cases, the intended effect may not be seen during the 5- to 7-day period for which the knockdown is effective. For example, compensatory/redundant mechanisms may get activated in the short term, and a consistent knockdown over a long period is required for the effect to be measurable. Several advances in plasmid- or viral-mediated siRNA or sh(short-hairpin)RNA delivery allow for long-term and/or inducible knockdown of the target gene. These have been extensively reviewed elsewhere (Devroe and Silver, 2004; Fewell and Schmitt, 2006; Ghosh *et al.*, 2006; Morris and Rossi, 2006; Pardridge, 2007; Vorhies and Nemunaitis, 2007).

2.2. Inducible expression of PKC constructs with tetracycline-inducible gene expression system

Inducible systems have been used to provide quantitative control of exogenous genes in mammalian cells and transgenic mice and plants. Through the use of the *E. coli* lactose and tetracycline (Tet) resistance operons, several transcriptional control systems have been developed to function in eukaryotic cells (Gossen *et al.*, 1995). The Tet-inducible systems were originally created by Bujard and Gossen to provide tight control of inducible genes in mammalian cells (Gossen and Bujard, 1992). Depending on the response to tetracycline, two versions of the Tet-inducible system exist. In the Tet-Off system, gene expression is turned on in the absence of tetracycline, and, conversely, in the Tet-On system gene expression is turned on in the presence of tetracycline. We have focused on the Tet-On system for our purpose in this chapter, although the Tet-Off system has been extensively reviewed (Alexander *et al.*, 2007; Bindels and van den Brekel, 2005; Handler, 2002; McGuire *et al.*, 2004; Sun *et al.*, 2007). The components of Tet-On inducible systems include (1) regulatory plasmid, which encodes the tetracycline repressor (TetR), (2) response plasmid, which contains the gene of interest under the control of a CMV promoter and tetracycline operator sites (*TetO*), and (3) effector substance, either Tet or its derivative, doxycycline (Dox). Since Gossen and Bujard's tetracycline-inducible gene expression systems were developed over a decade ago, several modifications have been made to improve the control of gene expression. These newly designed Tet-based systems are summarized in Table 8.2 and have been focused on eliminating background leakage, increasing sensitivity to tetracycline or Dox, and increasing inducibility of the gene of interest.

In this section, we have illustrated the usefulness of the Tet-On method by use of the T-REx system (Invitrogen Corp., Carlsbad, CA), developed by Yao and coworkers (Yao *et al.*, 1998). In this system, the gene of interest is repressed unless induced by the effector substance Tet or Dox. Because our attempts to generate long-term stable clones overexpressing PKCs involved in apoptosis were unsuccessful as a result of cytotoxicity issues, the repressed expression, and conditional inducibility of genes by use of the T-REx system is extremely advantageous. This system uses two Tet operator 2 sequences (*TetO$_2$*) in the expression plasmid pcDNA^{TM}5/TO that produce high levels of gene expression under the control of a strong CMV promoter. Gene expression is repressed when homodimers of the Tet repressors encoded by the regulatory vector pcDNA6/TR$^©$ bind to *TetO$_2$*. A total of four molecules of TetR binding to the two *TetO$_2$* confer strong repression of gene transcription. When Tet or Dox is added, it binds to the Tet-repressors, causing them to undergo a conformational change. Subsequently, the repressors are released from their bound state resulting in induction of gene expression (Fig. 8.2).

In the following section, we have described how the response plasmids containing full-length, regulatory domain (RD) and catalytic domain (CD) PKCδ-EGFP were generated, the general method of developing stable clones expressing these genes under the control of the regulatory plasmid components, and important considerations when to use Tet-inducible systems. Similar strategies could be used to generate other PKC isozyme constructs.

2.2.1. Cloning of full-length, regulatory domain (RD) and catalytic domain (CD) PKCδ into expression vector pcDNA^{TM}5/TO

1. Full-length PKCδ that we previously cloned into pcDNA3 was cloned into pcDNA^{TM}5/TO by restriction enzyme digestion by use of *BamHI* and *XhoI*.
2. Full-length PKCδ was PCR amplified from pcDNA^{TM}5/TO by use of forward primer 5′-GGGGTACCCCAATTTTATGGCACCGTTCCT GCG-3′ containing a *KpnI* site and reverse primer 5′-CTCACCATGGT GGCGACTGGATCCGGTTTATCT-3′ containing a *BamHI* site. The PCR product was ligated into pcDNA^{TM}5/EGFP (previously generated by Jie Huang) by use of *KpnI* and *BamHI* restriction enzymes to generate PKCδ-EGFP/pcDNA^{TM}5/TO (Fig. 8.3).
3. To generate RD-PKCδ-EGFP/pcDNA^{TM}5/TO, RD-PKCδ was PCR amplified from PKCδ-EGFP/pcDNA^{TM}5/TO by use of forward primer containing the *KpnI* site mentioned in step 2 and reverse primer 5′-CAT TTTCGCGGATCCGGTTTGCTGCCTTTGCCCAGGA-3′ containing a *BamHI* site. The resulting PCR product (RD-PKCδ) was ligated into pcDNA5/EGFP plasmid that was digested with *KpnI* and *BamHI*.

Table 8.2 Modifications in select tet-based systems

Type of system	Modification	Function	Reference
Tet Off/ Tet-On	pTRE-Tight (modified TRE w/seven repeats of a 36 bp sequence and 19 bp *tetO* upstream of the minimal CMV promoter. pTRE-Tight-DsRed (contains gene encoding red fluorescent protein variant isolated from *Discosoma sp*)	to minimize background expression levels of toxic proteins in cell lines	(Clontech, 2003)
pTet-tTS	encodes transcriptional silencer (TS) that blocks transcription of genes under the control of TRE in the absence of Dox.	for regulated expression of toxic genes, extremely low basal expression	(Freundlieb *et al.*, 1999)
Tet-On	rtTA-M2 transactivator driven by the minimal thymidine kinase (Tk) promoter. Luciferase driven by either pTRE, pTRE-tight, pTk-tetO in opposite directions separated y 5 kb human p53 intron. Internal ribosomal entry site (IRES) sequence placed 5' to rtTA-M2. Kanamycin, neomycin resistance.	to facilitate te eneration of conditional transgenic animals	(Backman[a] *et al.*, 2004)
pBI	Bidirectional vectors. TRE flanked on either side by two minimal CMV promoters driving the expression of a reporter gene in one direction and gene of interest in the other.	Simultaneous induction of reporter gene and gene of interest.	(Baron[b] *et al.*, 1995)
ptTA2, ptTA3, ptTA4	VP16 minimal domain vectors. Contain modified VP16 activation domains and each vector allows protein expression over different induction ranges.	Knock in/knock out experiments *in vivo*	(Baron *et al.*, 1997)

[a] Backman, C.M., Zhang, Y., Hoffer, B.J., and Tomac, A.C. (2004) Tetracycline-inducible expression systems for the generation of transgenic animals: A comparision of various inducible systems carried in a single vector. *J. Neurosci Methods* **139**, 257–262.
[b] Baron, U., Freundlieb, S., Gossen, M., and Bujard, H. (1995). Co-regulation of two gene activities by tetracycline via a bidirectional promoter. *Nucleic Acids Res.* **23**, 3605–3606.

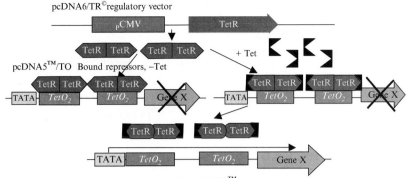

pcDNA6/TR©regulatory vector

Adapted from T-REx™ system: A tetracycline-regulated expression system for mammalian cells, manual, version E, invitrogen life technologies, 2001.

Figure 8.2 Schematic of T-REX System. The regulatory plasmid, pcDNA6/TR©, encodes the tetracyline repressor (TetR). Tet repressors form homodimers and bind to two tetracyline operator 2 sequences (TetO₂) on the pcDNA5 expression plasmid in the absence of tetracycline and represses gene expression. In the presence of tetracycline, the antibiotic binds to the repressors still bound to the operator sequences, causing a conformatinal change and releases the repressors, thus activating gene expression.

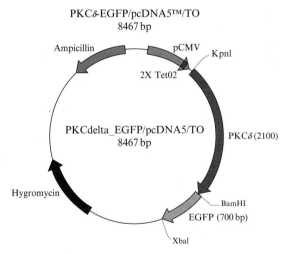

Figure 8.3 Schematic representation of the expression plasmid PKCδEGFP/pcDNA™5/TO. The plasmid contains a strong CMV promoter and the tetracycline operator (2× TetO₂) controlling PKCδ-EGFP expression. The plasmid also contains both hygromycin and ampillicin cassettes for the generation of stable clones. PKCδ-EGFP was cloned into pcDNA™5/TO using KpnI, BamHI, and XbaI restriction enzymes.

4. CD-PKCδ-EGFP/pcDNA^{TM}5/TO was generated by PCR amplification of the catalytic domain from PKCδ-EGFP/pcDNA^{TM}5/TO by use of forward primer 5′- GG<u>GGTACC</u>ATGAAAGGCAGCTTCGGGAAG-3′ containing a *KpnI* site and reverse primer 5′-GGGGC<u>TCTAGA</u>GCTTCTTACTTGTACAGCTCGT-3′ of GFP containing an *XbaI* site. The resulting PCR product (CD-PKCδ) and PKCδ/EGFP/pcDNA^{TM}5/TO were digested with *KpnI* and *BamHI* and ligated to create the CD-PKCδ-EGFP/pcDNA^{TM}5/TO construct.

2.2.2. Creating a stable cell line expressing the Tet repressor

For long-term inducible expression of the gene of interest, a stable cell line that constitutively expresses the Tet repressor should be generated. This cell line is then used to generate a second cell line that is transfected with the expression plasmids to express PKCδ-EGFP, RD-PKCδ-EGFP, and CD-PKCδ-EGFP described in the previous section. The following protocol was adapted from the T-REx System Manual Version E (Invitrogen, 2001).

1. Determine the minimum concentration of antibiotic (blasticidin) required to kill the untransfected cell line by performing concentration-response tests.
   a. Plate cells to approximately 25% confluency.
   b. The following day, treat cells with different concentrations of blasticidin (0 to 10 μg/ml).
   c. Feed cells with selective media every 3 to 4 days and observe percentage of surviving cells.
   d. Count viable cells to determine the appropriate concentration of anti-biotic that kills cells within 1 to 2 weeks after addition of the blasticidin.
2. Transfect the cell line with pcDNA6/TR© by use of desired protocol. Linearization of pcDNA6/TR© may be necessary to increase the integration of the vector in a way that does not disrupt the Tet repressor gene or other elements.
3. Add fresh medium to cells 24 h after transfection.
4. Split cells to 25% confluency into fresh medium containing the minimal concentration of blasticidin determined in step 1.
5. Feed cells with selective medium every 3 to 4 days until colonies appear.
6. Pick and expand at least 20 colonies and test for expression by transiently transfecting with the positive control plasmid expressing β-galactosidase (pcDNA^{TM}5/TO/LacZ). Screen for clones that show the lowest basal expression and the highest levels of β-galactosidase after the addition of Tet. β-Galactosidase expression can be detected by staining with X-gal (β-gal Staining Kit, Invitrogen, Catalog #K1465-01) or by detecting the β-galactosidase fusion protein by Western blot analysis (Anti-β-galactosidase antiserum, Invitrogen, Catalog #R901-25).

−Tet   +Tet

RD-PKCδ

Figure 8.4   Inducible expression of regulatory domain (RD) of PKCδ HeLa cells were transfected with the regulatory domain of PKCδ in tetracycline-inducible system. The expression of RD-PKCδ was induced by treatment with 1 μg/ml tetracycline. The regulatory domain of PKCδ was detected using monoclonal antibody to PKCδ raised against the regulatory domain.

7. Use the stable Tet repressor expressing cell line to isolate a stable clone expressing the gene of interest by transfecting with the pcDNA^{TM}5/TO expression plasmid containing the gene of interest.
8. Follow steps 1 to 5 with hygromycin to select colonies and maintain cells in medium containing blasticidin.
9. Pick and expand at least 20 colonies to test for Tet-regulated gene expression. Figure 8.4 illustrates the inducible expression of PKCδ regulatory domain in the presence of Tet.

2.2.3. Leakiness of system

Transient cotransfection of regulatory and response plasmids can reduce the time taken to generate Tet-inducible gene regulation. This method, however, greatly increases the chance of cointegration of the plasmids into the chromosome. Cointegration events can cause enhancer elements from the strong CMV promoter on the regulatory plasmid to bind and induce basal expression of the gene. Consecutive transfections to generate double stable cell lines expressing the regulatory and response plasmids can prevent cointegration and reduce basal expression of the gene. Countermeasures used to eliminate leaky basal expression in Tet-based systems have recently been summarized (Alexander *et al.*, 2007).

An inherent shortcoming of the Tet-based system is inappropriate binding of regulatory proteins to the Tet operator sequences under basal conditions (Berens and Hillen, 2003; Sun *et al.*, 2007). More tightly controlled regulatory systems can be achieved when modified or advanced versions of Tet-dependent transcriptional activators (Urlinger *et al.*, 2000), Tet-controlled transcriptional silencer (tTS) (Freundlieb *et al.*, 1999), and a minimal promoter with a modified Tet response element (pTRE-Tight) (Clontech, 2003) are used in systems described in Table 8.2.

2.2.4. Tetracycline versus doxycycline

The Tet-Off System is responsive to both Tet and Dox; however, the Tet-On system is only responsive to Dox (Gossen *et al.*, 1995). Dox has a longer half-life than Tet (24 vs 12 h), and this antibiotic can be used at low concentrations to avoid cytotoxic effects.

2.2.5. Purchasing stable cell lines expressing the tet repressor

The ultimate goal in successfully setting up a Tet–inducible system is the generation of double-stable cell lines expressing both the response plasmid and the regulatory plasmid. Companies such as Clontech and Invitrogen have developed premade cell lines that stably express the regulatory plasmid. One can significantly reduce the time needed to establish a stable cell line expressing the Tet regulatory proteins by purchasing a stable cell line. PKCη and PKCδ gene expression was successfully regulated by use of the MCF7 Tet-Off cell line purchased from Clontech to study the function of these proteins in cell cycle control and cell growth (Fima *et al.*, 2001). If a premade cell line stably expressing the regulatory plasmid is used, it is important to cotransfect a selectable marker plasmid such as pTK-Hygro (Clontech, Mountain View, CA.) with the response plasmid containing the gene of interest to select double-stable clones.

2.2.6. Loss of regulation

Over time, stable cell lines can lose their responsiveness to Tet or Dox. This phenomenon may be due to frequent passage that may cause the viral promoter to become methylated or switched off. To circumvent this problem, one recommendation would be to freeze stocks of cells at different stages. One group (Rennel and Gerwins, 2002) reported that in Tet-Off systems, where addition of Tet or Dox suppresses gene expression, there was considerable variability in induction after removal of Dox that could not be explained by clonal variability or the nature of the gene. They found that Dox bound nonspecifically to cells and extracellular matrix that was slowly released after removal from the culture media. The released Dox was sufficient to turn off gene expression. After replating and washing cells 3 h later to remove the released Dox, rapid and high induction of the gene was detected. Another reason for loss of regulation could be due to contamination of serum with Tet. Tet-reduced serum should be used to culture cells in Tet-On systems to reduce basal expression of the gene.

2.2.7. Toxicity of the VP16 activation domain

In the Tet-On and Tet-Off systems originally created by Bujard and Gossen, the TetR is fused to the VP16 activation domain of the herpes simplex virus to create a Tet-transactivator. Because the transcription of the Tet-transactivator is driven by a strong tissue-specific promoter, cytotoxicity of the Tet-transactivator protein can hinder the generation of stable cell lines and may result in negative pleiotrophic effects *in vivo*. Clontech has developed VP16 Minimal Domain Vectors to address this problem (See Table 8.2). The VP16 domain in these vectors is modified so that it contains just the minimal functional amino acids necessary for a tolerated level of transactivator over different induction ranges (Baron *et al.*, 1997).

ACKNOWLEDGMENTS

The authors wish to acknowledge Dr. Baohua Sun for the cloning of PKCδ and Ms. Jie Huang for the generation of the PKCδ Tet-inducible system. This work was supported by grants CA71727 and CA85682 from the National Cancer Institute.

REFERENCES

Alexander, H. K., Booy, E. P., Xiao, W., Ezzati, P., Baust, H., and Los, M. (2007). Selected technologies to control genes and their products for experimental and clinical purposes. *Arch. Immunol. Ther. Exp. (Warsz).* **55,** 139–149.

Baron, U., Gossen, M., and Bujard, H. (1997). Tetracycline-controlled transcription in eukaryotes: Novel transactivators with graded transactivation potential. *Nucleic Acids Res.* **25,** 2723–2729.

Basu, A. (1993). The potential of protein kinase C as a target for anticancer treatment. *Pharmacol. Ther.* **59,** 257–280.

Basu, A. (2003). Involvement of PKC-d in DNA damage-induced apoptosis. *J. Cell Mol. Med.* **7,** 341–350.

Basu, A., Lu, D., Sun, B., Moor, A. N., Akkaraju, G. R., and Huang, J. (2002). Proteolytic activation of protein kinase C-epsilon by caspase-mediated processing and transduction of antiapoptotic signals. *J. Biol. Chem.* **277,** 41850–41856.

Basu, A., and Sivaprasad, U. (2007). Protein kinase Cepsilon makes the life and death decision. *Cell Signal.* **19,** 1633–1642.

Berens, C., and Hillen, W. (2003). Gene regulation by tetracyclines. Constraints of resistance regulation in bacteria shape TetR for application in eukaryotes. *Eur. J. Biochem.* **270,** 3109–3121.

Bindels, E. M., and van den Brekel, M. W. (2005). Development of a conditional mouse model for head and neck squamous cell carcinoma. *Adv. Otorhinolaryngol.* **62,** 1–11.

Brazas, R. M., and Hagstrom, J. E. (2005). Delivery of small interfering RNA to mammalian cells in culture by using cationic lipid/polymer-based transfection reagents. *Methods Enzymol.* **392,** 112–124.

Brodie, C., and Blumberg, P. M. (2003). Regulation of cell apoptosis by protein kinase c d. *Apoptosis* **8,** 19–27.

Clontech, pTR E-Tight,Vectors. (2003). Clontechniques. **XVIII,** 10–11.

Datta, R., Kojima, H., Yoshida, K., and Kufe, D. (1997). Caspase-3-mediated cleavage of protein kinase C q in induction of apoptosis. *J. Biol. Chem.* **272,** 20317–20320.

Devroe, E., and Silver, P. A. (2004). Therapeutic potential of retroviral RNAi vectors. *Expert Opin. Biol. Ther.* **4,** 319–327.

Emoto, Y., Manome, Y., Meinhardt, G., Kisaki, H., Kharbanda, S., Robertson, M., Ghayur, T., Wong, W. W., Kamen, R., Weichselbaum, R., and Kufe, D. (1995). Proteolytic activation of protein kinase C d by an ICE-like protease in apoptotic cells. *EMBO J.* **14,** 6148–6156.

Fewell, G. D., and Schmitt, K. (2006). Vector-based RNAi approaches for stable, inducible and genome-wide screens. *Drug Discov. Today.* **11,** 975–982.

Fima, E., Shtutman, M., Libros, P., Missel, A., Shahaf, G., Kahana, G., and Livneh, E. (2001). PKCeta enhances cell cycle progression, the expression of G1 cyclins and p21 in MCF-7 cells. *Oncogene* **20,** 6794–6804.

Fire, A., Xu, S., Montgomery, M. K., Kostas, S. A., Driver, S. E., and Mello, C. C. (1998). Potent and specific genetic interference by double-stranded RNA in Caenorhabditis elegans. *Nature* **391,** 806–811.

Freundlieb, S., Schirra-Muller, C., and Bujard, H. (1999). A tetracycline controlled activation/repression system with increased potential for gene transfer into mammalian cells. *J. Gene Med.* **1,** 4–12.

Ghosh, S. S., Gopinath, P., and Ramesh, A. (2006). Adenoviral vectors: A promising tool for gene therapy. *Appl. Biochem. Biotechnol.* **133,** 9–29.

Gossen, M., and Bujard, H. (1992). Tight control of gene expression in mammalian cells by tetracycline-responsive promoters. *Proc. Natl. Acad. Sci. USA* **89,** 5547–5551.

Gossen, M., Freundlieb, S., Bender, G., Muller, G., Hillen, W., and Bujard, H. (1995). Transcriptional activation by tetracyclines in mammalian cells. *Science* **268,** 1766–1769.

Griner, E. M., and Kazanietz, M. G. (2007). Protein kinase C and other diacylglycerol effectors in cancer. *Nat. Rev. Cancer.* **7,** 281–294.

Handler, A. M. (2002). Prospects for using genetic transformation for improved SIT and new biocontrol methods. *Genetica* **116,** 137–149.

Hannon, G. J., and Rossi, J. J. (2004). Unlocking the potential of the human genome with RNA interference. *Nature* **431,** 371–378.

House, C., and Kemp, B. E. (1987). Protein kinase C contains a pseudosubstrate prototope in its regulatory domain. *Science* **238,** 1726–1728.

Huppi, K., Martin, S. E., and Caplen, N. J. (2005). Defining and assaying RNAi in mammalian cells. *Mol. Cell.* **17,** 1–10.

Kim, V. N. (2003). RNA interference in functional genomics and medicine. *J. Korean Med. Sci.* **18,** 309–318.

Makowske, M., and Rosen, O. M. (1989). Complete activation of protein kinase C by an antipeptide antibody directed against the pseudosubstrate prototope. *J. Biol. Chem.* **264,** 16155–16159.

McGuire, S. E., Roman, G., and Davis, R. L. (2004). Gene expression systems in *Drosophila*: A synthesis of time and space. *Trends Genet.* **20,** 384–391.

Meister, G., and Tuschl, T. (2004). Mechanisms of gene silencing by double-stranded RNA. *Nature* **431,** 343–349.

Mizuno, K., Noda, K., Araki, T., Imaoka, T., Kobayashi, Y., Akita, Y., Shimonaka, M., Kishi, S., and Ohno, S. (1997). The proteolytic cleavage of protein kinase C isotypes, which generates kinase and regulatory fragments, correlates with Fas-mediated and 12-O-tetradecanoyl-phorbol-13-acetate–induced apoptosis. *Eur. J. Biochem.* **250,** 7–18.

Mohanty, S., Huang, J., and Basu, A. (2005). Enhancement of cisplatin sensitivity of cisplatin-resistant human cervical carcinoma cells by bryostatin 1. *Clin. Cancer Res.* **11,** 6730–6737.

Morris, K. V., and Rossi, J. J. (2006). Lentiviral-mediated delivery of siRNAs for antiviral therapy. *Gene Ther.* **13,** 553–558.

Newton, A. C. (2003). Regulation of the ABC kinases by phosphorylation: Protein kinase C as a paradigm. *Biochem. J.* **370,** 361–371.

Pardridge, W. M. (2007). shRNA and siRNA delivery to the brain. *Adv. Drug Deliv. Rev.* **59,** 141–152.

Parker, P. J., and Parkinson, S. J. (2001). AGC protein kinase phosphorylation and protein kinase C. *Biochem. Soc. Trans.* **29,** 860–863.

Peek, A. S., and Behlke, M. A. (2007). Design of active small interfering RNAs. *Curr. Opin. Mol. Ther.* **9,** 110–118.

Rennel, E., and Gerwins, P. (2002). How to make tetracycline-regulated transgene expression go on and off. *Anal. Biochem.* **309,** 79–84.

Reynolds, A., Leake, D., Boese, Q., Scaringe, S., Marshall, W. S., and Khvorova, A. (2004). Rational siRNA design for RNA interference. *Nat. Biotechnol.* **22,** 326–330.

Schultz, A., Jonsson, J. I., and Larsson, C. (2003). The regulatory domain of protein kinase Ctheta localises to the Golgi complex and induces apoptosis in neuroblastoma and Jurkat cells. *Cell Death Differ.* **10,** 662–675.

Sivaprasad, U., Shankar, E., and Basu, A. (2007). Downregulation of Bid is associated with PKCepsilon-mediated TRAIL resistance. *Cell Death Differ.* **14,** 851–860.

Smith, L., Chen, L., Reyland, M. E., DeVries, T. A., Talanian, R. V., Omura, S., and Smith, J. B. (2000). Activation of atypical protein kinase Cz by caspase processing and degradation by the ubiquitin-proteasome system. *J. Biol. Chem.* **275,** 40620–40627.

Soltoff, S. P. (2001). Rottlerin is a mitochondrial uncoupler that decreases cellular ATP levels and indirectly blocks protein kinase Cd tyrosine phosphorylation. *J. Biol. Chem.* **276,** 37986–37992.

Straka, M., and Boese, Q. (2005). Why RATIONAL siRNA Pooling is SMART. *Dharmacon Technical Notes.* (http://www.dharmacon.com/docs/article_pooling.pdf).

Sun, Y., Chen, X., and Xiao, D. (2007). Tetracycline-inducible expression systems: New strategies and practices in the transgenic mouse modeling. *Acta Biochim. Biophys. Sin. (Shanghai).* **39,** 235–246.

Urlinger, S., Baron, U., Thellmann, M., Hasan, M. T., Bujard, H., and Hillen, W. (2000). Exploring the sequence space for tetracycline-dependent transcriptional activators: Novel mutations yield expanded range and sensitivity. *Proc. Natl. Acad. Sci. USA* **97,** 7963–7968.

Vorhies, J. S., and Nemunaitis, J. (2007). Nonviral delivery vehicles for use in short hairpin RNA-based cancer therapies. *Expert Rev. Anticancer Ther.* **7,** 373–382.

Yao, F., Svensjo, T., Winkler, T., Lu, M., Eriksson, C., and Eriksson, E. (1998). Tetracycline repressor, tetR, rather than the tetR-mammalian cell transcription factor fusion derivatives, regulates inducible gene expression in mammalian cells. *Hum. Gene Ther.* **9,** 1939–1950.

CHAPTER NINE

STUDYING p53-DEPENDENT CELL DEATH *IN VITRO* AND *IN VIVO*

Kageaki Kuribayashi, Niklas Finnberg, *and* Wafik S. El-Deiry

Contents

1. INTRODUCTION

The p53 pathway is inactivated in most human tumors. It is inactivated directly as a result of mutations in the p53-gene, indirectly by binding of the p53-protein to viral or cellular proteins, or as a consequence of alterations in proteins regulating its functions (Vogelstein *et al.*, 2000). P53 function is usually switched off, although when the cells get exposed to stress such as

University of Pennsylvania School of Medicine, Philadelphia, Pennsylvania

Methods in Enzymology, Volume 446
ISSN 0076-6879, DOI: 10.1016/S0076-6879(08)01609-1

DNA damage induced by ionizing radiation, ultraviolet rays, activation of oncogenic signaling, hypoxia, or nucleotide depletion, p53 accumulates in the nucleus in a tetrameric form (Bode and Dong, 2004). On activation, p53 mediates a growth-suppressive effect on cells by blocking the cell cycle or trigger programmed cell death (apoptosis) primarily by binding to particular DNA sequences and activating transcription of e.g., cell cycle arrest or pro-apoptotic genes (El-Deiry; Kurybayashi and El-Deiry, 2007). In this chapter, we describe how to study p53-dependent cell death *in vitro* and *in vivo*.

2. STUDYING P53-DEPENDENT CELL DEATH *IN VITRO*

Essentially, studying p53-dependent cell death is comparing the apo-ptotic response between p53 wild-type cells or tissues to that of p53 null or mutants. Basically, there are two ways to study how this depends on p53 *in vitro*. One is to introduce wild-type p53 in p53-null or mutant cells, and the other is to silence p53 in the wild-type cells. There are benefits and disadvantages regarding both methods. It is an advantage of introducing wild-type p53 into the null or mutant cells that direct effects of p53 can be studied. However, because of the growth suppressive effect of p53, the response could not be followed up for a long time. Creating a stable cell line in which the p53 gene is silenced by a short interfering RNA technique solves the problem. However, efficiency of the gene silencing is usually incomplete, and activating the p53 pathway requires additional stimuli, such as cytotoxic drugs, that usually activate the p53 pathway, as well as the other pathways simultaneously. The use of a specific cell line such as HCT116 p53 $^{-/-}$ generated by homolo-gous recombination created by Dr. Vogelstein's laboratory is useful. Which strategy to choose should be decided by the aim of each experiment.

3. OVEREXPRESSION OF P53 BY ADENOVIRUS (AD)

3.1. Construction of the adenoviral vectors

We have created Ad vectors that overexpress p53 (Wang *et al.*, 2003). Details about the construction of adenoviral vectors is beyond the scoop of this chapter and therefore only briefly described here. For Ad/GFP-p53 construction, green fluorescent protein (GFP) was fused to the N-terminus of p53, and the open reading frame of the fused protein was inserted into pAdTrack-CMV vector. For Ad/His-p53 and Ad/Flag-p53 construc-tion, His-p53 and Flag-p53 were inserted into pAdTrack-CMV vector, which expresses the protein under the control of CMV promoter. The pAdTrack-CMV constructs were recombined with pAdEasy-1 (Stratagene) in *Escherichia coli* BJ5183 cells to get recombined Ad plasmid. The pAd-Track-CMV that only expresses GFP without p53 expression was used as a control vector. The plasmids were linearized by Pac I digestion and transfected to 293 cells to obtain Ads.

3.2. Cloning the adenovirus

It is not mandatory to clone adenovirus every time. However, it should be taken into account that the process of amplifying the virus leads to increased percentages of recombined mutant viruses, which are deficient of replication or expressing the protein.

Things to be prepared:

A. 2% methylcellulose (MC) in sterile PBS, autoclave and store in 10-ml tubes at 4 °C.

1. On day 1, seed 5×10^5 293 cells/well in 6-well plate so 50 to 70% confluence can be obtained on the next day.
2. Dilute viral stocks in 1-ml volumes over a 10-fold series from 10^{-5} to 10^{-9} in culture media.
3. Add 1 ml of each dilution to each well and incubate at 37 °C for 2 h.
4. During incubation, warm the tubes containing 2% MC in boiling water. After the MC has melt, keep the tubes in 42 °C waterbath.
5. After incubating Ad with 293 cells, remove the media from the wells and overlay 2 ml of MC on 293 cells. It should be gently overlayed from the sidewall of the wells so as not to disrupt the monolayer of the cells.
6. Incubate the plate at 37 °C until the plaques become visible. It usually takes 3 to 10 days.
7. Pipette the plaque in 250 μl culture media, and the virus will be amplified in the next step.

3.3. Amplification of the adenovirus

First step of amplification (a plaque to 6-well)

1. Seed 293 cells in 6-well plate so the cells will achieve 50 to 70% confluency on the next day.
2. Infect the cells by adding adenoviral stock that is prepared in step 3.2.7) at 37 °C for 2 h. The amount of the medium during the infection process should be as small as possible just enough to cover the cells to obtain higher concentration of the virus.
3. After 2 h of infection, add culture media to the dish and incubate at 37 °C until cytopathic effect (CPE) appears, which is the time for harvesting the virus. CPE can be observed as rounding up and detaching the cells from the plate. At this stage, the nucleus will occupy a major part of the cell.

When the primary virus stock has higher titer, CPE will appear approximately 3 days and when that is low, it will take up to 10 days.

4. After CPE is observed, harvest the adenovirus-producing cells by cell scraper. If the cells were detached from the plate already, make sure to recover the floating cells by pipetting.

Second step of amplification (6-well to T75 flask)

5. Seed 293 cells in T75 flask so the cells will achieve 50 to70% confluency on the next day.
6. Infect the cells by adding Ad-infected 293 cells harvested in step 3.3.4), and incubate at 37 °C until CPE appears. It usually takes 3 days.
7. After CPE is observed, harvest the adenovirus-producing cells by cell scraper. If the cells were detached from the plate already, make sure to recover the floating cells by pipetting.

Third step of amplification (T75 flask to 6 T150 flasks)

8. Seed 293 cells in six T150 flasks so the cells will achieve 50 to 70% confluency in each flask on the next day.
9. Infect the cells by adding Ad-infected 293 cells harvested in step 3.3.7), and incubate at 37 °C until CPE appears. It usually takes 3 days.
10. After CPE is observed, harvest the adenovirus-producing cells by cell scraper. If the cells were detached from the plate already, make sure to recover the floating cells by pipetting. Resuspend the cells in PBS and pellet the cells by centrifugation. Resuspend the cell pellet in 5 ml of PBS and store at −80 °C.

3.4. Purification of the Ad by CsCl banding

Reagents to be prepared

A. CsCl solutions:
CsCl 1.25 g/ml = 36.16 g CsCl + 100 ml PBS
CsCl 1.35 g/ml = 51.20 g CsCl + 100 ml PBS
CsCl 1.40 g/ml = 62.00 g CsCl + 100 ml PBS

B. $2 \times$ Glycerol storage buffer
10 mM Tris-HCl, pH 8.0, 100 mM NaCl, 1 mM MgCl$_2$, 0.1% BSA, 50% glycerol
Sterilize the buffer by filtration.

1. Thaw the cells prepared in step 3.3.10) in 37 °C waterbath for 5 min and vortex for 10 sec.
2. Freeze the cells by putting the tube in liquid nitrogen for 5 min.
3. Repeat the above steps for four times.

4. Centrifuge the sample at 4000 rpm for 5 min at 4 °C to pellet the cell debris and recover the supernatant.

5. We use plastic centrifuge tubes (14 × 89 mm) for SW41 swinging bucket rotor in the following steps. Add 2.5 ml 1.4 g/ml CsCl in the tube, overlay 2.5 ml 1.2 g/ml CsCl on the top, and then add supernatant recovered in step 3.4.4) on the top (Fig. 9.1).

6. Centrifuge the tube at 32,000 rpm for 1 h at RT.

7. In the tube, from the top, there are PBS phase, interface of the PBS and the CsCl, and a single band of Ad (Fig. 9.1). Occasionally, there are two bands in CsCl. The virus is found in lower larger band. The upper smaller band is an empty-shell virus.

8. Gently remove the layers in the tube to the level of the virus band; be careful not to disrupt the band of Ad.

9. To recover the band of Ad, use a 3-ml syringe with 18-gauge needle and aspirate the band.

10. As an option, dialyze the Ad-containing solution to remove CsCl. Add equal volume of 2 × glycerol storage buffer to the Ad-containing solution and store at −20 °C.

3.5. Infecting the target cells

The primary viral stock is now ready to infect target cells as described in 3.3 amplification of the adenovirus. The infection efficiency differs from cell line to cell line. Because our adenovirus expresses GFP, the infected cells

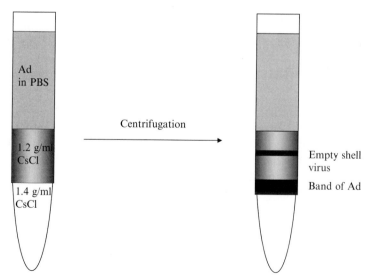

Figure 9.1 Purification of the Ad by CsCl banding.

can be monitored by fluorescence microscopy. We use the smallest multiplicity of infection (MOI) of the Ad that induces nearly 100% infection. P53 expression can be achieved within 24 h, and usually cell death starts to become apparent in 24 to 48 h after the infection.

4. SILENCING THE P53 GENE BY RETROVIRAL VECTOR

4.1. Construction of the retroviral vectors

We have several retroviral vectors that contain short hairpin RNA (shRNA) against p53. We confirmed that micro30 RNA sequences from Openbiosystems (V2HS_93615) works fine. These sequences are amplified by PCR and cloned into the *BamHI* and *EcoRI* site of pSIREN-RetroQ-Zs Green or PURO vector (Clontech). However, in some cell types, single retroviral infection is not enough to achieve efficient knock down. In these cases, the use of different sequences or adding second retroviral knockdown with different sequences is helpful. We use gactccagtggtaatctac as alternative 19mer sequences against p53 in these cases.

4.2. Obtaining retroviral soup

We use Phoenix cells for a packaging cell line. Because the efficiency of gene silencing greatly depends on its virus titer, we obtained a clone from parental Phoenix cells that produce highest virus titer.

In some cell types, the retrovirus infection is toxic and highly infected cells are difficult to grow, resulting in inefficient gene silencing. In these cases, we transform the cells with the vector by lipofection and clone highly knocked down cells.

1. On day 1, seed 2×10^5 Phoenix cells/well in 6-well plate so approximately 70% confluency can be obtained on the next day.
2. On day 2, transfect 6 μg of pSIREN-RetroQ vector that contains the sequence against p53 or control sequences by Lipofectamine 2000 (Invitrogen), according to the manufacture's instruction.
3. On day 4 or 5, harvest the supernatant that contains retrovirus. At this step, the retrovirus soup can be stored at $-80\,^{\circ}$C; however, because freezing and thawing of the retrovirus soup will decrease the virus titer, it is ideal to proceed directly to the next step. Add polybrene to the supernatant to the final concentration of 8 μg/ml. Add sodium bicarbonate if the pH of the medium is too acidic. Filter the retrovirus soup by 0.45-μm filter; this retrovirus soup can be used to infect target cells.

4.3. Infecting the target cells

1. Seed the cells for gene silencing in 6-well plate on the day before infection so 50–70% confluency can be achieved on the next day. As an option, use phosphate-free buffer, because it will increase the expression of receptor for retrovirus.
2. Remove the medium and add retrovirus soup obtained in step 4.2.3) Centrifuge the plate at 1500g for 60 min and incubate for another 1 h to overnight. Incubate at 32 °C as the virus is more stable than at 37 °C. Change the medium to complete media after the incubation.
3. Gene expression (GFP) can be seen approximately 24 h after the infection. Selection by puromycin can be started 24 to 48 h after the infection. If the Zs-Green is used as a selection marker, wait for 4 to 5 days before the selection so there are enough cells to be sorted.
4. Gene silencing can be confirmed by Western blot following the common procedure. We also recommend analyzing not only p53 but also its target gene such as p21 to confirm the efficiency of gene silencing.

5. STUDYING THE APOPTOTIC RESPONSE

To compare the apoptotic response between p53 wild-type and its counterpart, we often use sub-G1 analysis. Because p53-dependent cell death usually relies on transcription of its target genes, we incubate the cells under stresses for few hours to few days. Description of the methods can be found later in this chapter.

6. STUDYING P53-DEPENDENT APOPTOSIS *IN VIVO*

6.1. p53-Deficient mouse models and DNA damage

p53-knockout mice were first generated in 1992 (Donehower *et al.*, 1992) and have been extensively used for cancer studies, because they, in a homozygous state, develop lymphomas and soft tissue sarcomas within 3 to 5 months of age. However, use of the p53-deficient model has not been restricted to cancer studies. Important insights have been achieved into p53 function in cell-cycle control, response to DNA damage, apoptosis, hypoxia, oncogenic stimuli, embryonic development, and immunity (Attardi *et al.*, 1996; Donehower, 19961; 1996b). Recent p53-deficient mouse models also make use of temporally controlled and/or tissue-specific ablation of the p53 gene and the possibility to reversible reactivate the p53-gene

(Christophorou *et al.*, 2005; Marino *et al.*, 2000; Ventura *et al.*, 2007). This opens up a number of possibilities to study the function of p53 in restricted tissues or cells without interference because of systemic loss of p53.

This part of the chapter will focus on the use of p53-deficient mice as an essential tool to study p53-dependent apoptosis *in vivo* and discuss some methods that are used. We use ionizing irradiation to trigger DNA damage, because many hurdles in metabolism, pharmacodynamics, and kinetics can be overcome that are otherwise encountered when genotoxic agents are injected into animals. However, in certain instances, such compounds may be desirable to use because they may hold clinical relevance and/or target radiation-resistant tissues to DNA damaged-induced apoptosis.

Clearly, p53-deficiency has a profound effect on diminishing the apoptotic response after DNA damage induced by ionizing radiation in lymphoid organs (e.g., spleen, thymus, and bone marrow as well as in the GI tract) (Fei *et al.*, 2002). Furthermore, loss of p53 protects mice from doses of ionizing radiation that trigger lethal bone marrow aplasia (a hematopoietic syndrome), and this correlates well with p53's ability to protect cells from apoptosis in hematopoietic organs (bone marrow and spleen in mice) (Komarova *et al.*, 2004).

However, p53 does not protect from higher doses of radiation (>15 Gy) that trigger lethality through the erosion of the epithelium and inflammation in the gut of mice, rather p53-deficiency seems to increase sensitivity to this syndrome (Komarova *et al.*, 2004). A possible explanation is that surviving cells express p53 to induce cell cycle arrest and DNA damage repair and may subsequently yield viable progeny. Thus, although the acute apoptotic response in the gut is related to p53, the relevance, in contrast to the hematopoietic syndrome, in protecting from lethality because of irradiation-induced gastrointestinal damage is unclear.

6.2. Detection of apoptosis *in vivo*

Apoptosis typically is described as a sequence of morphologic events that can be easily detected histologically. The initial characterization of apoptosis was on the basis of *in vivo* microscopic observations by the father of modern pathology Rudolph Virchow (1821–1902). He recognized that cell death was a fundamental process in maintaining tissue homeostasis. Although appreciating that the appearance of cell death could be variable, Rudolph Virchow referred to apoptotic cells, and all other dying or dead cells, as necrotic cells. It was Professor John Kerr and colleagues who suggested because not all dead cells looked the same this could mean that the mechanism for cell death could be different (Kerr, 1971). Still, more than a century after Rudolph Virchow's death, morphologic hallmarks of apoptosis as described by Kerr in conventionally prepared tissue sections remains the "gold standard" in characterizing the process (Table 9.1). Apoptosis typically

Table 9.1 Some morphologic hallmarks that distinguish apoptotic cells from necrotic *in vivo*

Apoptosis	Necrosis
Scattered individual cells	Often a group of confluent cells
Cell shrinkage, loss of cell–cell contact, narrow empty halo	Swelling of cells
The forming of apoptotic bodies	Inflammation
Phagocytosis by neighboring cells and macrophages	

involves scattered individual cells as opposed to necrosis, which can involve confluent groups of cells. Apoptotic cells shrink and lose contact with neighboring cells; therefore, they are often surrounded by a narrow empty halo. Necrotic cells have a tendency to swell. In an apoptotic cell the chromatin is condensed and packed into smooth, round, or curved profiles situated in close opposition to the nuclear membrane. These cells shrink into a dense single apoptotic body and the cell emits buds that contain nuclear fragments surrounded by narrow rim of cytoplasm. These apoptotic bodies may remain in the tissue or become phagocytosed by macrophages or neighboring cells.

The *in vivo* morphology of apoptosis is identical regardless of species, cell or tissue type, and the appearance is not influenced by cause of therapeutic drugs, ionizing radiation, hypoxia, or pathologic conditions. Apoptosis may occur rapidly, and apoptotic cells may be quickly removed; therefore, successful detection of apoptosis relies on proper timing. A fundamental understanding of the kinetics of induction of apoptosis in any given tissue is important. Apoptosis can be distinguished from necrosis in sections of tissue fixed in buffered formalin or paraformaldehyde, processed by conventional methods, and stained by hematoxylin and eosin (H&E).

Identification of apoptosis in tissue sections has been facilitated by a number of immunohistochemical methods. One method described by Gavrieli *et al.* (1992), in which DNA breaks in apoptotic nuclei are marked by dUTP-biotin transferred to the free $3'$ end of cleaved DNA by nick-end labeling, it is often referred to "TUNEL" (Terminal dUTP-biotin deoxynucleotidyl transferase nick-end labeling). The method has become the standard technique for the study of apoptosis in tissue sections. TUNEL staining of tissue with hematoxylin as counterstain allows one to assess the morphologic features used with H&E together with the parameter of a positive peroxidase reaction. It is important to assess the morphology in combination with TUNEL positivity, because necrotic cells, autolytic cells, and debris may all show positive reactions (Grasl-Kraupp *et al.*, 1995).

TUNEL labeling can also be done with fluorescent FITC (green) in combination with propidium iodide as a counterstain that labels nuclei (red). However, many of the morphologic criteria for apoptotic cells will be difficult to assess with this type of staining. More recently, a wealth of markers for immunohistochemical detection of apoptosis has become available (e.g.,. antibodies toward activated caspases and other cleaved caspase substrates such as PARP).

6.3. Quantitating apoptotic responses

Apoptosis can be quantified in tissues with simple methods like H&E-staining and immunohistochemical methods. At higher doses of ionizing radiation (>1 Gy) this becomes difficult in the hematopoietic and lymphoid organs because of the massive amount of apoptosis one encounters. A better approach is to isolate cells from these organs and quantify apoptosis with regard to apoptosis marker and flow cytometry. We routinely use Sub-G1 DNA content as a marker for apoptosis, because DNA fragmentation is easily detectable in cells from these organs after ionizing radiation. However, other markers (not discussed in this chapter) frequently used for immunohistochemistry can be used (e.g., active-caspase-3 [BD], TUNEL [Chemicon] or the pan-caspase assay FLICA [Promega] [Fluorescent Labeling of Caspase Activity]). The apoptotic response in the small intestine and colon can be assessed in paraformaldehyde-fixed and paraffin-embedded tissue section slides prepared routine histological methods by H&E staining. It is also very helpful if the tissue section slides are stained for TUNEL and active caspase-3, preferentially, with peroxidase (ABC, Vector Laboratories) and DAB as substrate (Vector Laboratories). The number of apoptotic cells per crypt can then be counted under a light microscope. Entire half-crypts should be counted so the crypt position of the apoptotic cell can be recorded. Counting between 50 and 100 crypts typically provides the researcher with a good idea of the extent of apoptosis in the organ.

6.4. Material and methods

6.4.1. Animals
Six- to eight-week old female B6;129S2-Trp53^{tm1Tyj}/J (Jackson Laboratories; p53 $^{-/-}$) mice and 6- to 8-week old female C57Bl/6 (Jackson Laboratories) control animals.

6.4.2. Fixation procedures for optimal immunohistochemical detection
Prepare 4% paraformaldehyde (PFA) by diluting a $10 \times$ stock of PBS into 70% of the total volume and adding the appropriate amount of PFA. Heat to $70°$ and keep the temperature and mix solution for 10 min. Add one drop

of 2 *N* sodium hydroxide (the solution should clear) and chill to room temperature. Adjust pH to 7.2.

For routine histological study of most tissues, fixation by immersing the tissues in 4% PFA and fixing over night at 4 °C is sufficient. Spleens should be cut lengthwise before fixation, whereas the thymus and lymph nodes can be immersed whole. The small intestine and colon should be cleared from luminal content. This can be done by either pushing the content out gently with the backside of scalpel or flushing the content out with PBS followed by 4%PFA. It is imperative that the organs be removed immediately after the sacrifice of the animal. The gut can be cross-sectioned or fixed as a "Swiss-role"; ideally, the later allows assessment of the entire organ at once. For some organs, such as liver and brain, whole-body transcardial perfusion with fixative (not described here) of the animal will remove all blood and initiate fixation immediately something that may be crucial for accurate immunohistochemistry.

6.4.3. Tunel staining

TUNEL staining is made by use of either the ApopTag® Plus Fluorescein In Situ Apoptosis Detection Kit or ApopTag® Plus Peroxidase In Situ Apoptosis Detection Kit (Chemicon International, Millipore) according to the manufacturer's instruction, except in the case of the use of ApopTag® Plus Peroxidase *In Situ* Apoptosis Detection Kit hematoxylin, which is used as counterstain in favor of methyl green.

6.4.4. Immunohistochemical/immunofluorescent detection of phosphor-ATM, p53, p21, Bax, cleaved caspase-3 and cleaved caspase-9

Deparaffinize and hydrate sections:

1. Prewarm sections at 65 ° for 15 min.
2. Incubate sections in two washes of xylene for 2 min each.
3. Incubate sections in two washes of 100% ethanol for 2 min each.
4. Incubate sections in two washes of 95% ethanol for 1 min each.
5. Incubate sections in one wash of 80% ethanol for 1 min.
6. Incubate sections in two washes of 70% ethanol for 1 min.
7. Incubate sections in PBS for 1 min.
8. Unmask antigen on sections.
9. Bring slides to boil in 10 m*M* sodium citrate buffer, pH 6.0, then maintain at a subboiling temperature for 10 min. This can be done by submerging slide racks in a 2-L plastic beaker and microwave for 11 min at full effect.
10. Take slides out of the microwave oven and let cool on bench top for 20 min.

11. Wash three times with distilled water for 5 min or, preferably, rinse under slowly running warm water for 5 min.
12. Block endogenous peroxidases (skip this step if doing immuno-fluorescence).
13. Incubate slides in 7.5% of H_2O_2 in distilled water for 15 min.
14. Wash sections twice in TBS with 1% Tween-20 (TBST) for 5 min each.
15. Block sections 1 h the following blocking buffers: 5% donkey serum in TBST for immunofluoresence or 5% goat serum in TBST for immunoperoxidase (ABC).
16. Prepare the primary antibodies (Table 9.2) in appropriate blocking buffer while waiting.
17. Add primary antibodies and incubate overnight in 4 °C.
18. Wash sections in TBST two times 5 min each.
19. Prepare appropriate secondary antibody (Table 9.3) in blocking buffer.
20. Protect secondary antibody solution and slides for immunoflouresence from light from now on.
21. For immunoperoxidase staining, prepare the avidin–biotin complex by use of the ABC-HRP kit (Vector Laboratories) by mixing 10 μl of A with 10 μl of B per ml of blocking buffer (i.e., 5% goat serum in TBST). Let this sit for at least 30 min.
22. Incubate slides with secondary antibodies for 30 min at 37 °C.
23. Wash sections in TBST two times 5 min each.
24. For immunofluorescence
    a. Transfer slides to distilled water and wash twice for 5 min each.
    b. Stain nuclei (counterstain) slides in DAPI (0.5 μg/ml, Molecular Probes) for 5 min.

Table 9.2 Primary antibodies for immunohistochemistry

Antigen	Antibody	Manufacturer	Source	Dilution
Bax	P–19	Santa Cruz	Rabbit	1:200 (IF)
CC9 (asp353)	#9504	Cell Signal	Rabbit	1:100 (IF)
CC3 (asp175)	#9661	Cell Signal	Rabbit	1:500 (IP)
P21	F–5	Santa Cruz	Mouse	1:200 (IP)
P53	CM5	Vector Laboratories Inc	Rabbit	1:2,000 (IP)
p-ATM	#7C10D8	Rockland Immunochemical Inc	Mouse	1:200 (IF)

IF, Immunofluorescence; IP, immunoperoxidase (by use of the ABC kit from Vector Laboratories).
CC9 (asp353), cleaved caspase-9 (asp353); CC3 (asp175), cleaved caspase-3 (asp175); p-ATM, phospho (ser 1981)-ATM

Table 9.3 Secondary antibodies for immunohistochemistry

Antigen Antibody	Manufacturer	Source	Dilution
Rabbit IgG ImmunoPure Goat Anti-Rabbit IgG (H + L), biotin conjugated	Pierce	Goat	1:400 (IP)
Mouse IgG ImmunoPure goat anti-mouse IgG (H + L), (min × BvHnHs Sr Prot), biotin conjugated	Pierce	Goat	1:400 (IP)
Rabbit IgG Cyanine Cy3, anti-rabbit IgG (H + L)	Jackson ImmunoResearch	Donkey	1:400 (IF)
Mouse IgG Cyanine Cy3, anti-mouse IgG (H + L)	Jackson ImmunoResearch	Donkey	1:400 (IF)

IF, Immunofluorescence; IP, immunoperoxidase (by use of the ABC kit from Vector Laboratories).

    c. Wash two times in water for 5 min.

    d. Mount in fluorescent mounting media (e.g., Kirkegaard glycerol-based mounting media).

    e. View under fluorescent microscope. Store at 4 °C.

25. For immunoperoxidase

    a. Incubate slides with ABC for 30 min at 37 °C.

    b. Wash slides two times in TTBS for 5 min each.

    c. Prepare DAB by use of the DAB substrate kit (Vector) according to manufacturer's instructions.

    d. Develop signal under light microscope for positive control tissue and record time until satisfactory color development is reached. Incubate the rest of the slides accordingly. Stop reaction by washing in distilled water.

    e. Counterstain with hematoxylin, dehydrate to xylene, and mount in xylene-based mounting media.

6.4.5. Flow cytometric analysis of sub-G1 content

Animals were euthanized, and the bone marrow (femur and tibia), spleen, and thymus were isolated. The bone marrows were flushed with RPMI-1640 supplemented with 5% fetal bovine serum and 1% penicillin and streptomycin through a 50-μm cell strainer (Fisher). The spleen and thymus were cut and gently disrupted with a 10-ml syringe plunger in supplemented RPMI-1640, and the cell suspensions were filtered through a 100-μm mesh. Red blood cells were removed by incubating the cell suspensions in ACK-lysis buffer (0.15 M NH$_4$Cl, 1.0 mM KHCO$_3$, 0.1 mM Na$_2$EDTA, pH 7.2) for 5 min. RPMI-1640 was added, and cells were spun down, washed, fixed in ice-cold 70% ethanol during vortexing and incubated for 30 min at 4 °C. The final cell pellet was resuspended in 300 μl of PBS with 1% FBS supplemented with 50 μg/ml RNAse A and 25 μg/ml propidium iodide (Molecular Probes). Cells were incubated at room temperature for 1 h and analyzed for sub-G$_1$ DNA content with an Epics Elite flow cytometer (Beckman Coulter).

REFERENCES

Attardi, L. D., Lowe, S. W., Brugarolas, J., and Jacks, T. (1996). Transcriptional activation by p53, but not induction of the p21 gene, is essential for oncogene-mediated apoptosis. *EMBO J.* **15,** 3693–3701.

Bode, A. M., and Dong, Z. (2004). Post-translational modification of p53 in tumorigenesis. *Nat. Rev. Cancer* **4,** 793–805.

Christophorou, M. A., Martin-Zanca, D., Soucek, L., Lawlor, E. R., Brown-Swigart, L., Verschuren, E. W., and Evan, G. I. (2005). Temporal dissection of p53 function *in vitro* and *in vivo. Nat. Genet* **37,** 718–726.

Donehower, L. A. (1996a). Effects of p53 mutation on tumor progression: Recent insights from mouse tumor models. *Biochim. Biophys. Acta.* **1242,** 171–176.

Donehower, L. A. (1996b). The p53-deficient mouse: A model for basic and applied cancer studies. *Semin. Cancer Biol.* **7,** 269–278.

Donehower, L. A., Harvey, M., Slagle, B. L., McArthur, M. J., Montgomery, C. A., Jr., Butel, J. S., and Bradley, A. (1992). Mice deficient for p53 are developmentally normal but susceptible to spontaneous tumours. *Nature* **356,** 215–221.

El-Deiry, W. S. (2003). The role of p53 in chemosensitivity and radiosensitivity. *Oncogene* **22,** 7486–7495.

Fei, P., Bernhard, E. J., and El-Deiry, W. S. (2002). Tissue-specific induction of p53 targets *in vivo. Cancer Res.* **62,** 7316–7327.

Gavrieli, Y., Sherman, Y., and Ben-Sasson, S. A. (1992). Identification of programmed cell death *in situ* via specific labeling of nuclear DNA fragmentation. *J. Cell Biol.* **119,** 493–501.

Grasl-Kraupp, B., Ruttkay-Nedecky, B., Koudelka, H., Bukowska, K., Bursch, W., and Schulte-Hermann, R. (1995). *In situ* detection of fragmented DNA (TUNEL assay) fails to discriminate among apoptosis, necrosis, and autolytic cell death: A cautionary note. *Hepatology* **21,** 1465–1468.

Kerr, J. F. (1971). Shrinkage necrosis: A distinct mode of cellular death. *J. Pathol.* **105,** 13–20.

Komarova, E. A., Kondratov, R. V., Wang, K., Christov, K., Golovkina, T. V., Goldblum, J. R., and Gudkov, A. V. (2004). Dual effect of p53 on radiation sensitivity *in vivo*: p53 promotes hematopoietic injury, but protects from gastro-intestinal syndrome in mice. *Oncogene* **23,** 3265–3271.

Kuribayashi, K., and El-Deiry, W. S. (2007). Regulation of programmed cell death by the p53 pathway. *In* "Programmed Cell Death in Cancer Progression and Therapy." (R. Khorsravi-Far and E White, eds.), Springer, 201–221.

Marino, S., Vooijs, M., van Der, G. H., Jonkers, J., and Berns, A. (2000). Induction of medulloblastomas in p53-null mutant mice by somatic inactivation of Rb in the external granular layer cells of the cerebellum. *Genes Dev.* **14,** 994–1004.

Ozoren, N., and El-Deiry, W. S. (2003). Cell surface death receptor signaling in normal and cancer cells. *Semin. Cancer Biol.* **13,** 135–147.

Ventura, A., Kirsch, D. G., McLaughlin, M. E., Tuveson, D. A., Grimm, J., Lintault, L., Newman, J., Reczek, E. E., Weissleder, R., and Jacks, T. (2007). Restoration of p53 function leads to tumour regression *in vivo. Nature* **445,** 661–665.

Vogelstein, B., Lane, D., and Levine, A. J. (2000). Surfing the p53 network. *Nature* **408,** 307–310.

NF-κB as a Determinant of Distinct Cell Death Pathways

Irene L. Ch'en,* Stephen M. Hedrick,* *and* Alexander Hoffmann[†]

Contents

Abstract

The NF-κB signaling system has important and distinct roles in determining cell fate decisions, such as cell proliferation and cell death. Specifically, recent evidence indicates that NF-B regulates several types of programmed cell death, such as apoptosis, necroptosis, necrosis, as well as cellular senescence, but its precise role in these is not fully understood. Distinguishing these cell fates experimentally is therefore important, and several techniques are available to researchers. We summarize experimental strategies and protocols that reveal changes in nuclear morphology and cell shrinkage, exposure of phosphatidyl-serine, compromised membrane integrity, DNA fragmentation, and altered

* Division of Biological Sciences and the Department of Cellular and Molecular Medicine, University of California, San Diego, La Jolla, California
† Signaling Systems Laboratory, Department of Chemistry and Biochemistry, University of California, San Diego, La Jolla, California

Methods in Enzymology, Volume 446
ISSN 0076-6879, DOI: 10.1016/S0076-6879(08)01610-8

mitochondrial membrane potential. Together, these may discriminate distinct cell death pathways and lead to a better understanding of the underlying regulatory mechanisms.

1. INTRODUCTION

Starting with the genetics of *Clostridium elegans*, the molecular mechanisms underlying developmentally programmed cell death have been characterized over the past 20 years or so. In humans, negative selection of T cells by means of the death receptor Fas provided a similar starting point. The speed and success of these studies may have given the impression that this type of cell death—apoptosis—is the most important, the most common, or even the only one that fits the description of programmed cell death. However, other forms of programmed cell death, such as necrosis and cellular senescence, have regained prominence in a number of physiologic and pathologic settings.

Our understanding of NF-κB's role in cell death has similarly evolved. Although this transcription factor was first identified as a B-cell developmental regulator and was then associated more broadly with inflammation and immune responses, knockout mice deficient for the ubiquitous family member RelA were found to be embryonic lethal because of massive apoptotic cell death in fetal liver hepatocytes (Beg *et al.*, 1995). That finding established NF-κB as a major anti-apoptotic or "pro-survival" regulator. In recent years, NF-κB, which neither exists in *C. elegans* nor plays a major role in Fas-induced apoptosis, has been increasingly implicated in regulating a variety of cell death pathways, and not always as a pro-survival transcription factor (Fig. 10.1).

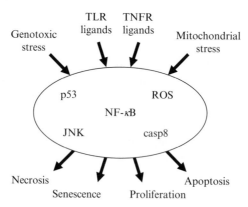

Figure 10.1   Signaling networks involving NF-κB that control cell fate. Diverse signals impinging on cells activate different combinations of signal transducers, only some of which are shown in the schematic. The network of interactions produces different cellular responses. NF-κB has been implicated in the control of cell proliferation, apoptosis, necrosis, and senescence.

Genetic ablation of NF-κB/RelA led to embryonic lethality because of massive apoptosis in the embryonic liver brought about by embryonic tumor necrosis factor (TNF) signaling (Alcamo *et al.*, 2001). TNF was previously known to kill cultured cells sensitized by treatment with the ribosome inhibitor cycloheximide—the new results suggested that NF-κB was responsible for this *de novo* gene expression (Beg and Baltimore, 1996). Subsequent screens for NF-κB target genes that act as anti-apoptotic regulators produced several candidates, including Bcl2 family member Bfl/A1, FLIP, A20, cIAPs, ferritin heavy chain (FHC), and Gadd45ß, (Papa *et al.*, 2004), but it remains unclear which are the relevant targets in specific physiologic contexts. However, blockade of NF-κB activity, either genetically by IκB super-repressor expression or RNAi, or pharmacologically, renders many cell types sensitive to apoptosis in response to diverse stimuli. Cancer cells, in particular, seem to lose their chemoresistance when elevated NF-κB activity is reduced by either of these strategies.

However, in other experimental scenarios, NF-κB was shown to play a pro-death role. The first such report showed that Sindbis virus caused cell death in an NF-κB–dependent manner (Lin *et al.*, 1998). More recently, UV irradiation was shown to signal through PKCδ, whose expression is NF-κB–dependent, toward a death pathway involving JNK (Liu *et al.*, 2006). These studies identified homeostatic NF-κB activity, and not stimulus-induced NF-κB activity, as being responsible for determining cell death sensitivity. Another study used a panel of IKK knockout cells and distinguished between apoptotic and necrotic cell death; whereas NF-κB–deficient cells were sensitive to apoptosis, chemical inhibition of caspases increased the amount of necrotic cell death (May and Madge, 2007). Furthermore, chronic NF-κB activity is also associated with cell senescence or aging (Adler *et al.*, 2007). Senescent cells are not dead, but they are unable to proliferate or exit G0, and physical manipulation (trypsinization) will often cause cell death.

Thus, NF-κB seems to be involved in a host of cell fate decisions that often fall within the cell death category. Genes known to be important regulators of these processes have been identified as NF-κB targets, including the anti-apoptotic regulators (mentioned earlier), anti-reactive oxygen species (ROS), regulators (ferritin heavy chain, FHC; superoxide dismutase, SOD; catalase), and cell cycle regulators (p21, cyclin D). The role of NF-κB is clearly cell type specific, as well as stimulus specific. TNF Receptor and Toll like Receptor superfamilies, as well as metabolic and genotoxic stress agents, engage a number of different signaling pathways. On the basis of recent publications, it seems that the role of NF-κB may, in part, be determined by the status and activity of the JNK pathway, as well as the p53 regulatory system (Gurova *et al.*, 2005; Ryan *et al.*, 2000). However, much more needs to be done to produce a predictive model. Therefore, distinguishing between the different types of cell death is an important component to understanding NF-κB's role in determining specific death pathways.

2. METHODS

2.1. Electron microscopy

Both apoptotic and necrotic cell death can be determined by their unique physiologic hallmarks. One recognized characteristic of cells undergoing apoptosis includes change in nuclear morphology, such as chromatin condensation (pyknosis) followed by chromosomal DNA fragmentation (karyorrhexis). Cell shrinkage and the formation of apoptotic bodies that contain cellular material all occur before the loss of plasma membrane integrity (Kerr *et al.*, 1972). In contrast, the morphologic characteristics of necrotic cells do not display these criteria that define apoptosis. Necrotic cells show early plasma membrane rupture and the presence of cellular debris (Kroemer and Martin, 2005). Visualization by electron microscopy is a very clear way to distinguish between these two types of cell death.

Carefully suspend cells in fresh cell culture medium. Add an equal volume of Karnovsky's fixative (1.5% glutaraldehyde, 3% paraformaldehyde, 5% sucrose in 0.1 M cacodylate buffer, pH 7.4) and incubate at room temperature for 15 min. Wash cells three times in 0.1 M cacodylate buffer with slow speed centrifugation. Postfix in 1% OsO_4 in 0.1 M cacodylate buffer for 1 h at room temperature. Wash three times in cacodylate buffer followed by one wash in water. Carefully resuspend in 10% ethanol and stain *en bloc* in 1% uranyl acetate in 10% ethanol for 1 h in the dark. Dehydrate with a series of ethanol washes (25%, 50%, 75%, 95%), incubating 10 to 20 min each. Finally, wash two times in 100% ethanol. Dehydrate in a 1:1 mixture of propylene oxide and 100% ethanol for 20 to 30 min, followed by 100% propylene oxide. Treat samples 2 to 4 h each with increasing mixtures of resin (1:3, 1:1) and leave overnight in 3:1 resin. The next day, transfer samples into 100% resin for a few hours. Repeat twice with fresh 100% resin before finally transferring samples to embedding capsules with fresh resin. Polymerize overnight at 60 °C. Cut ultrathin sections with an ultramicrotome and contrast with 1% uranyl acetate and lead nitrate. Examine sections with a transmission electron microscope. (Protocol designed by K. Kudlicka.)

2.2. Membrane health

2.2.1. Distinguishing between apoptosis and necrosis

Lipids in biological membranes are asymmetrically distributed on either side of the bilayer as an active process, depending on the origin of lipid synthesis and enzymatic maintenance (Daleke, 2003). Cells undergoing apoptosis lose membrane asymmetry, and this has been used as a distinctive marker to discriminate apoptotic cells from other forms of cell death. Membrane integrity (i.e., the exclusion of normally membrane-impermeable molecules) is maintained through most of the apoptotic process, whereas necrotic cells rapidly lose integrity of the plasma membrane (Golstein and Kroemer, 2007). These categorical differences can be useful in distinguishing between the two death processes. The caveats are that, ultimately, all dying cells lose membrane integrity, and as necrotic cells lose membrane integrity, they can appear to have lost asymmetry as probes gain access to the intracellular side. Specifically, Annexin V binding is not a feature restricted to apoptosis, because cells undergoing necrosis can also appear positive (Lecoeur *et al.*, 2001). Likewise, it has been shown that early apoptotic and early necrotic cells are $7AAD^{lo}Annexin\ V^+PI^-$, whereas both late apoptotic and necrotic cells are $7AAD^{hi}Annexin\ V^+PI^+$ (Lecoeur *et al.*, 2002). However, others have published the ability to distinguish between apoptotic and necrotic cells with the simultaneous staining of Annexin V-fluorescein isothiocyanate (FITC) and propidium iodide (Matteucci *et al.*, 1999). With these caveats in mind, cell death can be analyzed productively by taking advantage of these mechanistic distinctions.

2.2.2. Detection of extracellular phosphatidylserine

In a viable cell, phosphatidylserine is found on the cytosolic surface of the plasma membrane. Cells undergoing apoptosis lose membrane asymmetry, exposing phosphatidylserine on the extracellular surface (Daleke, 2003). The translocation of this phospholipid marks the cells for removal by phagocytosis through macrophages (Henson *et al.*, 2001).

To detect exposed phosphatidylserine by Annexin V binding (Vermes *et al.*, 1995), cells can be stained with phycoerythrin (PE)- or allophycocyanin (APC)-labeled Annexin V (Caltag Laboratories, Invitrogen, cat. no. ANNEXINV04, ANNEXIN05). Because Annexin V binds more specifically to negatively charged phosphatidylserine in the presence of calcium, staining should be performed in Annexin V binding buffer (140 mM NaCl, 2.5 mM CaCl$_2$, 10 mM HEPES). Wash 0.5 to 1×10^6 cultured cells with PBS by centrifuging at 1400 rpm for 5 min, resuspend in 100 μl of Annexin V diluted 1:150 in binding buffer, and incubate for 15 min in the dark at room temperature. Cells do not need to be washed. Add 100 to 150 μl of binding buffer to each sample and analyze by flow cytometry within an hour. It is important to note that it may be difficult to determine the exact percentage of death (Annexin V positivity) from freshly isolated cells, because dead and dying cells are quickly phagocytosed *in vivo*.

2.2.3. Cell viability stains, membrane integrity

7AAD (7-amino-actinomycin D, Molecular Probes, Invitrogen, Eugene, OR, cat. no. A1310) is a membrane-excluded dye that becomes highly fluorescent upon DNA intercalation and thus stains membrane-compromised cells. It can be excited with an argon laser (488 nm), although the maximum excitation wavelength is 546 nm. It fluoresces red (647 nm) showing minimum overlap with fluorescein or phycoerythrin. It is excluded by viable cells but can penetrate the cell membrane of those that are dead or dying. Wash 0.5 to 1 $\times$ 10^6 cells in FACS buffer (1% FCS, 1 $\times$ PBS, 0.1% NaN$_3$), centrifuge at 1400 rpm for 5 min, and stain with 7AAD at 10 μg/ml for 15 min at room temperature. Wash and resuspend in FACS buffer before collection on the flow cytometer.

Another compound that measures cell viability is propidium iodide (PI, Sigma, St. Louis, MO, cat. no. P4170). Similar to 7AAD, its fluorescence is substantially enhanced by DNA intercalation, and it is also excluded from viable cells. One significant drawback is that it fluoresces between channels (517 nm), creating background that is difficult to compensate. However, if one chooses to use PI to identify dead cells, it should be added to cells in FACS buffer at a final concentration of 50 μg/ml just before collection by flow cytometry.

Membrane integrity can also be determined by staining cells with trypan blue. Viable cells do not uptake trypan blue; however, the dye easily passes through the membrane of a dead or dying cell, resulting in a blue color. Determining the percentage of live cells excluding trypan blue is called the dye exclusion method. Make a 1:1 mixture of cells and 0.1% trypan blue. Place 10 μl of stained cells on a hemocytometer and allow cells time to settle. By use of a microscope, count the number of unstained live cells and the number of total cells (unstained and stained). Calculate the percentage of viable cells by dividing the number of unstained cells by the total number of cells and multiplying by 100.

3. DNA

3.1. Cell cycle (sub2n)

Another method of determining apoptotic cell death is by quantifying subdiploid levels of DNA, usually visualized by dye intercalation into double-stranded DNA. If cells are permeabilized, 7AAD and PI cause cells to fluoresce in proportion to the amount of DNA per cell (Fig. 10.2). G0/G1 cells have 2n DNA, S/G2 cells have >2n DNA, and apoptotic cells undergoing DNA fragmentation have subdiploid amounts of DNA. These dyes can thus be used to determine the proportion of cells in a population that are in each phase of the cell cycle.

Before staining with 7AAD, 0.5 to 1 $\times$ 10^6 cells need to be suspended in 0.5 ml 150 m*M* NaCl and kept on ice. Add 1.2 ml cold 100% ethanol dropwise while vortexing the cells. Incubate on ice for 30 min before washing cells with PBS. Follow by resuspending in 1% paraformaldehyde (PFA) and 0.01% Tween-20 in PBS. Incubate at room temperature for 30 min then

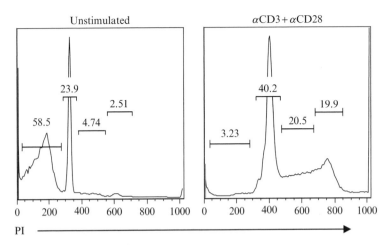

Figure 10.2 Detection of apoptotic death through measurement of subdiploid DNA. Purified wild-type T cells were cultured in media alone or with anti-CD3 and anti-CD28 for 72 h. Cells were fixed, permeabilized, and stained with propidium iodide. DNA content was measured by passing the cells through a flow cytometer and using a linear scale. Histograms were made by use of the FlowJo software (Tree Star, Ashland, OR).

wash cells with PBS. Centrifuge fixed cells at 1800 rpm. Incubate another 30 min at room temperature in 100 μl of 380 μM sodium citrate (diluted in PBS) containing 10 μg/ml 7AAD. Wash with PBS, resuspend in FACS buffer and analyze by flow cytometry by use of a linear scale (protocol designed by DR Beisner). To measure cell cycle by staining with propidium iodide, first fix cells with 0.5% PFA on ice for 20 min. Wash with PBS. Next, permeabilize the cells with 0.2% TritonX-100. Wash and resuspend in PI buffer (0.1% TritonX-100, 1 mM Tris pH 8.0, 0.1 mM EDTA, 0.1% Na citrate, 50 μg/ml PI) with 50 μg/ml RNase A added just before use. Stain cells on ice in the dark for at least 10 min. Wash and resuspend with FACS buffer. Collect on the flow cytometer by use of a linear scale.

3.2. Tunel

A defining characteristic of apoptosis is the cleavage of DNA by endonucleases, resulting in double- and single-stranded breaks (nicks). Terminal deoxynucleotidyl transferase dUTP Nick End Labeling (TUNEL) is a common method used to detect DNA fragmentation (Fig. 10.3).

The enzyme terminal deoxynucleotidyl transferase (TdT) identifies free 3′-OH ends in DNA and catalyzes the template-independent polymerization of deoxyribonucleotides (dUTP). Modified dUTPs (i.e., fluorescein-dUTP) allow for detection by flow cytometry or fluorescence microscopy (*In situ* cell death detection kit, TMR red, Roche, cat. no. 12156792910; FragEL DNA fragmentation detection kit, Fluorescent TdT enzyme,

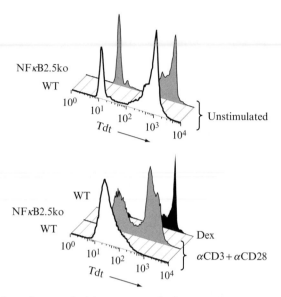

Figure 10.3 Lymphocytes lacking NF-κB die by apoptosis during proliferation. Wild-type (WT) and NFκB2.5ko (*nfkb1⁻/⁻crel⁻/⁻rela⁺/⁻*) T cells were purified from lymph nodes by negative selection. Cells were cultured in media alone or with anti-CD3 and anti-CD28 for 72 h. Wild-type cells were treated overnight with dexamethasone (Dex) as a positive control for TUNEL staining. Data were collected with a flow cytometer and analysis was done with the Flowjo software.

Calbiochem, cat. no. QIA39.) To allow for the surface staining of cells for flow cytometry, we have derived an alternative fixation and permeabilization protocol when the FragEL DNA fragmentation detection kit is used. Once 0.5 to 1×10^6 cells have been washed with FACS buffer, incubate with your desired antibodies conjugated to any fluorophore, except those than fall into the FL1 channel, for 15 min at room temperature. Wash with PBS and fix cells with 1% PFA in PBS for 10 to 20 min at room temperature. Wash and permeabilize with 0.2% Tween-20 in PBS for 10 to 20 min. Wash and resuspend cells in the equilibrium buffer provided in the FragEL DNA kit as described in the kit directions. Follow with TdT labeling for 1 h at 37 °C. The protocol included with the *in situ* cell death kit, TMR red, allows for surface staining before fixation and does not need to be altered. A decent positive control for TUNEL can be obtained by the use of cells cultured with dexamethasone (Sigma, cat. no. D-4902) or the anti-Fas antibody, Jo-2 (BD Pharmingen, cat. no. 554255), overnight.

3.3. Nucleosomal DNA laddering (DNA fragmentation)

The activation of endonucleases preferentially results in the degradation of the internucleosomal linker regions within genomic DNA, producing 180- to 200-base pair DNA fragments (Fig. 10.4). These nucleosomal DNA

	WT			NFκB2.5ko			
αCD3 + αCD28	–	+	–	–	+	–	M
Time (h)	0	48	48	0	48	48	1kB

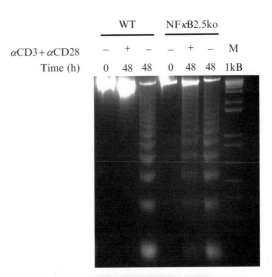

Figure 10.4   Proliferation of NF-κB–deficient T cells results in fragmented DNA. Purified wild-type and NFκB2.5ko ($nfkb1^{-/-}crel^{-/-}rela^{+/-}$) lymphocytes were stimulated with anti-CD3 and anti-CD28 for 48 h. As a positive control, lymphocytes were also cultured in media alone for 48 h. DNA was prepared as described and resolved on a 2% agarose gel in comparison with a size marker (M), the 1-kb plus ladder.

fragments containing different numbers of nucleosomes appear as a "ladder" by gel electrophoresis and are indicative of cells undergoing the late stages of apoptosis (Wyllie *et al.*, 1980).

To access DNA laddering, at least 2 to 5×10^6 cells are needed. Harvest, wash and lyse cells with 0.2% TritonX-100 in PBS. Incubate on ice for 10 min. Add RNAseA at 50 μg/ml and incubate for 1 h at 37 °C. Bring cell suspension to a final concentration of 0.5% SDS and 150 μg/ml proteinase K and incubate for 1 h at 50 °C. Add an equal volume of phenol/chloroform, vortex, and centrifuge at 14,000 rpm for 5 min. Recover the aqueous phase and add 0.1 volume of 3 *M* sodium acetate. Use two volumes of ethanol to precipitate the DNA at −20 °C for 1 h. Pellet the DNA by centrifugation, dry, and resuspend in TE buffer before resolving on a 2% agarose gel. Primary cells left in culture with no stimulation for at least 2 days or treated with dexamethasone overnight are sufficient positive laddering controls.

4. PROTEINS

Many signaling cascades are triggered during the events leading to programmed cell death. For example, the mitochondria could release cytochrome *c*, resulting in the activation of caspase-9 and the caspase-signaling cascade, ending in the activation of executioner caspases, such as caspases-3 and -7. Likewise, the extrinsic death pathway involving caspase-8 also leads to the activation of executioner caspases. Changes in protein levels of these key players can be measured by immunoblotting.

Prepare cytoplasmic extracts with lysis buffer containing protease and phosphatase inhibitors to prevent protein degradation. Separate 10 to 20 μg of protein by electrophoresis on a SDS-polyacrylamide gel (SDS-PAGE). Transfer to a PVDF membrane and block with 5% milk dissolved in TBST (10 mM Tris-Cl pH7.5, 150 mM NaCl, 0.25% Tween-20). With gentle agitation, incubate overnight at 4 °C or 2 h at room temperature with primary antibody, diluted 1:100 in 5% milk and 0.02% NaN$_3$. Caspase antibodies worth mentioning within the intrinsic and extrinsic death pathways are as follows: anti-Caspase-8 clone 1G12, 3B10 (Alexis Biochemicals, cat. no. 804-447-C100, 804-448-C100), cleaved anti-Caspase-3 (BD Pharmingen, cat. no. 559565), pro- and cleaved- anti-Caspase-3 (Cell Signaling Technology, Danvers, MA, cat. no. 9665), and anti-Caspase-9 (Cell Signaling Technology, cat. no. 9504). A few antibodies that detect components in the death receptor (noncanonical) and mitochondrial (canonical) death pathways that lead to apoptosis are as follows: anti-Bim (Sigma, cat. no. B7929), anti-Bax (Cell Signaling, cat. no. 2772), anti-Bcl-xL (Cell Signaling, cat. no. 2762), anti-Bcl-2 (BD Pharmingen, cat. no. 554218), and anti-cytochrome c (Cell Signaling Technology, cat. no. 4272). Wash membrane three times with TBST to remove unbound primary antibody, followed by a 1 h incubation with a horseradish peroxidase (HRP)–conjugated secondary antibody to the appropriate species in 5% milk (no NaN$_3$). Wash membrane and visualize on autoradiography film by use of the ECL system.

5. METABOLIC HEALTH

5.1. Mitochondrial membrane potential

Change in the mitochondrial transmembrane potential is one major cause of programmed cell death (Green and Kroemer, 1998). Measuring such a transformation in the cell is one method used to determine the mechanism involved in apoptotic and necrotic death. An increase in membrane permeability results from the mitochondrial permeability transition (MPT) that alters the electrochemical gradient across the membrane, opening the mitochondrial transition pores.

One very useful tool is the JC-1 Mitochondrial Membrane Potential Detection Kit from Molecular Probes (Invitrogen, cat. no. M34152). Similar to TMRE, DiOC$_6$, and rhodamine 123, JC-1 is a fluorescent cationic dye. However, rather than measuring a loss of fluorescence, shifts in mitochondrial polarization can be specifically detected by the color change of JC-1. In viable cells, the JC-1 dye accumulates in the mitochondria as an aggregate (red). When the mitochondrial potential of the cell collapses, the dye remains in the cytoplasm as a monomer (green). Analysis can be done by flow cytometry or fluorescence microscopy; however, titrations should be done to ensure the collection of accurate data with the least amount of background staining.

5.2. Reactive oxygen species

Under normal physiologic conditions, reactive oxygen species (ROS) are by-products generated in the mitochondria that are quickly removed by antioxidant enzymes. However, when disrupted, such as in instances of environmental stress, the accumulation of ROS can mediate programmed cell death in many cell types (Fiers *et al.*, 1999). Intracellular ROS can be identified by the oxidation of a reduced, cell permeable probe that is naturally nonfluorescent but becomes fluorescent when its acetate groups are cleaved by intracellular esterases.

The ROS detection agent we currently use is the carboxy derivative of fluorescein, carboxy-H_2DCFDA (5-(and-6)-carboxy-$2',7'$- dichlorohydro-fluorescein diacetate, Molecular Probes, cat. no. C-400); 0.5 to 1×10^6 cells are incubated for 30 min at 37 °C in FACS buffer with 5 μM carboxy-H_2DCFDA. Surface marker staining in other fluorophores can be done simultaneously. Samples are washed and centrifuged at 1500 rpm for 5 min. The presence of ROS activity is then measured by flow cytometry. As a positive control, cells can be treated with 100 μM H_2O_2 for 20 h, although it is best to do a titration as each cell type can differ. A few known inhibitors of ROS activity are the reducing agent N-acetylcysteine (NAC) and the ROS scavenger butylated hydroxyanisole (BHA).

6. SENESCENCE DETECTION

Cellular senescence involves an irreversible entering of the G0 phase of the cell cycle concomitant with morphologic changes of the cell. Senescent cells are flat and often damaged when physically manipulated (e.g., trypsinization). Cellular senescence can be brought about by mitochondrial or genotoxic stresses or chronically elevated NF-κB activity, but it is also part of the normal proliferation limit of primary fibroblasts obtained from mouse embryos or human sources. Whereas mouse fibroblasts' proliferative capacity is limited by the elevated oxygen levels of cell culture, human fibroblasts are driven into senescence by shortening telomeres.

One hallmark of senescent adherent cells (such as fibroblasts) is that they are flat. In simple bright-field microscopy, senescent fibroblasts are several times bigger than proliferating fibroblasts and show less birefringence and contrast. In addition, an endogenous ß-galactosidase with an acidic activity preference is expressed at elevated levels in senescent cells. This senescence-associated ß-galactosidase, or SA-ß-gal, can be revealed by use of X-gal in acidic buffer conditions.

Starting with adherent cells attached to a tissue culture plate, wash cells with PBS at least twice, then fix with 2% (v/v) formaldehyde and 0.2% (w/v) glutaraldehyde in PBS. Wash again with PBS and add 1 ml per 30-mm plate of 5 mM potassium ferrocyanide ($K_3Fe(CN)_6$), 2 mM $MgCl_2$, 20 mM citric acid, 40 mM sodium phosphate dibasic (Na_2HPO_4) at pH 6.0 and 1 mg/ml X-gal. Incubate for 2 to 24 h. Blue staining indicates SA-ß-gal expression and is used as a marker for senescence.

Other hallmarks of senescent cells are elevated levels of the cell cycle inhibitor p16. Within a population, p16 levels can be assayed by immunoblot, but standard immunohistochemistry can also be used to reveal p16 levels in individual cells.

REFERENCES

Adler, A. S., Sinha, S., Kawahara, T. L., Zhang, J. Y., Segal, E., and Chang, H. Y. (2007). Motif module map reveals enforcement of aging by continual NF-{kappa}B activity. *Genes Dev.*

Alcamo, E., Mizgerd, J. P., Horwitz, B. H., Bronson, R., Beg, A. A., Scott, M., Doerschuk, C. M., Hynes, R. O., and Baltimore, D. (2001). Targeted mutation of TNF receptor I rescues the RelA-deficient mouse and reveals a critical role for NF-kappa B in leukocyte recruitment. *J. Immunol.* **167,** 1592–1600.

Beg, A. A., and Baltimore, D. (1996). An essential role for NF-kappa B in preventing TNF-alpha-induced cell death. *Science* **274,** 782–784.

Beg, A. A., Sha, W. C., Bronson, R. T., Ghosh, S., and Baltimore, D. (1995). Embryonic lethality and liver degeneration in mice lacking the RelA component of NF-kappa B. *Nature* **376,** 167–170.

Daleke, D. L. (2003). Regulation of transbilayer plasma membrane phospholipid asymmetry. *J. Lipid Res.* **44,** 233–242.

Fiers, W., Beyaert, R., Declercq, W., and Vandenabeele, P. (1999). More than one way to die: Apoptosis, necrosis and reactive oxygen damage. *Oncogene* **18,** 7719–7730.

Golstein, P., and Kroemer, G. (2007). Cell death by necrosis: Towards a molecular definition. *Trends Biochem. Sci.* **32,** 37–43.

Green, D., and Kroemer, G. (1998). The central executioners of apoptosis: Caspases or mitochondria? *Trends Cell Biol.* **8,** 267–271.

Gurova, K. V., Hill, J. E., Guo, C., Prokvolit, A., Burdelya, L. G., Samoylova, E., Khodyakova, A. V., Ganapathi, R., Ganapathi, M., Tararova, N. D., et al. (2005). Small molecules that reactivate p53 in renal cell carcinoma reveal a NF-kappa B–dependent mechanism of p53 suppression in tumors. *Proc. Natl. Acad. Sci. USA* **102,** 17448–17453.

Henson, P. M., Bratton, D. L., and Fadok, V. A. (2001). Apoptotic cell removal. *Curr. Biol.* **11,** R795–R805.

Kerr, J. F., Wyllie, A. H., and Currie, A. R. (1972). Apoptosis: A basic biological phenomenon with wide-ranging implications in tissue kinetics. *Br. J. Cancer* **26,** 239–257.

Kroemer, G., and Martin, S. J. (2005). Caspase-independent cell death. *Nat. Med.* **11,** 725–730.

Lecoeur, H., de Oliveira-Pinto, L. M., and Gougeon, M. L. (2002). Multiparametric flow cytometric analysis of biochemical and functional events associated with apoptosis and oncosis using the 7-aminoactinomycin D assay. *J. Immunol. Methods* **265,** 81–96.

Lecoeur, H., Fevrier, M., Garcia, S., Riviere, Y., and Gougeon, M. L. (2001). A novel flow cytometric assay for quantitation and multiparametric characterization of cell-mediated cytotoxicity. *J. Immunol. Methods* **253,** 177–187.

Lin, K. I., DiDonato, J. A., Hoffmann, A., Hardwick, J. M., and Ratan, R. R. (1998). Suppression of steady-state, but not stimulus-induced NF-kappa B activity inhibits alphavirus-induced apoptosis. *J. Cell Biol.* **141,** 1479–1487.

Liu, J., Yang, D., Minemoto, Y., Leitges, M., Rosner, M. R., and Lin, A. (2006). NF-kappa B is required for UV-induced JNK activation via induction of PKC delta. *Mol. Cell* **21,** 467–480.

Matteucci, C., Grelli, S., De Smaele, E., Fontana, C., and Mastino, A. (1999). Identification of nuclei from apoptotic, necrotic, and viable lymphoid cells by using multiparameter flow cytometry. *Cytometry* **35,** 145–153.

May, M. J., and Madge, L. A. (2007). Caspase inhibition sensitizes inhibitor of NF-kappa B kinase beta-deficient fibroblasts to caspase-independent cell death via the generation of reactive oxygen species. *J. Biol. Chem.* **282,** 16105–16116.

Papa, S., Zazzeroni, F., Pham, C. G., Bubici, C., and Franzoso, G. (2004). Linking JNK signaling to NF-kappa B: A key to survival. *J. Cell Sci.* **117,** 5197–5208.

Ryan, K. M., Ernst, M. K., Rice, N. R., and Vousden, K. H. (2000). Role of NF-kappa B in p53-mediated programmed cell death. *Nature* **404,** 892–897.

Vermes, I., Haanen, C., Steffens-Nakken, H., and Reutelingsperger, C. (1995). A novel assay for apoptosis. Flow cytometric detection of phosphatidylserine expression on early apoptotic cells using fluorescein labelled Annexin V. *J. Immunol. Methods* **184,** 39–51.

Wyllie, A. H., Kerr, J. F., and Currie, A. R. (1980). Cell death: The significance of apoptosis. *Int. Rev. Cytol.* **68,** 251–306.

CHAPTER ELEVEN

PURIFICATION AND BIOASSAY OF HEDGEHOG LIGANDS FOR THE STUDY OF CELL DEATH AND SURVIVAL

Pilar Martinez-Chinchilla *and* Natalia A. Riobo

Contents

Abstract

The Hedgehog (Hh) family of secreted ligands—composed of Sonic Hedgehog (Shh), Indian Hedgehog (Ihh), and Desert Hedgehog (Dhh)—possesses many roles during embryonic development, adult homeostasis, and cancer. The specific functions of the Hh proteins are intertwined with their requirement as survival factors in Hh-responsive cells. However, studies designed to dissect the anti-apoptotic role of Hhs have been hindered by the lack of simple approaches to purify large quantities of recombinant ligands in the average laboratory setting because of the natural modifications of these proteins with palmitic acid and cholesterol. In this chapter, we provide a comprehensive protocol for the expression of Shh, Ihh, and Dhh in *Escherichia coli* as fusion proteins with calmodulin-binding peptide to allow easy and rapid purification. The ligands are engineered with a new N-terminus containing two isoleucine

Department of Biochemistry and Molecular Biology and Kimmel Cancer Center, Thomas Jefferson University

Methods in Enzymology, Volume 446
ISSN 0076-6879, DOI: 10.1016/S0076-6879(08)01611-X

residues to provide an essential hydrophobic interphase for achieving high biologic activity. The protocol includes a detailed description of a method for determination of the specific activity of the generated proteins by use of a cell culture–based luciferase approach.

1. INTRODUCTION

1.1. Overview of the hedgehog signaling pathway

The *hedgehog* gene was initially identified in a *Drosophila* mutational screening for genes that regulate the differences between the anterior and posterior parts of individual body segments (known as "segment polarity genes"). In particular, the *hedgehog* mutant larva presented a fusion of the denticle-containing stripes and loss of stripes of naked cuticle, giving it the appearance of a "hedgehog" (Nusslein-Volhard and Wieschaus, 1980; Mohler, 1988).

The vertebrate family of secreted Hedgehog proteins is composed of three highly homologous isoforms encoded by separate genes: Sonic Hedgehog (Shh), Indian Hedgehog (Ihh), and Desert Hedgehog (Dhh). By far the most studied isoform is Shh, followed by Ihh and Dhh, the latter being the most similar to the fly homolog. The three isoforms have a distinct and partially overlapping expression pattern. In midgestation embryos, Shh is expressed by the notochord, neural tube floor plate, zone of polarizing activity (ZPA) of the limb buds, and additional restricted locations (Echelard *et al.*, 1993; Harfe *et al.*, 2004). Ihh is more abundant in the primitive gut where it patterns the growth of several endoderm-derived organs and is also expressed in the developing skeleton and yolk sac where it has a prominent role in angiogenesis (Byrd *et al.*, 2002; Colnot *et al.*, 2005). Dhh is restricted in expression to Schwann cell and Sertoli cell progenitors (Bitgood and McMahon, 1995).

Shh initiates a group of signals by binding to a receptor complex composed of Patched-1 (Marigo *et al.*, 1996; Stone *et al.*, 1996) and either CDO or BCO family, both members of the Ig family of single transmembrane proteins (Tenzen *et al.*, 2006; Zhang *et al.*, 2006). In the absence of Shh, Patched-1 acts catalytically to repress the constitutive activity of Smoothened (Smo), a G protein–coupled receptor (Taipale, 2002). Binding of Shh to the receptor complex inhibits Patched-1 activity, resulting in derepression of Smo (Fig. 11.1). Mammalian Smo has been recently shown to engage heterotrimeric G proteins of the Gi subfamily through the intracellular loops and to initiate another signal of unclear nature that was mapped to a small stretch of residues of the long cytoplasmic C-terminal tail (Riobo *et al.*, 2006a; Varjosalo, 2006). At least these two signals originated at Smo are required for the stabilization and nuclear accumulation of the

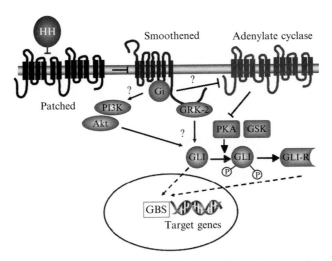

Figure 11.1 Current model of the mammalian Hedgehog signaling pathway. Binding of a Hedgehog ligand (Hh) to Patched results in derepression of Smoothened, which signals through heterotrimeric G_i proteins (G_i) and G protein–coupled receptor regulator kinase-2 (GRK-2). G_i is then believed to activate PI3K and Akt and inhibit adenylate cyclase, both events leading to stabilization of the GLI family of transcription factors (GLI). In the absence of Hh, GLIs are targeted for partial proteasomal degradation by PKA and GSK-3β–mediated phosphorylation. Both full-length GLIs and GLI-R (a fragment spared of degradation) compete for GLI binding sites (GBS) in the promoter regions of target genes to promote and repress their expression, respectively.

full-length forms of the Gli transcription factors (Riobo *et al.*, 2006a). The Gli2 and Gli3 variants are processed into a transcriptional repressor in the absence of Smo activation through phosphorylation by protein kinase A, glycogen synthase kinase 3, and casein kinase I followed by partial proteasomal degradation (Riobo and Manning, 2007). Smo-dependent signals, which include phosphoinositide 3-kinase and Akt, prevent their phosphorylation and result in stimulated transcription of Gli-target genes (such as *gli1*, *patched1*, *bcl2*, etc) (Riobo *et al.*, 2006b,c).

The use of powerful genetic models, spontaneous mutations, and the availability of specific pathway inhibitors such as cyclopamine and its derivatives have been instrumental in defining numerous functions of the Hh ligands during embryonic development, adult homeostasis, and disease. During vertebrate development, Hh has functions as a morphogen in the dorsoventral patterning of the neural tube, in the anterior-posterior patterning of the limbs, as a mitogen for neuronal precursors, and as a promoter for tissue remodeling processes like vasculogenesis, angiogenesis, and branching morphogenesis, and as a prosurvival factor in all those tissues (McMahon *et al.*, 2003).

In adult animals, Shh has shown a critical role for stem cell maintenance and tissue regeneration because of normal cell replacement or on injury

(Beachy *et al.*, 2004). When dysregulated, Hh signaling promotes tumor formation and growth and inhibits apoptosis so strongly that the blockade of the Hh pathway with specific pharmacologic agents suffices to increase cell death and in many cases leads to a significant tumor regression (Taipale, 2001).

1.2. Biosynthetic modifications of the secreted hedgehog proteins

The three Hh proteins are synthesized as precursor proteins of ~45 kDa, which enter the secretory pathway and undergo a particular set of posttranslational modifications (Mann and Beachy, 2004). First, the precursors are cleaved to generate the ~19-kDa N-terminal signaling fragments. The proteolytic process is catalyzed by the C-terminal half of the precursor. The enzymatic mechanism of this cleavage involves the formation of an internal thioester intermediate that is then cleaved by a nucleophilic substitution. Surprisingly, the attacking nucleophile for this substitution is cholesterol, resulting in an N-terminal fragment (N-Hh) covalently attached at the C-terminus to cholesterol. Subsequently, N-Hh is palmitoylated at its N-terminus generating N-Hhp, the Hh derivative with the highest potency. This dually lipidated molecule forms a multimeric complex through the hydrophobic moieties that is soluble in aqueous environments and allows signaling at a distance from the source (Chen *et al.*, 2004).

It is evident from the atypical modifications of Hh proteins that engineering of recombinant active proteins is not trivial. The cholesterol modification is not necessary for binding to Patched but modifies Hh trafficking, affecting the activity range in physiologic settings (Dawber *et al.*, 2005; Li *et al.*, 2006); however, cholesterol is an essential substrate for the cleavage reaction to generate the N-terminal Hh fragment (Guy, 2000). Therefore, the cholesterol modification can be omitted if we synthesize N-Hh by replacing the cleavage site with a premature stop codon. Nonetheless, the N-terminal modification has a large impact on the potency of N-Hh (Dawber *et al.*, 2005). The specific requirement for palmitate was investigated by replacing it with several other fatty acids or hydrophobic compounds or hydrophobic residues (Taylor *et al.*, 2001). It was concluded that it is the hydrophobicity of the modification rather than its identity as palmitate what increases the potency of N-Hhp. Therefore, we decided to generate recombinant N-Hh carrying two additional Ile residues at its N-terminus to increase its hydrophobicity, which has an intermediate potency between the natural N-Hhp and the unmodified recombinant N-Hh. This same approach is used by R&D to produce commercial Hh ligands.

2. HEDGEHOG AND THE REGULATION OF CELL DEATH

2.1. Hedgehog-dependent survival in the nervous system

The anti-apoptotic activity of Shh in the developing nervous system has been extensively documented. Seminal experiments in developing chick embryos demonstrated that removal of the notochord and floor plate, two natural sources of Shh, induces massive apoptosis of neuroepithelial cells of the neural tube, as well as of the myogenic and chondrogenic somitic lineages (Charrier *et al.*, 2001; Teillet *et al.*, 1998). Moreover, re-addition of ectopic Shh to the embryo through transplantation of Shh-producing cells completely rescued the cell death process and the truncation of the animal (Charrier *et al.*, 2001). Validation of this paradigm was also provided by the generation of Shh–null mice, which also showed massive cell death in the developing spinal cord (Chiang *et al.*, 1996; Litingtung and Chiang, 2000). Moreover, not only the neural tube but also branchial arch structures of Shh mutant mice are affected (Ahlgren and Bronner-Fraser, 1999). Application of a blocking antibody against Shh (mAb5E1) also results in the apoptosis of neural crest cells. Shh has also been shown to promote survival of dopamine- and γ-aminobutyric acid (GABA)–producing neurons *in vitro* (Miao *et al.*, 1997). Along the same line, Borycki and colleagues demonstrated that Shh is required for the survival of sclerotome cells, motoneuron progenitors in the ventral neural tube, and presumptive trunk neural crest cells (Borycki *et al.*, 1999). All this evidence suggests that Shh displays a major anti-apoptotic activity in the developing and mature CNS. However, it was reported that ectopic application of Shh could also enhance apoptosis within the ventral neural tube, although this cell death may be a consequence of misspecification of ventral neuronal types (Oppenheim *et al.*, 1999).

During forebrain development, abnormal activation of the Hedgehog pathway by deletion of αE-catenin shortens the cell cycle and reduces apoptosis, resulting in cortical hyperplasia (Lien *et al.*, 2006).

Two additional observations support the view that Shh is a prosurvival molecule: first, downstream targets of Shh are clearly involved in cell survival (Akt and Bcl-2) (Bigelow *et al.*, 2004; Riobo *et al.*, 2006b); second, in the absence of Shh, Patched triggers apoptosis by transducing a death signal (Thibert *et al.*, 2003). In this context, Patched behaves as a dependence receptor, which induces caspase-triggered apoptosis in the absence of ligand, thus rendering the cells dependent on ligand availability for survival (Mehlen and Thibert, 2004). It has been proposed that Patched triggers apoptosis through the cleavage of its seventh intracellular domain to expose a cryptic proapoptotic domain. Indeed, expression of a dominant-negative Patched that cannot be cleaved in chick neural tube cells deprived of Shh is able to prevent cell death (Mehlen and Thibert, 2004). It is not clear how Shh blocks the proapoptotic activity of Patched. A truncated form of Patched is proapoptotic both in the absence and in the presence of Shh, suggesting that Shh binding may control Patched cleavage. It seems that the proapoptotic role of Patched is a recent evolutionary aquisition, because the cleavage site is present in mammals and avians but not in *Drosophila* (Guerrero and Ruiz i Altaba, 2003).

Later in development, when the cerebellum is formed, a burst of granule cell progenitor (GCP) proliferation occurs in the outer external granule layer, which is sustained mainly by Purkinje cell–derived Shh. Once GCPs move inwardly, they stop responding to Shh and begin differentiating and migrating toward the internal granule layer (Ruiz i Altaba *et al.*, 2002). Failure to interrupt Shh signals results in uncoordinated proliferation and differentiation of GCPs and is associated with medulloblastoma formation. Hedgehog antagonism was proposed to be mediated by REN, a protein that when overexpressed drives caspase-3 activation and TUNEL labeling of GPCs (Argenti *et al.*, 2005). The Hh antagonist cyclopamine reduces Bcl-2 expression levels and increases apoptosis in DAOY and UW228 medulloblastoma cells (Bar *et al.*, 2007). Apoptotic induction caused by cyclopamine can be partially rescued by ectopic expression of Gli1 or Bcl-2 (Bar *et al.*, 2007). This and other studies indicate that Bcl-2 is an important mediator of Hh anti-apoptotic activity in medulloblastoma.

2.2. Regulation of apoptosis in adult tissues by hedgehog

Several studies have revealed an anti-apoptotic role of Hh in the adult immune system on both T and B lymphocytes. Inhibition of Hh activity by specific inhibitors leads to cell cycle arrest and apoptosis in human acute lymphoblastic leukemia cells, suggesting that that the Hh pathway is critical for leukemia T cell growth and survival (Ji *et al.*, 2007). Moreover, Hh ligands secreted by the bone marrow, nodal, and splenic stromal cells function as survival factors for malignant B lymphoma and plasmacytoma cells isolated from humans with these malignancies (Dierks *et al.*, 2007). Inhibition of Hh signaling in lymphomas induces apoptosis through downregulation of Bcl-2, prevents expansion of lymphoma cells in syngeneic animals, and reduces tumor mass in mice with fully developed disease (Dierks *et al.*, 2007). Shh is one of the survival signals provided by follicular dendritic cells to prevent apoptosis in germinal center B cells: blockade of the Hh signaling reduces the survival and, consequently, antibody secretion of germinal center B cells. Moreover, Shh rescues germinal center B cells from apoptosis induced by Fas ligation (Sacedon *et al.*, 2005).

Solid tumors have also been reported to depend on active Hh signaling for survival. The Hh pathway is kept almost silent in normal adult tissues and is highly upregulated during tumorigenesis. This is the case of ovarian carcinoma cells, in which immunoreactivity for Shh, Dhh, Ptch, Smo, and Gli1 highly correlates with cell proliferation assessed by Ki-67. Blocking the Hh signal by use of either the Hh pathway inhibitor cyclopamine or Gli1 siRNA induced G_1 arrest and apoptosis along with downregulation of cyclin A and cyclin D1 and upregulation of p21 and p27 (Chen *et al.*, 2007). Another example of the anti-apoptotic role of Hh is the induction of cell death in established basal cell carcinomas by molecular or pharmacologic inhibition of the Hh pathway (Hutchin *et al.*, 2005).

Another paradigm that has evidenced a role of Shh in survival is tissue repair and regeneration. In an elegant study of myocardial infarction, intramyocardial injection of naked DNA encoding Shh was shown to reduce infarct size by promoting neovascularization and stem cell recruitment, reducing fibrosis, and by reducing apoptotic cell death of preexisting cardiomyocytes (Kusano *et al.*, 2005).

3. GENERATION OF RECOMBINANT HEDGEHOG LIGANDS

3.1. Rationale and overview of the experimental strategy

Given the extensive processing and lipid modifications occurring in natural Hedgehog ligands, of which only the N-terminal hydrophobicity is required for activity, we designed an engineered N-Shh made up of only the mature fragment (spanning amino acids 24 to 197) with no C-terminal domain and, therefore, no cholesterol modification at Gly197 and with two additional Ile residues N-terminal to Cys24 to provide increased hydrophobicity (Taylor *et al.*, 2001). In addition, we constructed it as a C-terminal fusion with a cleavable calmodulin-binding peptide (CBP) N-terminal tag to allow selective separation of the fusion protein from crude bacterial extracts. Effective purification of the fusion is achieved by affinity chromatography by use of a calmodulin (CaM) affinity resin in the presence of high calcium levels followed by cleavage of the CBP tag and removal of this fragment from the intact Ilex2-N-Shh.

3.2. Materials

The pCAL-n-EK vector (cat # 214310) from Stratagene was used to subclone the cDNA encoding Ilex2-N-Hh ligands. Either the Affinity Protein Expression and Purification Systems kit (cat # 204301 and # 204302) or the separated items, including CaM affinity resin (cat # 214303) and BL21 (Gold)DE3 cells (cat # 230132) were also obtained from Stratagene. Enterokinase was purchased from Novagen (cat # 69066-3), EKapture Agarose from Novagen (cat # 69068-3), *EarI*

(Eam1104I isoschizomer), *HindIII*, calf intestinal alkaline phosphatase, and T4 ligase were from New England Biolabs, isopropyl-β-D-thiogalactopyranoside (IPTG) was from Denville Scientific Inc. (cat # CI8280-13), LB broth was from Fisher (cat # BP1426-500), and LB agar was from Invitrogen (cat # 22700-025). Monoclonal anti-Shh/Ihh antibody (clone 5E1) is available from Developmental Studies Hybridoma Bank, anti-Dhh is from Santa Cruz Biotechnology (cat # sc-1193). Amicon filters of 10-kDa cutoff (cat # UFC901024) are from Millipore. NIH3T3 fibroblasts are from American Type Cell Collection (cat # CRL-1658). FuGene6 transfection reagent is from Roche (cat # 11815091001), DMEM and penicillin/streptomycin are from Invitrogen, iron–enriched calf serum from Sigma (cat # C8056), Dual Luciferase Assay System from Promega (cat # E1960). All other chemicals were from Sigma Aldrich.

3.3. Generation of recombinant plasmids

The 24 to 197 region of the Shh, Ihh, and Dhh cDNAs was amplified by PCR by use of forward primers containing an *Eam1104I/EarI* restriction site followed by two Ile codons and reverse primers containing a premature stop codon upstream of Gly197 followed by a *HindIII* restriction site. The full primer sequences are shown in Table 11.1. Low-stringency PCR was performed by five cycles by use of $T_{annealing} = 55\,^{\circ}C$ followed by high-stringency PCR at $T_{annealing} = 67\,^{\circ}C$ for 25 additional cycles.

The Hh recombinant sequences created by PCR were ligated into TA-TOPO (Invitrogen) and transformed into TOP10 competent cells (Invitrogen). Single ampicillin-resistant colonies were grown overnight in LB ampicillin broth and recombinant plasmids isolated with the Qiagen Miniprep Kit. Ten micrograms of each plasmid was digested with *EarI* and *HindIII* at 37 $^{\circ}$C for 1 h and the excised fragments were gel purified. The recipient vector pCAL-n-EK was similarly digested, column purified, and treated with calf intestinal alkaline phosphatase (CIP) to reduce background by religation. The ligation of the engineered Hh cDNA into pCAL-n-EK was performed at 18 $^{\circ}$C overnight in the presence of T4 DNA ligase and ATP, followed by transformation of TOP10 cells and plasmid Miniprep. All constructs were fully sequenced after this step. A diagram of the expression construct is provided in Fig. 11.2.

3.4. Purification protocol

3.4.1. Induction of fusion protein synthesis in *E. coli*

After transforming the pCAL-N-Hh constructs into BL21(Gold)DE3 cells following the manufacturer's guidelines, a well–isolated ampicillin-resistant colony is inoculated into 50 ml of LB broth with ampicillin (100 μg/ml) and expanded overnight with shaking at 225 to 250 rpm and 37 $^{\circ}$C.

Table 11.1   Primer Sequences for PCR Amplification of N-Hh

Hh ligand	cDNA origin	Primers
Sonic Hedgehog	pRK5-mShh (courtesy of P. A. Beachy, Stanford University)	5'- <u>CTCTTC</u> CAAG**ATTATT**GGGCCC GGCAG-3' 5'- <u>AAGCTT</u> **CTA**GCCGCCGGATT-3'
Indian Hedgehog	Pax4-mIhh (courtesy of H. Matthias, Diabetes Center, UCSF)	5'-<u>CTCTTC</u> CAAG**ATTATT** GGGCCGGGCCG-3' 5'-<u>AAGCTT</u>**CTA**GCCCGGCCGGAC-3'
Desert Hedgehog	pCMV-SPORT6 human clone 5169286 (from ATCC)	5'-<u>CTCTTC</u> CAAG**ATTATT** GGGCCGGGCCG-3' 5'-<u>AAGCTT</u>**CTA**GCCCGCCCGGAC-3'

Variants with necessary modifications for later cloning into pCAL-n-EK. Restriction sites for *Eam1104I/EarI* and *HindIII* are underlined; Ile and stop codons are indicated in bold.

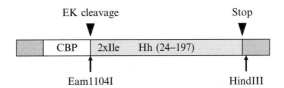

Figure 11.2   Design of pCAL-N-Hh constructs. The cloning vector pCAL-n-EK contains a calmodulin-binding peptide (CBP) coding sequence followed by an entero-kinase (EK) cleavage site that is preserved on *Eam1104I* digestion and ligation with the insert. Hh cDNA fragments encoding aa 24 to 197 were cloned between *Eam1104I* and *HindIII* restriction sites. Inducible expression in *E. coli* is controlled by the bacteriophage T7 gene 10 promoter and a leader sequence, which confers high-level expression in the presence of IPTG.

The next morning this culture is diluted 20 times into 1 L LB broth containing no antibiotics and incubated at 30 °C for 3 h with shaking. We optimized the induction time, isopropyl β-D-1-thiogalactopyranoside (IPTG) concentration and temperature for optimal induction of soluble

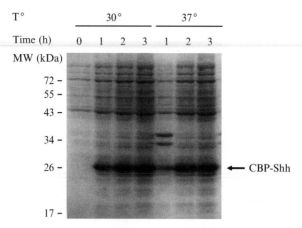

Figure 11.3 Effect of induction time and temperature on expression of soluble CBP-Shh. BL21(Gold)DE3 cells transformed with pCAL-N-Shh were incubated for the indicated time frames in the presence of 1 mM IPTG at either 30 °C or 37 °C. Clarified bacterial lysates (soluble proteins) were separated in a 12% SDS-PAGE and stained with Coomassie blue.

CBP-N-Hh proteins (Fig. 11.3). The optimal conditions were found to be almost identical for the three Hh isoforms, so we will describe a general protocol that can be used with any N-Hh ligand.

3.4.2. Affinity purification of CBP-N-Hh proteins

After the induction phase is completed, cells are recovered by centrifugation at 4000g for 20 min. To release the soluble proteins, the pellet is resuspended in binding buffer (50 mM Tris/HCl, pH 8; 150 mM NaCl, 10 mM β-mercaptoethanol, 1 mM magnesium acetate, 1 mM imidazole, 2 mM CaCl$_2$). The high concentration of CaCl$_2$ will later facilitate the binding of the CBP peptide to the CaM resin. The suspended cells are incubated with lysozyme (0.2 mg/ml) with gentle rocking for 15 min at 4 °C. To ensure total bacterial lysis, five rounds of sonication are performed for 30 sec at setting 4, with 30-sec intervals. After sonication, intact cells are separated by centrifugation at 16,000g for 15 min at 4 °C to recover the soluble fraction (lysate).

The previous supernatant is applied to 2-ml bed volume of CaM resin previously equilibrated in binding buffer in a chromatographic column and let run by gravity in a cold room. After loading the sample, the resin is washed with 5 to 10 column volumes of binding buffer to efficiently remove any unbound material. The washes usually contain a significant amount of CBP-N-Hh that does not bind the resin probably because of partial denaturation, because increasing the bed volume does not increase the retention of the fusion protein (Fig. 11.4). The CaM column is further washed with 5 to 10 column volumes of preelution buffer (50 mM

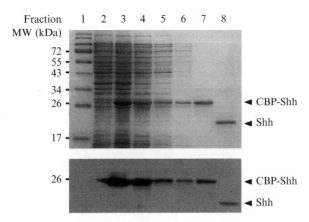

Figure 11.4 Coomassie blue staining (upper gel) and Western blot (lower gel) of the Shh purification fractions. Lane 1, molecular weight markers; lane 2, noninduced BL21 (Gold)DE3 cells transformed with pCAL-N-Shh; lane 3, BL21(Gold)DE3 cells after 3 h IPTG induction at 30 °C; lane 4, bacterial lysate; lane 5, flow-through; lane 6, concentrated preelution wash; lane 7, elution; lane 8, free N-Shh after CBP cleavage with EK.

Tris-HCl; pH 8.0, 10 mM β-mercaptoethanol, 150 mM NaCl). The lack of calcium lowers the affinity of CaM-binding proteins for CaM; however, we have found that CBP–N–Hh fusions remain associated to the resin, likely because of strong electrostatic interactions, allowing the use of this additional step to remove most of the low CaM-affinity proteins. Although some fusion is lost during this step, the purity of the final preparation is greatly improved. Finally, the CBP–N–Hh fusion protein is eluted from the column with 1 volume of elution buffer (50 mM Tris-HCl; pH 8.0, 10 mM β-mercaptoethanol, 1 M NaCl, 2 mM EGTA).

3.4.3. Removal of the CBP tag

The eluted CBP–N–Hh is concentrated by centrifugation on Amicon Ultra filters (10-kDa cutoff) and the buffer exchanged for enterokinase (EK) cleavage buffer (50 mM Tris-HCl; pH 8.0, 50 mM NaCl, 2 mM CaCl$_2$, 0.1% Tween-20). The CBP tag is then removed by incubation with 40 U EK/mg protein during 4 h at room temperature. To remove the released CBP peptide and EK, the reaction is consequently incubated for 30 min at 4 °C with 0.5 ml CaM resin per mg protein and 1 ml EKapture. Both CaM resin per mg protein and EKapture agarose need to be previously equilibrated in EK buffer with 200 mM NaCl. After removal of the resin-bound CBP and EK by centrifugation, the supernatant is filtered through a endotoxin removal column, and after protein determination, the cryoprotective agents BSA and glycerol are added to a final concentration of 0.1 and 25%, respectively. Aliquots should be stored at −80° to minimize loss of biologic activity.

It is highly recommended that a small aliquot of each purification step be saved to assess the purification efficiency by SDS-PAGE followed by both Coomassie blue staining and Western blot. Problematic steps can then be tracked down and corrected. Representative SDS-PAGE of total protein and a Western blot to illustrate the purification of N-Shh is shown in Fig. 11.4. We use an anti-Shh/Ihh monoclonal antibody 5E1 at 1:1,000 and anti-Dhh polyclonal at 1:1,000 dilution.

3.5. Bioassay to determine activity of the purified ligands

To determine the ability of the purified recombinant proteins to activate the Hedgehog canonical pathway, we measure their capability to activate Gli-mediated transcription in the Hh-responsive NIH3T3 fibroblasts (ATCC). To this end, we monitor the activity of a *Firefly*-luciferase reporter plasmid (p8XGBS-Luc, described in Sasaki *et al.*, 1997) under the control of a minimal γ-crystalline promoter and eight tandem copies of the HNF-3β gene's Gli-binding site (TGGGTGGTC) and normalize the transfection efficiency with a thymidine kinase promoter–driven *Renilla*-luciferase reporter (pRL-TK, from Promega). The luciferase activity is proportional and extremely sensitive to small variations in the activation of the Hh pathway.

NIH3T3 cells are maintained in high-glucose DMEM supplemented with 10% iron-enriched calf serum and penicillin/streptomycin. To perform the Hh activity assay, the cells are seeded in the same medium in a

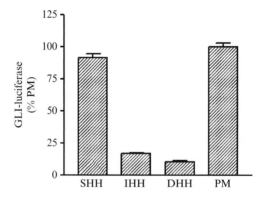

Figure 11.5 Evaluation of activity of the three Hedgehog isoforms. NIH3T3 cells transfected with p8XGBS-luc and pRL-TK were stimulated during 24 h with 5 μg/ml Shh, 8 μg/ml Ihh, 35 μg/ml Dhh, or 5 μM purmorphamine as a positive control. Purified Ihh and Dhh are considerably less active than Shh, as previously reported in other models. Bars represent the mean ±SEM of the *Firefly* to *Renilla* activity ratio of three independent experiments and are expressed as the percentage with respect to purmorphamine.

48-well plate. When they reach 70 to 90% confluency, the cells are transfected with a mixture containing 0.5 μl Fugene6 reagent (Roche), 70 ng p8XGBS-Luc, 17 ng pRL-TK, and 85 ng empty pcDNA vector per well, following the manufacturer's instructions. The addition of DNA mass in the form of pcDNA vector simply increases transfection efficiency. After the culture reaches a very high confluency, and this is a critical step, the medium is changed to DMEM with 0.5% calf serum containing the purified Hh ligands or vehicle (EK buffer with BSA and glycerol). We always include a positive control consisting of 5 μM purmorphamine (Calbiochem), a synthetic Smoothened agonist. After 24 to 48 h, the cells are carefully washed with phosphate buffer and lysed for luciferase activity determination by use of the Dual-Luciferase Reporter Assay System (Promega). Typical results are shown in Fig. 11.5. Note the pecking order of potencies for the Hh ligands on this assay (Shh>Ihh>>Dhh).

REFERENCES

Ahlgren, S. C., and Bronner-Fraser, M. (1999). Inhibition of sonic hedgehog signaling *in vivo* results in craniofacial neural crest cell death. *Curr. Biol.* **9,** 1304–1314.

Argenti, B., Gallo, R., Di Marcotullio, L., Ferretti, E., Napolitano, M., Canterini, S., De Smaele, E., Greco, A., Fiorenza, M. T., Maroder, M., Screpanti, I., Alesse, E., *et al.* (2005). Hedgehog antagonist REN(KCTD11) regulates proliferation and apoptosis of developing granule cell progenitors. *J. Neurosci.* **25,** 8338–8346.

Bar, E. E., Chaudhry, A., Farah, M. H., and Eberhart, C. G. (2007). Hedgehog signaling promotes medulloblastoma survival via BclII. *Am. J. Pathol.* **170,** 347–355.

Beachy, P. A., Karhadkar, S. S., and Berman, D. M. (2004). Tissue repair and stem cell renewal in carcinogenesis. *Nature* **432,** 324–331.

Bigelow, R. L., Chari, N. S., Unden, A. B., Spurgers, K. B., Lee, S., Roop, D. R., Toftgard, R., and McDonnell, T. J. (2004). Transcriptional regulation of bcl-2 mediated by the sonic hedgehog signaling pathway through gli-1. *J. Biol. Chem.* **279,** 1197–1205.

Bitgood, M. J., and McMahon, A. P. (1995). Hedgehog and Bmp genes are coexpressed at many diverse sites of cell–cell interaction in the mouse embryo. *Dev. Biol.* **172,** 126–138.

Borycki, A. G., Brunk, B., Tajbakhsh, S., Buckingham, M., Chiang, C., and Emerson, C. P., Jr. (1999). Sonic hedgehog controls epaxial muscle determination through Myf5 activation. *Development* **126,** 4053–4063.

Byrd, N., Becker, S., Maye, P., Narasimhaiah, R., St-Jacques, B., Zhang, X., McMahon, J., McMahon, A., and Grabel, L. (2002). Hedgehog is required for murine yolk sac angiogenesis. *Development* **129,** 361–372.

Charrier, J. B., Lapointe, F., Le Douarin, N. M., and Teillet, M. A. (2001). Anti-apoptotic role of Sonic hedgehog protein at the early stages of nervous system organogenesis. *Development* **128,** 4011–4020.

Chen, M. H., Li, Y. J., Kawakami, T., Xu, S. M., and Chuang, P. T. (2004). Palmitoylation is required for the production of a soluble multimeric Hedgehog protein complex and long-range signaling in vertebrates. *Genes Dev.* **18,** 641–659.

Chen, X., Horiuchi, A., Kikuchi, N., Osada, R., Yoshida, J., Shiozawa, T., and Konishi, I. (2007). Hedgehog signal pathway is activated in ovarian carcinomas, correlating with cell proliferation: It's inhibition leads to growth suppression and apoptosis. *Cancer Sci.* **98,** 68–76.

Chiang, C., Litingtung, Y., Lee, E., Young, K. E., Corden, J. L., Westphal, H., and Beachy, P. A. (1996). Cyclopia and defective axial patterning in mice lacking Sonic hedgehog gene function. *Nature* **383,** 407–413.

Colnot, C., de la Fuente, L., Huang, S., Hu, D., Lu, C., St-Jacques, B., and Helms, J. A. (2005). Indian hedgehog synchronizes skeletal angiogenesis and perichondrial maturation with cartilage development. *Development* **132,** 1057–1067.

Dawber, R. J., Hebbes, S., Herpers, B., Docquier, F., and van den Heuvel, M. (2005). Differential range and activity of various forms of the Hedgehog protein. *BMC Dev. Biol.* **5,** 21.

Dierks, C., Grbic, J., Zirlik, K., Beigi, R., Englund, N. P., Guo, G. R., Veelken, H., Engelhardt, M., Mertelsmann, R., Kelleher, J. F., Schultz, P., and Warmuth, M. (2007). Essential role of stromally induced hedgehog signaling in B-cell malignancies. *Nat. Med.* **13,** 944–951.

Echelard, Y., Epstein, D. J., St-Jacques, B., Shen, L., Mohler, J., McMahon, J. A., and McMahon, A. P. (1993). Sonic hedgehog, a member of a family of putative signaling molecules, is implicated in the regulation of CNS polarity. *Cell* **75,** 1417–1430.

Guerrero, I., and Ruiz i Altaba, A. (2003). Longing for ligand: Hedgehog, patched, and cell death. *Science* **301,** 774–776.

Guy, R. K. (2000). Inhibition of sonic hedgehog autoprocessing in cultured mammalian cells by sterol deprivation. *Proc. Natl. Acad. Sci. USA* **97,** 7307–7312.

Harfe, B. D., Scherz, P. J., Nissim, S., Tian, H., McMahon, A. P., and Tabin, C. J. (2004). Evidence for an expansion-based temporal Shh gradient in specifying vertebrate digit identities. *Cell* **118,** 517–528.

Hutchin, M. E., Kariapper, M. S., Grachtchouk, M., Wang, A., Wei, L., Cummings, D., Liu, J., Michael, L. E., Glick, A., and Dlugosz, A. A. (2005). Sustained Hedgehog signaling is required for basal cell carcinoma proliferation and survival: Conditional skin tumorigenesis recapitulates the hair growth cycle. *Genes Dev.* **19,** 214–223.

Ji, Z., Mei, F. C., Johnson, B. H., Thompson, E. B., and Cheng, X. (2007). PKA, not Epac, suppresses hedgehog activity and regulates glucocorticoid sensitivity in acute lympho-blastic leukemia cells. *J. Biol. Chem. Epub Sept* 25.

Kusano, K. F., Pola, R., Murayama, T., Curry, C., Kawamoto, A., Iwakura, A., Shintani, S., Ii, M., Asai, J., Tkebuchava, T., Thorne, T., Takenaka, H., Aikawa, R., Goukassian, D., von Samson, P., Hamada, H., Yoon, Y. S., Silver, M., Eaton, E., Ma, H., Heyd, L., Kearney, M., Munger, W., Porter, J. A., Kishore, R., and Losordo, D. W. (2005). Sonic hedgehog myocardial gene therapy: Tissue repair through transient reconstitution of embryonic signaling. *Nat. Med.* **11,** 1197–1204.

Li, Y., Zhang, H., Litingtung, Y., and Chiang, C. (2006). Cholesterol modification restricts the spread of Shh gradient in the limb bud. *Proc. Natl. Acad. Sci. USA* **103,** 6548–6553.

Lien, W. H., Klezovitch, O., Fernandez, T. E., Delrow, J., and Vasioukhin, V. (2006). AlphaE-catenin controls cerebral cortical size by regulating the hedgehog signaling pathway. *Science* **311,** 1609–1612.

Litingtung, Y., and Chiang, C. (2000). Specification of ventral neuron types is mediated by an antagonistic interaction between Shh and Gli3. *Nat. Neurosci.* **3,** 979–985.

Mann, R. K., and Beachy, P. A. (2004). Novel lipid modifications of secreted protein signals. *Annu. Rev. Biochem.* **73,** 891–923.

Marigo, V., Davey, R. A., Zuo, Y., Cunningham, J. M., and Tabin, C. J. (1996). Biochemical evidence that patched is the Hedgehog receptor. *Nature* **384,** 176–179.

McMahon, A. P., Ingham, P. W., and Tabin, C. J. (2003). Developmental roles and clinical significance of hedgehog signaling. *Curr. Top. Dev. Biol.* **53,** 1–114.

Mehlen, P., and Thibert, C. (2004). Dependence receptors: Between life and death. *Cell. Mol. Life Sci.* **61,** 1854–1866.

Miao, N., Wang, M., Ott, J. A., D'Alessandro, J. S., Woolf, T. M., Bumcrot, D. A., Mahanthappa, N. K., and Pang, K. (1997). Sonic hedgehog promotes the survival of specific CNS neuron populations and protects these cells from toxic insult *in vitro*. *J. Neurosci.* **17**, 5891–5899.

Mohler, J. (1988). Requirements for hedgehog, a segmental polarity gene, in patterning larval and adult cuticle of Drosophila. *Genetics* **120**, 1061–1072.

Nusslein-Volhard, C., and Wieschaus, E. (1980). Mutations affecting segment number and polarity in *Drosophila*. *Nature* **287**, 795–801.

Oppenheim, R. W., Homma, S., Marti, E., Prevette, D., Wang, S., Yaginuma, H., and McMahon, A. P. (1999). Modulation of early but not later stages of programmed cell death in embryonic avian spinal cord by sonic hedgehog. *Mol. Cell. Neurosci.* **13**, 348–361.

Riobo, N. A., Saucy, B., Dilizio, C., and Manning, D. R. (2006a). Activation of hetero-trimeric G proteins by Smoothened. *Proc. Natl. Acad. Sci. USA* **103**, 12607–12612.

Riobo, N. A., Lu, K., Ai, X., Haines, G. M., and Emerson, C. P., Jr. (2006b). Phosphoi-nositide 3-kinase and Akt are essential for Sonic Hedgehog signaling. *Proc. Natl. Acad. Sci. USA* **103**, 4505–4510.

Riobo, N. A., Lu, K., and Emerson, C. P., Jr. (2006c). Hedgehog signal transduction: Signal integration and cross talk in development and cancer. *Cell Cycle* **5**, 1612–1615.

Riobo, N. A., and Manning, D. R. (2007). Pathways of signal transduction employed by vertebrate Hedgehogs. *Biochem. J.* **403**, 369–379.

Ruiz i Altaba, A., Palma, V., and Dahmane, N. (2002). Hedgehog-Gli signalling and the growth of the brain. *Nat. Rev. Neurosci.* **3**, 24–33.

Sacedón, R., Díez, B., Nuñez, V., Hernández-López, C., Gutierrez-Frías, C., Cejalvo, T., Outram, S. V., Crompton, T., Zapata, A. G., Vicente, A., and Varas, A. (2005). Sonic hedgehog is produced by follicular dendritic cells and protects germinal center B cells from apoptosis. *J. Immunol.* **174**, 1456–1461.

Sasaki, H., Hui, C., Nakafuku, M., and Kondoh, H. (1997). A binding site for Gli proteins is essential for HNF-3beta floor plate enhancer activity in transgenics and can respond to Shh *in vitro*. *Development* **124**, 1313–1322.

Stone, D. M., Hynes, M., Armanini, M., Swanson, T. A., Gu, Q., Johnson, R. L., Scott, M. P., Pennica, D., Goddard, A., Phillips, H., Noll, M., Hooper, J. E., de Sauvage, F, and Rosenthal, A. (1996). The tumour-suppressor gene patched encodes a candidate receptor for Sonic hedgehog. *Nature* **384**, 129–134.

Taipale, J., and Beachy, P. A. (2001). The Hedgehog and Wnt signalling pathways in cancer. *Nature* **411**, 349–354.

Taipale, J., Cooper, M. K., Maiti, T., and Beachy, P. A. (2002). Patched acts catalytically to suppress the activity of Smoothened. *Nature* **418**, 892–897.

Taylor, F. R., Wen, D., Garber, E. A., Carmillo, A. N., Baker, D. P., Arduini, R. M., Williams, K. P., Weinreb, P. H., Rayhorn, P., Hronowski, X., Whitty, A., Day, E. S., Boriack-Sjodin, A., Shapiro, R. I., Galdes, A., and Pepinsky, R. B. (2001). Enhanced potency of human Sonic hedgehog by hydrophobic modification. *Biochemistry* **40**, 4359–4371.

Teillet, M., Watanabe, Y., Jeffs, P., Duprez, D., Lapointe, F., and Le Douarin, N. M. (1998). Sonic hedgehog is required for survival of both myogenic and chondrogenic somitic lineages. *Development* **125**, 2019–2030.

Tenzen, T., Allen, B. L., Cole, F., Kang, J. S., Krauss, R. S., and McMahon, A. P. (2006). The cell surface membrane proteins Cdo and Boc are components and targets of the Hedgehog signaling pathway and feedback network in mice. *Dev. Cell.* **10**, 647–656.

Thibert, C., Teillet, M. A., Lapointe, F., Mazelin, L., Le Douarin, N. M., and Mehlen, P. (2003). Inhibition of neuroepithelial patched-induced apoptosis by sonic hedgehog. *Science* **301**, 843–846.

Varjosalo, M., Li, S. P., and Taipale, J. (2006). Divergence of hedgehog signal transduction mechanism between *Drosophila* and mammals. *Dev. Cell.* **10,** 177–186.

Zhang, W., Kang, J. S., Cole, F., Yi, M. J., and Krauss, R. S. (2006). Cdo functions at multiple points in the Sonic Hedgehog pathway, and Cdo-deficient mice accurately model human holoprosencephaly. *Dev. Cell* **10,** 657–665.

CHAPTER TWELVE

Experimental Approaches to Investigate the Proteasomal Degradation Pathways Involved in Regulation of Apoptosis

Alan Tseng,*,† Hiroyuki Inuzuka,*,† Daming Gao,* Amrik Singh,* *and* Wenyi Wei*

Contents

Abstract

Ubiquitin-mediated proteolysis plays a major role in a variety of cellular functions, including cell metabolism, cell cycle progression, cellular response to DNA damage, and programmed cell death. In most cases, the crucial regulators involved in the control of these diverse cellular functions are modified by specific E3 ubiquitin ligases through the attachment of multiple ubiquitin molecules, a signal that triggers the subsequent destruction by the 26S proteasome complex.

* Department of Pathology, Beth Israel Deaconess Medical Center, Harvard Medical School, Boston, Massachusetts
† Equal contributors

Methods in Enzymology, Volume 446
ISSN 0076-6879, DOI: 10.1016/S0076-6879(08)01612-1

Recent studies revealed that the proteasomal degradation pathway regulates the cellular apoptosis process on multiple levels. Thus, a better understanding of the molecular mechanisms that underlie the ubiquitination and destruction of these specific regulators of apoptosis will provide us with insight on how apoptosis is properly controlled in normal cells and how tumor cells evade the apoptosis pathways. This chapter provides an overview of the common methods used to examine whether a target protein is ubiquitinated, as well as the protocols to examine how a putative E3 ligase controls the destruction of the target protein.

1. INTRODUCTION

Recent studies demonstrated that many of the key regulators of various cellular functions are subject to the ubiquitin-mediated protein destruction pathway (Hoeller *et al.*, 2006). In this pathway, a protein complex composed of an activating enzyme (E1), a conjugating enzyme (E2), and a protein ligase enzyme (E3) will covalently attach multiple ubiquitin molecules to the target protein, which serves as a destruction signal recognized by the 26S proteasome complex (Hershko *et al.*, 1994; Orian *et al.*, 1995). This proteolysis process is highly specific and tightly regulated, because it requires the specific interaction between the E3 ligase and the target protein. In many cases, site-specific posttranslational modifications, including phosphorylation (Harper *et al.*, 2002; Wei *et al.*, 2005) and hydroxylation (Ivan *et al.*, 2001), are required for the substrate to be recognized by the specific E3 ligase. On the other hand, dysregulation of this destruction process has been observed in many cancer cells, leading to either overexpression of certain oncogenes or enhanced degradation of certain tumor suppressors (Crosetto *et al.*, 2006; Zhang *et al.*, 2004).

Proteasomal degradation has been shown to participate in regulating apoptosis pathways on multiple levels (Friedman and Xue, 2004; Wojcik, 2002; Zhang *et al.*, 2004). First, many proapoptotic proteins, including p53 (Fang *et al.*, 2000), FOXO1 (Huang *et al.*, 2005), FOXO3a, BIM, and caspases (Richter and Duckett, 2000), are regulated by ubiquitin-proteasome pathways. In normal dividing cells, the ubiquitin-dependent degradation of these proapoptotic proteins is constantly active to avoid the inadvertent activation of apoptosis. On the contrary, their destruction will be halted to allow for proper response to stresses such as DNA damage or oxidative stress. Furthermore, mutations leading to enhanced degradation of these proapoptotic proteins will favor the tumor cells to evade the apoptosis pathway. For example, amplification of Mdm2 leads to downregulation of p53 protein, whereas overexpression of Skp2 protein leads to a reduction of FOXO1, both of which prevent the efficient induction of apoptosis upon physiologic stress. However, for cells that are programmed for cellular death, the proteasome pathway also leads to rapid degradation of anti-apoptotic proteins, such as Mcl-1 (Nijhawan *et al.*, 2003; Zhong *et al.*, 2005). Failure to degrade Mcl-1 leads to resistance to apoptosis and subsequent tumor progression. The third mechanism by which the proteasome pathway influences apoptosis is through regulating the degradation of IκB (Hattori *et al.*, 1999; Winston *et al.*, 1999), an inhibitor of the NF-κB pathway, which serves as a repressor for the induction of apoptosis. It is anticipated that more regulatory elements will be identified to further implicate the ubiquitin-proteasome pathway as an efficient means for controlling programmed cell death. This chapter aims to provide practical methods that can be used to address whether a protein of interest is regulated by the putative upstream E3 ligase through the ubiquitin-proteasome pathway.

2. GENERAL INSTRUCTIONS

The experimental procedures described herein are aimed to help researchers to determine whether a protein of interest is ubiquitinated by a putative E3 ligase. However, it is beyond the scope of this chapter to discuss approaches to identify the substrates for a specific E3 ligase, or conversely, to identify a putative E3 ligase for a protein of interest, although it typically involves the use of either IP/mass spectrometry or yeast two-hybrid methods. We then only consider the protein of interest a physiological substrate for the specific E3 ligase if we obtain positive results in the four following assays: (1) they interact *in vivo* and *in vitro*; (2) depletion of the E3 ligase protein by siRNA results in upregulation of the protein of interest; (3) overexpression of the E3 ligase protein leads to downregulation of the protein of interest; and (4) the putative E3 ligase can promote the ubiquitination of the interested protein *in vivo* and *in vitro*.

3. PROTOCOLS

3.1. Co-immunoprecipitation experiment to examine protein interactions *in vivo*

3.1.1. Materials

- 37 °C incubator with 5% CO_2
- DMEM + 10% FBS + penicillin/streptomycin
- 0.05% trypsin-EDTA
- OPTI-MEM (Invitrogen)
- MG132 (10 mM in stock solution, and 25 μM working concentration)
- Sterile RNase-free 1.5-ml Eppendorf Safe-lock tubes

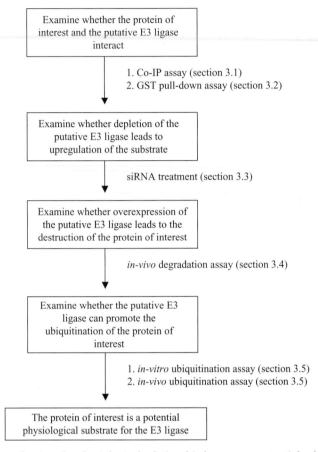

Schema for evaluating the physiological relationship between a potential substrate and its E3 ligase.

- Lipofectamine reagent (Invitrogen)
- PLUS reagent (Invitrogen)
- HeLa cells, U2OS cells, 293T cells
- P100 tissue culture dishes (Corning)
- 15-ml polypropylene conical tubes (Falcon)
- 50-ml polypropylene conical tubes (Falcon)
- EBC buffer (50 mM Tris, pH 7.5, 120 mM NaCl, 0.5% NP40)
- Complete MINI protease inhibitor cocktail tablets (Roche)
- 1 × sterile PBS
- Shaker (labnet)
- TBS (10 mM Tris, pH 8.0, 150 mM NaCl)
- Dry milk (grocery store)
- ECL detection reagent (Pierce, Millipore)

- NETN buffer (20 mM Tris, pH 8.0, 100 mM NaCl, 0.5% NP40, 1 mM EDTA)
- Protein-A sepharose beads (GE Healthcare)
- 3 × protein dye buffer (6.7% SDS, 33.3% glycerol, 300 mM DTT, bromophenol blue until orange)
- Protein assay dye reagent (Bio-rad)
- Purified BSA diluted to 1 mg/mL (NEB)
- DU800 Spectrophotometer (Beckman-Coulter)
- Immuno-Blot PVDF Membrane (Bio-rad)
- HRP-conjugated antibody (Sigma)

3.1.2. Procedures
3.1.2.1. Transient transfection

1. One day before transfection, trypsinize and resuspend the HeLa cells in DMEM + 10% FBS + penicillin/streptomycin medium. It is ideal to perform the transfection the next day when the cells are approximately 85% confluent.
2. Half an hour before starting the transfection, prewarm the OPTI-MEM in a 37 °C waterbath to reduce thermal shock to the cells.
3. Prepare one 15-ml conical tube for each sample and label each tube accordingly. Prepare one 50-ml conical tube as a master mix tube with a total volume of OPTI-MEM equal to 750 μl times the total number of samples.
4. Aliquot 750 μl OPTI-MEM to each tube and add the DNA to each aliquot. Vortex each tube briefly when all the DNA has been added.
5. Add 20 μl PLUS reagent to each tube. Vortex for 4 to 5 sec.
6. Add 30 μl/sample lipofectamine reagent to the master mix tube (total volume = 30 μl × total sample number).
7. Wait 15 min for the lipid/DNA complex to form.
8. Take 780 μl from the master mix tube and add it to each sample tube. Vortex for 5 to 6 sec.
9. Wait another 15 min. While waiting, take the dishes out and wash once with 10 ml warmed sterile 1 × PBS to remove the serum. Add 5 ml OPTI-MEM to each dish.
10. To each dish, add the respective sample in a drop-wise fashion, moving the pipette over the entire dish twice. Gently move the dish up and down and side to side to distribute evenly.
11. Incubate for 4 to 5 h at 37 °C. Change the medium to warmed 10% FBS DMEM (10 ml) after incubation. Make sure to aspirate with a new tip for every dish to prevent cross-contamination.
12. Because overexpression of the E3 ligase usually leads to the degradation of the substrate, to analyze the interaction, 36 h after transfection, block the proteasome degradation pathway by incubation in the presence of

25 μM of the specific proteasome inhibitor, MG132, for 6 h. Alternately, the samples can be treated with 10 μM MG132 overnight if 6 h treatment with 25 μM is insufficient.

13. Before harvesting the cells, prepare the lysis buffer by adding 1 tablet of the Complete Mini per 10 ml of EBC lysis buffer and prechill on ice.
14. Wash the sample dishes once with PBS.
15. Add 220 to 240 μl EBC + Complete Mini to each dish (for 70 to 80% confluent cells; for 95 to 100%, add 320 μl—want approximately 5 $\mu g/\mu l$ for the final protein concentration).
16. Scrape the dishes and transfer the lysates to microcentrifuge tubes on ice.
17. Place the tubes on a rocker in the cold room for 15 min.
18. Centrifuge the tubes at 14,000 rpm for 10 min.
19. Collect the supernatant and assay for the protein concentration.

3.1.2.2. Immunoprecipitation

1. On the basis of the measured protein concentration, calculate the volume needed for the amount of desired protein. We typically use 500 to 1000 μg total protein for each immunoprecipitation (IP).
2. Calculate the volume needed to bring the total volume per tube to 660 μl with EBC buffer.
3. Add 1μg antibody to each tube and incubate the mixture in the cold room for 3 to 4 h.
4. Take Protein A-sepharose beads (enough for 20 μl/sample) and wash the beads twice with NETN buffer by aspirating with a needle and syringe, and inverting the tubes repeatedly with fresh buffer. Resuspend with an equal amount of NETN buffer.
5. Add 40 μl total beads + buffer 50% slurry to each IP tube.
6. Return the tubes to the shaker in the cold room for approximately 1 h. In the meantime, prepare whole cell lysate (WCL) as input, by use of 5 to 10% protein (volume should not exceed 40 μl). Add 3 × protein dye buffer to the tubes. Boil the WCL samples at 100 °C for 5 min.
7. Wash the IP tubes 4 × by spinning down the tubes at a low speed (10,000 to 11,000 rpm) and aspirating with a syringe and needle. Invert the tubes with fresh NETN buffer added (800 to 1000 μl) to wash the tubes.
8. Cut 1000-μl pipette tips to prevent shearing of the beads, and use them to transfer each IP to a new tube (with 500 μl NETN buffer).
9. Use another 500 μl buffer to wash out the old tube and transfer the wash buffer to the new tube as well. Spin one final time and aspirate thoroughly (tip: hold the open needle hole against the wall of the tube and move the tip all the way to the bottom of the tube).
10. Check to see that the amount of beads is equal across all samples.

11. Add 25 μl 2× (equal volume 1× and 3× dye) loading dye and vortex. Boil at 100 °C for 5 min.

3.1.2.3. Western blot analysis

1. Load the samples on 10% SDS–PAGE mini-gels and perform electrophoresis at 120 to 130 V for 1 to 1.5 h.
2. Transfer the separated proteins onto the PVDF membrane by use of the Bio-Rad gel transfer apparatus at 100 V for 2 h on ice.
3. Block the membrane with TBST buffer in the presence of 5% dry milk for 30 to 45 min. Incubate with the appropriate primary antibody in the cold room overnight.
4. Wash the membrane four times (10 to 15 min for each wash) with TBST buffer and incubate the membrane with HRP-conjugated antibody (in TBST + 5% dry milk) for 1 h at room temperature.
5. Wash the membrane four times (10 to 15 min for each wash) with TBST buffer. Develop the membranes with ECL detection reagents according to the manufacturer's instructions.

3.2. GST pull-down experiment to examine protein Interactions *in vitro*

3.2.1. Materials

- BL21 *E. coli* cells
- pGEX GST-fusion protein bacteria expression vectors
- P100 bacteria culture dishes (Fisher)
- Sterile RNase-free 1.5-ml Eppendorf Safe-lock tubes
- 100 mM IPTG stock solution
- LB medium with ampicillin (100 μg/ml final concentration)
- Sonicator
- 15-ml polypropylene conical tubes (Falcon)
- 50-ml polypropylene conical tubes (Falcon)
- EBC buffer (50 mM Tris, pH 7.5, 120 mM NaCl, 0.5% NP40)
- Complete MINI protease inhibitor cocktail tablets (Roche)
- 1 × sterile PBS
- Shaker (labnet)
- TBS (10 mM Tris, pH 8.0, 150 mM NaCl)
- Dry milk (grocery store)
- ECL detection reagent (Pierce, Millipore)
- NETN buffer (20 mM Tris, pH 8.0, 100 mM NaCl, 0.5% NP40, 1 mM EDTA)
- Glutathione Sepharose 4B (GE Healthcare)

- 3 × protein dye buffer (6.7% SDS, 33.3% glycerol, 300 mM DTT, bromophenol blue until orange)
- TNT T7 quick coupled IVT kit (Promega)
- ^{35}S–Met
- 3-mm chromatography paper (Whatman)
- Gel drier apparatus (Bio-rad)
- Gel code blue reagent (Pierce)
- MS film (Kodak)
- Protein assay dye reagent (Bio-rad)
- Purified BSA diluted to 1 mg/ml (NEB)
- DU800 Spectrophotometer (Beckman-Coulter)

3.2.2. Procedures
3.2.2.1. Purification of GST protein from bacteria

1. Transform the pGEX GST fusion protein expression vector into BL21 bacteria by use of heat-shock methods. Pick a single colony of the BL21 transformant to inoculate in 4 ml of LB medium and shake overnight.
2. The next morning, dilute the culture 1:100-fold into prewarmed LB medium (100 to 400 ml) containing fresh 100 μg/ml ampicillin. Grow at 37 °C until the A600 reading reaches 0.6 to 0.9.
3. At this point, add in 100 mM IPTG to a final concentration of 0.1 mM and continue the incubation for an additional 3 to 4 h.
4. Harvest the bacteria by centrifugation at 7000g for 15 min and discard the supernatant. Place the pellet on ice.
5. Dissolve the pellet in 15 to 20 ml PBS solution (add 1 Complete MINI tablet per 15 ml of cold PBS solution to block the protease activity) in a 50-ml conical tube.
6. Disrupt the bacteria by sonication. After each burst of sonication (30 sec), incubate the tube on ice for 1 min to cool the solution. It should take six to nine rounds of sonication to result in a nice disruption of the bacteria, at which point the milky bacteria solution becomes more opaque.
7. Centrifuge at 12,000g for 15 min to get rid of the bacteria debris and transfer the supernatant into a fresh tube.
8. Add in NETN buffer to a final volume of 30 ml and add in 1 to 2 ml of prewashed GST-sepharose beads slurry. Incubate in the cold room for 3 to 4 h.
9. Wash intensively with NETN five times, then use an equal volume of NETN buffer to dissolve the GST-beads.
10. Aliquot into 1.5-ml Eppendorf tubes and store at 4 °C. Take 10 μl of beads out, mix with 5 μl 3 × lysis buffer and boil at 100 °C for 5 min before loading into a SDS-PAGE gel. Use a serial dilution of BSA protein with known concentrations to assess the protein concentration of the recovered GST-fusion protein.

3.2.2.2. In vitro translation (IVT)

11. Thaw the TNT T7 quick coupled IVT kit retic lysate master mix, ^{35}S-Met, and plasmids with T7 promoter on ice.
12. For each IVT reaction, prepare an RNase-free 1.5-ml Eppendorf tube, and label it accordingly.
13. Add in 20 μl TNT T7 quick coupled IVT retic lysate master mix, 2 μl RNase-free water, 2 μl ^{35}S-Met, and 1 μl plasmid DNA encoding the protein of interest.
14. Incubate at 37 °C for 1 to 2 h and afterwards store the reaction at -80 °C.

3.2.2.3. GST pull-down assay

15. For each reaction, prepare a prechilled 1.5-ml Eppendorf tube. Label the tube properly and add in 600 μl cold NETN buffer. Then add in 5 to 10 μl IVT reaction product with or without purified GST-fusion protein. Purified GST protein is used as a negative control.
16. Incubate reactions in the cold room for 3 to 5 h. During this period, prepare 5 to 10% of the IVT product used for the pull-down assay as input. Add 3× dye buffer to the input tubes. Boil the input samples at 100 °C for 5 min.
17. Follow the procedure from steps 26 to 31 in the Co-IP protocol for washing of the beads and resolving the samples in SDS-Page gel.
18. Cut the separation gel out, wash in distilled water for 30 to 60 min, and add in gel code reagent (Pierce) to stain the GST and GST-fusion protein.
19. After incubation at room temperature for 30 to 45 min, wash the gels with distilled water for 60 min to destain the gel.
20. Place the properly stained gel on a piece of 3-mm Whatman paper and cover the gel in plastic wrap. Place the Whatman paper with gel on the gel drier and dry at 75 °C for 1 to 2 h (per gel).
21. After drying, place a piece of MS film (Kodak) on top of the dried gel in the dark room. Store the film at -80 °C for 10 to 15 h before developing.

3.3. siRNA experiments to deplete the endogenous E3 ligase protein

3.3.1. Materials

- 37 °C incubator with 5%/10% CO_2
- 0.05% trypsin-EDTA
- DMEM + 10% FBS (no penicillin/streptomycin)
- DMEM + 30% FBS (no penicillin/streptomycin)
- OPTI-MEM (Invitrogen)

- Oligofectamine reagent (Invitrogen)
- siRNA oligos (Dharmacon)
- Sterile RNase-free 1.5-ml Eppendorf Safe-lock tubes
- HeLa cells, U2OS cells
- P60 tissue culture dishes (Corning)
- 15-ml polypropylene conical tubes (Falcon)
- 50-ml polypropylene conical tubes (Falcon)
- 1× Sterile PBS
- EBC buffer (50 mM Tris, pH 7.5, 120 mM NaCl, 0.5% NP40)
- Complete MINI protease inhibitor cocktail tablets (Roche)
- ECL detection reagent (Pierce, Millipore)
- 3 × protein dye buffer (6.7% SDS, 33.3% glycerol, 300 mM DTT, bromophenol blue until orange)
- Protein assay dye reagent (Bio-rad)
- Purified BSA diluted to 1 mg/ml (NEB)
- DU800 Spectrophotometer (Beckman-Coulter)
- Immuno-Blot PVDF Membrane (Bio-rad)
- HRP-conjugated antibody (Sigma)

3.3.2. Procedure

Note: The following protocol is for experiments that use P60 tissue culture dishes. For 6-well plates, it is recommended to use half the stated amounts.

1. One day before transfection, trypsinize and resuspend the HeLa cells or U2OS cells in antibiotic-free DMEM + 10% FBS medium. It is important to note that culturing cells in antibiotic-free medium will greatly enhance the transfection efficiency. The goal is to target approximately 20 to 25% confluency on the day of transfection. Typically we plate 2.2×10^5 HeLa cells and 3.0×10^5 U2OS cells at approximately 1:00 PM and the cells will be at the expected confluency approximately 9 to 10:00 AM the next day.

2. Before transfection, prewarm the OPTI-MEM in a 37 °C waterbath such that at the time of transfection the medium is around room temperature.

3. Prepare a 1.5-ml microcentrifuge tube (sterile and RNase free, open only in tissue culture hood) for each sample, and one for a master mix tube.

4. To each sample tube, add 356 μl OPTI-MEM and 14 μl oligo. Mix by gently vortexing.

5. Add 22 μl prewarmed OPTI-MEM and 8 μl Oligofectamine reagent to the master mix tube (per sample).

6. Let the mixture sit at room temperature for 10 min. In the meantime, label each plate to be transfected.

9. Wait another 15 min. While waiting, take the dishes out and wash 1× with warmed PBS to remove the serum. Add 2.0 ml OPTI-MEM to each dish.

10. To each dish, add the respective sample in a drop-wise fashion, moving the pipette over the entire dish twice. Move the dish up and down and side to side to distribute evenly.

11. Incubate for 4 to 5 h at 37 °C. Change the medium to warmed 10% FBS DMEM (5 ml) after incubation. Make sure to aspirate with a new tip for every dish to prevent cross-contamination.

12. To assess whether the decreased substrate abundance in response to overexpression of the E3 ligase is through the ubiquitin-proteasome pathway, 24 to 30 h after transfection, block the proteasome degradation pathway by incubation in the presence of 10 μM of the specific proteasome inhibitor, MG132, for 10 to 15 h.

13. 36 to 48 h after transfection, prepare the lysis buffer by adding 1 tablet Complete Mini per 10 ml EBC lysis buffer and prechill on ice.

14. Wash the dishes once with PBS.

15. Add 110 to 120 μl EBC + Complete Mini to each dish (for 70 to 80% confluent cells; for 95 to 100%, add 160 μl: want approximately 5 $\mu g/\mu l$ final concentration).

16. Scrape the dishes and transfer the lysates to microcentrifuge tubes on ice.

17. Place the tubes on a rocker in the cold room for 15 min.

18. Centrifuge the tubes at 14,000 rpm for 10 min.

19. Collect the supernatant and assay for protein concentration.

20. Follow the procedure from steps 31 to 35 in the Co-IP protocol to perform Western blot analysis.

It is also important to show that the decrease in the target protein abundance is due to a shorter half-life (quicker protein turnover). The following procedure can be used to measure changes in protein half-life.

3.4.2.2. Cycloheximide treatment to measure change in the protein half-life

1. Perform the transient transfection experiment as discussed previously, but scale the reactions up to 100-mm dishes by multiplying all reagent volumes by 2.5; 20 h after transfection, split the transfected cells into five 60-cm dishes.

2. Let the cells recover overnight, and the next morning start the cycloheximide treatment, which blocks protein synthesis, thus allowing the turnover rate of the already synthesized proteins to be measured.

3. One hour before the start of the experiment, dissolve cycloheximide in 100% EtOH (prepare this fresh, 50 mg in 500 μl = 100 mg/ml, 1000× stock).

4. Set aside the time = 0 plates, and decide on the time course, which will vary for different proteins. We typically use 30, 60, 120, and 240 min as a starting time scale.
5. Add 100 μg/ml cycloheximide to each dish in chronologic order, and immediately shake the dishes to make sure that the drug is distributed evenly. Return the plates to the incubator. After the drug is added to the first time point dishes, start a timer to count down the time points.
6. Wash the time = 0 plates with 1 $\times$ PBS and harvest the plate with 140 to 180 μl EBC + PI and place on a rotator in the cold room for 15 min and spin down for 10 min at 13,200 rpm.
7. For the sample time course, when 30 min has passed, wash and harvest the plates for time = 30 min.
8. Harvest all subsequent time points in a similar manner.
9. Measure the protein concentration and apply 20 to 40 μg for SDS-PAGE for Western blot analysis as described earlier.

3.5. Protocols for *in vivo* and *in vitro* ubiquitination assays to examine whether the putative E3 Ligase can promote the ubiquitination of the protein of interest *in vivo* and *in vitro*

3.5.1. *In vivo* ubiquitination

In vivo ubiquitination assay is an approach based on transient transfection and coimmunoprecipitation assays. Thus, the following procedures are very similar to that described in the assays to detect protein interaction. The purpose of this assay is to determine whether when coexpressed *in vivo* in the presence or absence of the ubiquitin molecule, the putative E3 ligase can promote the ubiquitination of the target protein. After recovering the target protein by immunoprecipitation procedure, the ubiquitination is detected by Western blot analysis as a ladder or smear of ubiquitinated protein species that migrate slower on the SDS-PAGE than the recovered target protein. When designing the cotransfection experiment, various negative controls should be included to make sure that the resulting target protein ubiquitination is solely due to the addition of the ubiquitin and the putative E3 ligase.

3.5.1.1. Materials

- 37 °C incubator with 5% CO_2
- 0.05% trypsin–EDTA
- DMEM + 10% FBS + penicillin/streptomycinOPTI-MEM (Invitrogen)
- MG132 (10 mM in stock solution, and 25 μM working concentration)
- Sterile RNase-free 1.5-ml Eppendorf Safe-lock tubes
- Lipofectamine reagent (Invitrogen)

- PLUS reagent (Invitrogen)
- HeLa cells, U2OS cells, 293T cells
- P100 tissue culture dishes (Corning)
- 15-ml polypropylene conical tubes (Falcon)
- 50-ml polypropylene conical tubes (Falcon)
- EBC buffer (50 mM Tris, pH 7.5, 120 mM NaCl, 0.5% NP40)
- Complete MINI protease inhibitor cocktail tablets (Roche)
- 1× sterile PBS
- Shaker (labnet)
- TBS (Tris 10 mM, pH 8.0, 150 mM NaCl)
- Dry milk (grocery store)
- ECL detection reagent (Pierce)
- NETN buffer (20 mM Tris, pH 8.0, 100 mM NaCl, 0.5% NP40, 1 mM EDTA)
- Protein-A sepharose beads (GE Healthcare)
- 3 × protein dye buffer (6.7% SDS, 33.3% glycerol, 300 mM DTT, bromophenol blue until orange)
- Protein assay dye reagent (Bio-rad)
- Purified BSA diluted to 1 mg/mL (NEB)
- DU800 Spectrophotometer (Beckman–Coulter)
- Immuno-Blot PVDF Membrane (Bio-rad)
- HRP-conjugated antibody (Sigma)

3.5.1.2. Procedures
3.5.1.2.1. Transient transfection
Follow the protocol in the Co-IP section for transient transfection (steps 1 to 19).

3.5.1.2.2. Immunoprecipitation

1. On the basis of the measured protein concentration, calculate the volume needed for the amount of desired protein. We typically use 500 to 1000 μg total protein for each immunoprecipitation.
2. Calculate the volume needed to bring the total volume to 660 μl with EBC buffer.
3. Add 1μg antibody targeting the substrate to each tube and incubate the mixture in the cold room for 3 to 4 h.
4. Take Protein A-sepharose beads (enough for 20 μl/sample) and wash the beads twice with NETN buffer by aspirating with a needle and syringe, and inverting the tubes repeatedly with fresh buffer. Resuspend with an equal amount of NETN buffer.
5. Add 40 μl total beads + buffer slurry to each tube.
6. Return the tubes to the shaker in the cold room for approximately 1 h. In the meantime, prepare whole cell lysate (WCL) as input, by use of

5 to 10% protein (total volume should not exceed 40 μl). Add 3 $\times$ dye buffer to the tubes. Boil the WCL samples at 100 °C for 5 min.

7. Wash the IP tubes 4$\times$ by spinning down the tubes at a low speed (10,000 to 11,000 rpm) and aspirating with a syringe and needle. Invert the tubes with fresh NETN buffer added (800 to 1000 μl) to wash the tubes.

8. Cut 1000-μl pipette tips to prevent shearing of the beads, and use them to transfer each IP to a new tube (with 500 μl NETN buffer).

9. Use another 500 μl buffer to wash out the old tube and transfer to the new tube as well. Spin one final time and aspirate thoroughly (tip: hold the open needle hole against the wall of the tube and move the tip all the way to the bottom of the tube).

10. Check to see the amount of beads is equal across all samples.

11. Add 25 μl 2 $\times$ loading dye and vortex. Boil at 100 °C for 5 min.

12. Follow the procedure from steps 31 to 35 in the Co–IP protocol to perform Western blot analysis.

3.5.2. *In vitro* ubiquitination

In vitro ubiquitination of a particular substrate is a useful technique to confirm observed ubiquitination *in vivo*. Once the molecular factors of a particular pathway that regulate ubiquitination of a protein have been identified, a reconstituted system consisting of only recombinant enzymes, substrate, ubiquitin, and an ATP-regenerating system can be used to promote *in vitro* ubiquitination. *In vitro* ubiquitination assays are excellent tools for demonstrating the specificity of a particular E3 ligase complex for ubiquitination of a substrate. *In vitro* ubiquitination assays can be performed by use of either an *in vitro* translated protein or a purified protein as a substrate.

3.5.2.1. *Materials*

- BL21 *E. coli* cells
- pGEX GST-fusion protein bacteria expression vectors
- P100 bacteria culture dishes (Fisher)
- Sterile RNase-free 1.5-ml Eppendorf Safe-lock tubes
- 100 mM IPTG stock solution
- LB medium with ampicillin (100 μg/ml final concentration)
- Sonicator
- 15-ml polypropylene conical tubes (Falcon)
- 50-ml polypropylene conical tubes (Falcon)
- 1% SDS (v/v)
- Dissociation dilution buffer (1% Triton X-100, 0.5% deoxycholate, 120 mM NaCl, 50 mM HEPES, pH 7.2)
- Complete MINI protease inhibitor cocktail tablets (Roche)

- 10 × reaction buffer (300 mM HEPES, pH 7.2, 20 mM ATP, 50 mM MgCl$_2$, 2 mM DTT)
- 1 × sterile PBS
- Shaker (labnet)
- TBS (10 mM Tris, pH 8.0, 150 mM NaCl)
- Dry milk (grocery store)
- ECL detection reagent (Pierce, Millipore)
- NETN buffer (20 mM Tris, pH 8.0, 100 mM NaCl, 0.5% NP40, 1 mM EDTA)
- Protein-A sepharose beads (GE Healthcare)
- Glutathione Sepharose 4B (GE Healthcare)
- 3 × protein dye buffer (6.7% SDS, 33.3% glycerol, 300 mM DTT, bromophenol blue until orange)
- Protein assay dye reagent (Bio-rad)
- Purified BSA diluted to 1 mg/ml (NEB)
- DU800 Spectrophotometer (Beckman-Coulter)
- Immuno-Blot PVDF Membrane (Bio-rad)
- HRP-conjugated antibody (Sigma)

3.5.2.2. Procedures

1. Purify the desired E3 ligase from appropriate mammalian cell lysates for *in vitro* ubiquitination assay by first boiling the lysates in 1% SDS for 6 min at 95 °C.
2. Dilute lysates 1:10 in the dissociation dilution buffer containing the Complete MINI protease inhibitor cocktail.
3. Immunopurify the desired E3 ligase from lysates with the appropriate IP antibody according to aforementioned procedures.
4. Purify the putative substrate to be ubiquitinated from *E. coli* as GST fusion proteins according to the methods described previously.
5. Perform *in vitro* ubiquitination reactions in 50-μl volumes (adjust to total volume by use of 1 × PBS) containing the following components as indicated: 2 μg FLAG-Ub, 0.2 μg E1, 1 μg E2 (all from Boston Biochem, Cambridge, MA), 0.1 μg desired purified E3 ligase, 1 μg substrate as a GST fusion protein and 5 μl 10 × reaction buffer.
6. Incubate reactions at 30 °C for 1 h with agitation at 750 rpm. Reactions should be performed in quadruplicate and after incubation, combine the quadruplicate reactions into a final volume of 200 μl.
7. Reserve 50 μl for immunoblotting and add 1% SDS to the remaining 150 μl, which should be heated at 90 °C for 10 min.
8. Dilute these samples to a final volume of 5 ml with dissociation dilution buffer and add the appropriate primary antibody for immunoprecipitation of ubiquitinated substrate.

9. Rotate samples overnight at 4 °C. Add 50 μl protein A/G beads (50% slurry) for an additional 2 h. Wash immunoprecipitates three times with the dissociation dilution buffer, and prepare samples for immunoblot analysis as described previously. To detect ubiquitinated proteins, immunoblot membranes with an anti–FLAG-M2-HRP antibody (Sigma).

REFERENCES

Crosetto, N., Bienko, M., and Dikic, I. (2006). Ubiquitin hubs in oncogenic networks. *Mol. Cancer Res.* **4,** 899–904.

Fang, S., Jensen, J. P., Ludwig, R. L., Vousden, K. H., and Weissman, A. M. (2000). Mdm2 is a RING finger-dependent ubiquitin protein ligase for itself and p53. *J. Biol. Chem.* **275,** 8945–8951.

Friedman, J., and Xue, D. (2004). To live or die by the sword: The regulation of apoptosis by the proteasome. *Dev. Cell* **6,** 460–461.

Harper, J. W., Burton, J. L., and Solomon, M. J. (2002). The anaphase-promoting complex: It's not just for mitosis any more. *Genes Dev.* **16,** 2179–2206.

Hattori, K., Hatakeyama, S., Shirane, M., Matsumoto, M., and Nakayama, K. (1999). Molecular dissection of the interactions among IkappaBalpha, FWD1, and Skp1 required for ubiquitin-mediated proteolysis of IkappaBalpha. *J. Biol. Chem.* **274,** 29641–29647.

Hershko, A., Ganoth, D., Sudakin, V., Dahan, A., Cohen, L. H., Luca, F. C., Ruderman, J. V., and Eytan, E. (1994). Components of a system that ligates cyclin to ubiquitin and their regulation by the protein kinase cdc2. *J. Biol. Chem.* **269,** 4940–4946.

Hoeller, D., Hecker, C. M., and Dikic, I. (2006). Ubiquitin and ubiquitin-like proteins in cancer pathogenesis. *Nat. Rev. Cancer* **6,** 776–788.

Huang, H., Regan, K. M., Wang, F., Wang, D., Smith, D. I., van Deursen, J. M., and Tindall, D. J. (2005). Skp2 inhibits FOXO1 in tumor suppression through ubiquitin-mediated degradation. *Proc. Natl. Acad. Sci. USA* **102,** 1649–1654.

Ivan, M., Kondo, K., Yang, H., Kim, W., Valiando, J., Ohh, M., Salic, A., Asara, J. M., Lane, W. S., and Kaelin, W. G., Jr. (2001). HIFalpha targeted for VHL-mediated destruction by proline hydroxylation: Implications for O_2 sensing. *Science* **292,** 464–468.

Nijhawan, D., Fang, M., Traer, E., Zhong, Q., Gao, W., Du, F., and Wang, X. (2003). Elimination of Mcl-1 is required for the initiation of apoptosis following ultraviolet irradiation. *Genes Dev.* **17,** 1475–1486.

Orian, A., Whiteside, S., Israel, A., Stancovski, I., Schwartz, A. L., and Ciechanover, A. (1995). Ubiquitin-mediated processing of NF-kappa B transcriptional activator precursor p105. Reconstitution of a cell-free system and identification of the ubiquitin-carrier protein, E2, and a novel ubiquitin-protein ligase, E3, involved in conjugation. *J. Biol. Chem.* **270,** 21707–21714.

Richter, B. W., and Duckett, C. S. (2000). The IAP proteins: caspase inhibitors and beyond. *Sci. STKE* 2000, PE1.

Wei, W., Jin, J., Schlisio, S., Harper, J. W., and Kaelin, W. G., Jr. (2005). The v-Jun point mutation allows c-Jun to escape GSK3-dependent recognition and destruction by the Fbw7 ubiquitin ligase. *Cancer Cell* **8,** 25–33.

Winston, J. T., Strack, P., Beer-Romero, P., Chu, C. Y., Elledge, S. J., and Harper, J. W. (1999). The SCFbeta-TRCP-ubiquitin ligase complex associates specifically with phosphorylated destruction motifs in IkappaBalpha and beta-catenin and stimulates IkappaBalpha ubiquitination *in vitro*. *Genes Dev.* **13,** 270–283.

Wojcik, C. (2002). Regulation of apoptosis by the ubiquitin and proteasome pathway. *J. Cell Mol. Med.* **6,** 25–48.

Zhang, H. G., Wang, J., Yang, X., Hsu, H. C., and Mountz, J. D. (2004). Regulation of apoptosis proteins in cancer cells by ubiquitin. *Oncogene* **23,** 2009–2015.

Zhong, Q., Gao, W., Du, F., and Wang, X. (2005). Mule/ARF-BP1, a BH3-only E3 ubiquitin ligase, catalyzes the polyubiquitination of Mcl-1 and regulates apoptosis. *Cell* **121,** 1085–1095.

CHAPTER THIRTEEN

UBIQUITINATION MEDIATED BY INHIBITOR OF APOPTOSIS PROTEINS

Sun-Mi Park,*,‡ Shimin Hu,† Tae H. Lee,* *and* Xiaolu Yang†

Contents

Abstract

Inhibitor of apoptosis (IAP) proteins are a family of evolutionarily conserved proteins that regulate apoptosis as well as other cellular processes. The functions of many IAPs are defined by their RING domains, which possess E3 ubiquitin ligase activity and promote proteasomal degradation of an increasing number of target proteins. In this chapter, we describe the methods used in our laboratories to study the IAP's E3 activity.

1. INTRODUCTION

IAP proteins were initially found in baculoviruses for their ability to block apoptosis in host cells (Birnbaum *et al.*, 1994; Crook *et al.*, 1993). The cellular homologs of baculoviral IAP proteins were subsequently identified in a wide range of species (Deveraux and Reed, 1999; Salvesen and Duckett, 2002; Vaux and Silke, 2005). Some of these cellular IAPs live up to their reputation

* Department of Biology and Protein Network Research Center, Yonsei University, Seoul, South Korea
† Abramson Family Cancer Research Institute and Department of Cancer Biology, University of Pennsylvania School of Medicine, Philadelphia, Pennsylvania
‡ Current address: Ben May Department for Cancer Research, The University of Chicago, Chicago, IL 60637, USA

Methods in Enzymology, Volume 446
ISSN 0076-6879, DOI: 10.1016/S0076-6879(08)01613-3

and mainly function in apoptosis, whereas the rest play important roles in other cellular processes in addition to, or even instead of, apoptosis regulation. All IAPs contain the characteristic baculoviral IAP repeats (BIRs), which are often accompanied by other motifs, including the *really interesting novel gene* (RING) domain. At least eight IAPs are found in humans: XIAP, cIAP1, cIAP2, ILP2, ML-IAP, NAIP, Survivin, and Bruce, the first five of which have a RING domain. The functions of IAPs in both apoptotic and non-apoptotic processes are often determined by the RING domain, which possesses E3 ubiquitin ligase activity and mediates degradation of the IAPs themselves (Yang *et al.*, 2000), as well as an other proteins such as caspases (Suzuki *et al.*, 2001), TRAF2 (Li *et al.*, 2002), Smac (Hu and Yang, 2003), RIP (Park *et al.*, 2004), and Bcl10 (Hu *et al.*, 2006a,b). This RING domain of IAPs recruits E2 ubiquitin–conjugating enzymes, and the substrate specificity is determined by the direct binding through the BIR region. IAPs are thus single-molecule E3s as opposed to the cullin-RING family of E3s, which are a large complex with the RING and the substrate-binding domains residing on different subunits. Here we describe methods used in our laboratories to assay ubiquitination of IAPs and their substrates both *in vitro* and *in vivo*.

2. *IN VITRO* UBIQUITINATION ASSAY

Components used for *in vitro* ubiquitination assays include the ubiquitin-activating enzyme (E1), ubiquitin-conjugating enzyme (E2), ubiquitin ligase (E3), ubiquitin, and the substrate (Ciechanover, 1998). E1 is encoded by a single gene in many organisms. Its function is well conserved, and an E1 from one species works well with E2s and E3s from another species. Of the dozen or so mammalian E2s, the Ubc5H subfamily members are promiscuous, and they support IAP- and other RING-containing E3-mediated ubiquitination. The other E2s seem to support IAP-mediated ubiquitination with some selectivity (Hu and Yang, 2003).

Various expression systems have been used to generate components of the ubiquitination reaction (Beaudenon and Huibregtse, 2005; Hatfield *et al.*, 1990). However, expressing these components in bacteria provides a unique advantage, because bacteria lack the ubiquitin-conjugating machinery, thereby eliminating the interference of contaminated host proteins. We were able to express six-His-tagged yeast Uba1 (E1), human UbcH5A (E2) (Kamura *et al.*, 2000) (kind gifts from J. W. Conaway), and ubiquitin (Tan *et al.*, 1999) (a kind gift from Z. Q. Pan) in bacteria by use of pET plasmids (Novagen). cIAP2, XIAP, and substrates were expressed with pGEX plasmids (Amersham Biosciences). The RING domain of IAPs invariably resides at the carboxyl termini of IAPs. Because in some cases, an epitope fused at the C-terminus can interfere with ubiquitination mediated by a carboxyl-terminal RING (Salvat *et al.*, 2004), we recommend the tag be placed at the amino terminus of IAP. Although crude bacterial lysates of bacteria expressing the components can be used for the assay, these proteins can also readily be purified to apparent homogeneity.

2.1. Expression of ubiquitination assay components in bacteria

Procedures

- The plasmids are transformed into bacterial strain BL21(DE3), which has decreased protease activity.
- Pick a colony and inoculate in 25 ml LB media containing ampicillin (100 μg/ml) or kanamycin (30 μg/ml), depending on the plasmid that was transformed in the bacteria.
- Transfer the 25-ml culture into 500 ml of LB and incubate the bacteria until the culture has reached the OD_{600} of approximately 0.6.
- Induce protein expression by adding isopropyl-β-D-thiogalactoside (IPTG) to a final concentration of 1 mM (for 6× His–tagged proteins) and 0.1 to 0.2 mM (for GST fusion proteins). Culture the bacteria for an additional 2 to 4 h at 37°.
- Centrifuge the culture at 6000g for 15 min at 4°.
- The cell pellet can be stored at −20° at this step until use.

Purification of His-tagged proteins

2.1.1.1.1. Solutions and materials

Lysis buffer: 50 mM NaH$_2$PO$_4$, pH 8.0, 300 mM NaCl, 10 mM imidazole, 1 mg/ml lysozyme. Imidazole in the buffer reduces nonspecific binding to beads, and for purifying His-tagged proteins that exhibit high binding affinity, the concentration of imidazole can be increased up to 20 mM.

Wash buffer: 50 mM NaH$_2$PO$_4$, pH 8.0, 300 mM NaCl, 10 mM imidazole.

Elution buffer: 50 mM NaH$_2$PO$_4$, pH 8.0, 300 mM NaCl, 250 mM imidazole.

Nickel-nitrilotriacetic acid-agarose (NiNTA) beads (Qiagen).

2.1.1.1.2. Procedure

- The cell pellet is resuspended in 5 ml of lysis buffer.
- Cells are sonicated on ice and centrifuged at 10,000g for 20 min.
- Transfer supernatant to new 15-ml conical tube and add 500 μl of bed volume of NiNTA beads that have been washed twice with 50 mM NaH$_2$PO$_4$, pH 8.0.
- Rotate the tube at 4° for 1 to 2 h.
- Wash beads with wash buffer three times.

- Elute the proteins from beads by incubating with elution buffer for 15 min at room temperature. Repeat this step twice and pool the elutes.
- Imidazole is removed by dialyzing elutes against 50 mM NaH$_2$PO$_4$, pH 8.0, with 2 mM dithiothreitol (DTT).
- The protein concentration is determined by Bradford method. The protein is aliquoted, snap-frozen, and stored at $-80°$.
- Five to ten microliters of the elute is boiled with protein sample loading buffer for 5 min and analyzed on SDS-PAGE.
- The gel is stained with Coomassie blue (Fig. 13.1).

Purification of GST-tagged proteins from bacteria
2.1.1.2.1. Materials

Glutathione-Sepharose 4B (Amersham Biosciences).

2.1.1.2.2. Procedure

- Suspend cell pellet in 20 ml of cold PBS and sonicate them on ice.
- Centrifuge at 10,000g at 4° for 30 min.
- Transfer the supernatant to a new 50-ml conical tube and add Triton X-100 to final concentration of 1% and phenylmethylsulfonyl fluoride (PMSF) to 1 mM.
- The supernatant is then incubated with 200 μl bed volume of PBS-equilibrated glutathione-Sepharose 4B at 4° for 2 to 4 h in a rotator.

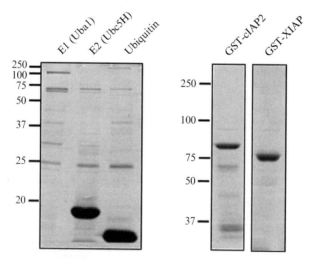

Figure 13.1 Purification of components for the *in vitro* ubiquitination assay. *Left panel,* His-tagged E1, E2, and ubiquitin were prepared as described in the text; 500 ng of E1, 10 μg of E2, and 10 μg of ubiquitin were resolved on 14% SDS-PAGE and gel was stained with Coomassie blue. *Right panel,* Purified GST-cIAP2 and -XIAP. Approximately 2 μg of GST-cIAP2 and XIAP was loaded on 8% SDS-PAGE and stained with Coomassie blue.

- The beads are collected by centrifugation at $1000g$ for 5 min at $4°$ and washed with PBS containing 1% Triton X-100 five times.
- Fusion proteins were purified by elution from glutathione sepharose beads by use of 5 mM glutathione.
- Protein was dialyzed overnight at $4°$ against buffer containing 50 mM Tris, pH 8.0.
- When it is necessary to cleave the GST from the protein, the beads are washed with specific protease buffer and resuspended in 300 μl of the cleavage buffer containing the appropriate amount of protease as recommended by the manufacturer.
- After 16 h to overnight incubation at $4°$, the supernatant is collected, snap frozen and stored at $-80°$ in small aliquots.

Production and isolation of proteins from mammalian cells IAPs and their substrates can be transiently expressed in HEK 293 cells. To facilitate purification, we use the Flag epitope tag (Tang *et al.*, 2004). The following method is for the expression and purification of Flag-tagged cIAP2.

2.1.1.3.1. Solutions and materials

Lysis buffer: 50 mM Tris, pH 7.4, 150 mM NaCl, 1 mM EDTA, 0.2 mM PMSF, 1.0% Triton X-100, protease inhibitor cocktail (Sigma), and 10 μM MG132 (Calbiochem). Proteasome inhibitors are used to prevent degradation of IAPs during the purification steps.

Elution buffer: 50 mM Tris, pH 7.4, 150 mM NaCl, 1 mM EDTA, 0.2 mM PMSF, protease inhibitor cocktail (Sigma), 10 μM MG132, and 0.1 mg/ml Flag peptide.

2×BBS: 50 mM N,N-bis(2-hydroxyethyl)-2-aminoethanesulfonic acid (Bes; Calbiochem), 280 mM NaCl, and 1.5 mM Na$_2$HPO$_4$, pH 6.98. The pH is crucial. It is recommended to test a few stocks with slightly different pHs or to use a previously tested stock as a reference.

Anti-flag mAb (M2)-Coupled agarose beads (Sigma)
2.1.1.4.1. Procedure

- HEK 293 cells are seeded in 10-cm dishes at 2×10^6 cells per plate the day before transfection.
- Cells are transfected with mammalian vector expressing the Flag-cIAP2 by use of calcium phosphate method; 10 μg of plasmid was added to mixture of 500 μl of 2× BSS, 65 μl of 2 M CaCl$_2$, and 435 μl H$_2$O. After 10 min of incubation, the CaPO$_4$-DNA precipitate is mixed well and added to the cell culture.

- 24 h later, cells are washed with cold PBS and lysed with lysis buffer.
- The cell lysates are centrifuged at 12,000g for 15min.
- Supernatant is incubated with M2 beads at 4° for 2 h.
- The beads are washed extensively with the lysis buffer at least six times. In the final washing step, the beads are washed with lysis buffer without detergent.
- Flag-cIAP2 is eluted from the beads with elution buffer at 4° for 30 min with gentle agitation.
- The eluted protein is analyzed by SDS-PAGE and Western blot by use of anti–Flag or anti–cIAP2 antibody.

Synthesis and purification of proteins with TNT transcription/translation system   An alternative way to generate IAPs and their substrates is to use the TNT transcription/translation system. This approach is convenient and especially suitable when a substrate is ubiquitinated when overexpressed in mammalian cells (e.g., RIP [Park *et al.*, 2004]).

2.1.1.5.1. *Solutions and materials*

Binding buffer: 50 mM Tris, pH 7.4, 150 mM NaCl, 1 mM EDTA, and 1.0% Triton X-100.
Elution buffer: 50 mM Tris, pH 7.4, 150 mM NaCl, and 0.1 mg/ml Flag peptide

SP6 TNT transcription/Translation system (Promega)
2.1.1.6.1. *Procedure*

- 1 μg of Flag-cIAP2 or Flag-RIP, both of which are cloned in the pRK5 plasmid, is added to 50 μl of the TNT Quick Master Mix containing 50 μM of either [^{35}S]-labeled or unlabeled methionine. The reaction is done at 30° for 60 to 90 min.
- 2.5 μl of the reaction mix is resolved on SDS-PAGE. The protein gel is transferred and subsequently blotted and probed with anti–Flag antibody to estimate the efficiency of the reaction. If [^{35}S]methionine is used in the reaction, the protein gel can also be dried and analyzed by autoradiography.
- Five reactions for each protein are performed and pooled.
- Binding buffer (3 ml) and 60 μl bed volume of anti–Flag M2 Ab-coupled agarose beads are added to the *in vitro* translated proteins and transferred to three Eppendorf tubes. The tubes are put on a rotator at 4° for 4 h.
- The beads are washed with binding buffer at least six times. Flag-cIAP2 or Flag-RIP is eluted with elution buffer. Alternately, the protein on the beads can be used directly in the *in vitro* ubiquitination reaction, although eluted proteins have given us more efficient ubiquitination.

- Estimate the amount of eluted proteins by Western blot analysis.

In vitro ubiquitination reaction
2.1.1.7.1. Solution and materials

Ubiquitination buffer: 50 mM Tris, pH 7.5, 2 mM ATP, 2.5 mM MgCl$_2$, 0.5 mM DTT, and 0.05% NP-40.

Antibodies for cIAP2 and XIAP (R&D systems)
2.1.1.8.1. Procedure.
The ubiquitination reaction is performed with 50 ng of E1(Uba1), 200 ng of E2 (UbcH5a), 0.8 μg of ubiquitin, 250 ng of E3 (GST-cIAP2 or XIAP), and substrate proteins in a total volume of 30 μl of ubiquitination buffer. Before performing *in vitro* ubiquitination assay with substrate, the purified components of the assay can be tested for activity by performing the assay in the absence of substrate. Functional enzymes should result in auto-ubiquitination of cIAP2 or XIAP as shown in Fig. 13.2. When bead-bound cIAP2 is used, beads are washed twice with ubiquitination reaction buffer before the reaction. The reaction is performed as described, except that the reaction is done with gentle agitation to keep beads in suspension. The reaction is performed at 30° for 1 h. SDS sample loading buffer is added to terminate the reaction and the sample is boiled and resolved on SDS-PAGE. The proteins are transferred to nitrocellulose membrane and blotted with antibody against the substrate and IAPs. Proper controls should be included, such as reactions that lack E1, E2, E3, or ubiquitin and reactions containing RING-deleted

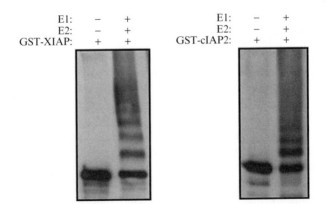

Figure 13.2 *In vitro* ubiquitination assay. Self-ubiquitination of cIAP2 and XIAP was performed in the presence of E1, E2, and ubiquitin as described in text. Samples were analyzed by immunoblotting with antibodies against cIAP2 and XIAP.

IAPs. An example of the *in vitro* ubiquitination reaction is shown in Fig. 13.2.

3. *In Vivo* Ubiquitination Assay

3.1. IAP-dependent ubiquitination

To assess IAP-mediated ubiquitination *in vivo*, an IAP and its putative substrate can be coexpressed in mammalian cells. The steady-state level and the ubiquitination of the substrates can then be assessed by immuno-precipitation and immunoblotting. The following is an example of assessing the cIAP2-mediated ubiquitination of Bcl10, which functions in signal transduction initiated by antigen receptor engagement.

3.1.1. Solution and material

NP-40 lysis buffer: 50 mM HEPES, 150 mM NaCl, 10% glycerol, 2 mM dithiothreitol, 1% Nonidet P-40, and 1 mM EDTA.
Complete Protease Inhibitor Cocktail (Roche).
Anti-Flag mAb M2 (Sigma).
Anti-HA and Bcl10 rabbit polyclonal antibodies (Santa Cruz Biotechnology).

Procedure

- Seed 2×10^6 HEK 293 cells in 10-cm plate the day before transfection.
- Cells are transfected with HA-Bcl10 plus Flag-cIAP2 or Flag-cIAP2Δ (which lacks the RING domain and serves as a negative control) by use of calcium phosphate precipitation method as described previously. The amount of DNA used for Flag-cIAP2 and Flag-cIAP2Δ is adjusted between 1 to 4 μg to achieve similar levels of expression.
- Sixteen to twenty-two hours after transfection, cells are washed once with cold PBS and lysed in 1 ml NP-40 lysis buffer containing protease inhibitors.
- Cell lysate is centrifuged at 12,000g for 15 min.
- Supernatant is incubated with anti-Flag M2 beads for 2 to 4 h at 4°.
- Beads are washed extensively (four to six times) with NP-40 lysis buffer.
- The immunoprecipitated proteins are resolved by SDS-PAGE and ana-lyzed by immunoblotting with anti-Flag and anti-HA antibodies. Bcl10 ubiquitination is indicated by the appearance of slower-migrating bands.
- The lysate can be analyzed by immunoblotting for the effect of cIAP2 and cIAP2Δ on the steady-state levels of Bcl10.

3.2. Ubiquitination of endogenous IAP targets

To confirm ubiquitination of endogenous IAP target proteins, the cell lysate is made under denaturing conditions to dissociate any interacting proteins before immunoprecipitation. The following procedure is to detect ubiquitination of endogenous Bcl10 in primary human T cells on T-cell receptor engagement.

3.2.1. Materials and solutions

Anti-Bcl10 antibody conjugated on beads: polyclonal anti-Bcl10 antibody (Santa Cruz) are conjugated to AminoLink Plus® Coupling gel (Pierce) according to the manufacturer's instructions.

Primary human CD4$^+$ T cells: Human peripheral blood lymphocytes are isolated from normal volunteer donors after apheresis and elutriation. CD4$^+$ T cells are purified by negative selection by use of magnetic beads (Dynal Biotech, Lake Success, NY) as previously described (Parry et al., 2003).

Anti-CD3/CD28 beads: Anti-human CD3 and anti-human CD28 are conjugated on magnetic beads as previously described (Parry et al., 2003).

Anti-ubiquitin antibody (Pierce).

Procedure

- Primary human CD4$^+$ T cells (20×10^6 at 5×10^6 cells/ml) are incubated with anti-CD3/CD28 beads (3 beads/cell). Culture them for 3 days in complete RPMI medium.
- Separate the beads with cells by use of magnetic bar/stand.
- Culture the cells for approximately 10 days. During this time, the cells will be expanded approximately 50- to 100-fold. Split cells regularly to keep the concentration of cells at ~5×10^6 cells/ml.
- When cells stop dividing, treat ~3×10^8 cells with anti-CD3/CD28 beads (3 beads/cell) for 30 min under normal culture conditions.
- The treated cells and 1×10^8 untreated control cells are lysed in NP-40 lysis buffer containing 1% SDS. Because of Bcl10 degradation in response to antigenic stimulation (Hu et al., 2006a,b), more treated than untreated cells are used to achieve the same level of Bcl10 in the lysates. Heat lysates for 5 min at 90° or greater.
- The lysates are diluted by a factor of 10 in buffer containing no SDS.
- After centrifugation, the lysates are incubated with anti-Bcl10 beads for 4 h to overnight at 4°.
- Wash beads four to six times with buffer.
- Bead-bound proteins are resolved on SDS-PAGE and transferred to nitrocellular membrane.

- The membrane is placed in distilled water between two pieces of filter paper and heated in a microwave oven for 2 min.
- The membrane is probed by use of anti–ubiquitin antibody.

ACKNOWLEDGMENTS

This work was supported by NIH grants CA88868 and CA108872 and a Leukemia & Lymphoma Society Scholar award to X. Y., and the Korea Research Foundation Grant (KRF-2006-005-J04502) and Brain Korea 21 project to T.H.L. The authors are grateful for critical reading of the manuscript by the members of the Yang laboratory.

REFERENCES

Beaudenon, S., and Huibregtse, J. M. (2005). High-level expression and purification of recombinant E1 enzyme. *Methods Enzymol.* **398,** 3–8.

Birnbaum, M. J., Clem, R. J., and Miller, L. K. (1994). An apoptosis-inhibiting gene from a nuclear polyhedrosis virus encoding a polypeptide with Cys/His sequence motifs. *J. Virol.* **68,** 2521–2528.

Ciechanover, A. (1998). The ubiquitin-proteasome pathway: On protein death and cell life. *EMBO J.* **17,** 7151–7160.

Crook, N. E., Clem, R. J., and Miller, L. K. (1993). An apoptosis-inhibiting baculovirus gene with a zinc finger-like motif. *J. Virol.* **67,** 2168–2174.

Deveraux, Q. L., and Reed, J. C. (1999). IAP family proteins–suppressors of apoptosis. *Genes Dev.* **13,** 239–252.

Hatfield, P. M., Callis, J., and Vierstra, R. D. (1990). Cloning of ubiquitin activating enzyme from wheat and expression of a functional protein in *Escherichia coli. J. Biol. Chem.* **265,** 15813–15817.

Hu, S., Alcivar, A., Qu, L., Tang, J., and Yang, X. (2006a). cIAP2 inhibits antigen receptor signaling by targeting Bcl10 for degradation. *Cell Cycle* **5,** 1438–1442.

Hu, S., Du, M. Q., Park, S. M., Alcivar, A., Qu, L., Gupta, S., Tang, J., Baens, M., Ye, H., Lee, T. H., Marynen, P., Riley, J. L., *et al.* (2006b). cIAP2 is a ubiquitin protein ligase for BCL10 and is dysregulated in mucosa-associated lymphoid tissue lymphomas. *J. Clin. Invest.* **116,** 174–181.

Hu, S., and Yang, X. (2003). Cellular inhibitor of apoptosis 1 and 2 are ubiquitin ligases for the apoptosis inducer Smac/DIABLO. *J. Biol. Chem.* **278,** 10055–10060.

Kamura, T., Sato, S., Iwai, K., Czyzyk-Krzeska, M., Conaway, R. C., and Conaway, J. W. (2000). Activation of HIF1alpha ubiquitination by a reconstituted von Hippel-Lindau (VHL) tumor suppressor complex. *Proc. Natl. Acad. Sci. USA* **97,** 10430–10435.

Li, X., Yang, Y., and Ashwell, J. D. (2002). TNF-RII and c-IAP1 mediate ubiquitination and degradation of Traf2. *Nature* **416,** 345–349.

Park, S. M., Yoon, J. B., and Lee, T. H. (2004). Receptor interacting protein is ubiquitinated by cellular inhibitor of apoptosis proteins (c-IAP1 and c-IAP2) *in vitro. FEBS Lett.* **566,** 151–156.

Parry, R. V., Rumbley, C. A., Vandenberghe, L. H., June, C. H., and Riley, J. L. (2003). CD28 and inducible costimulatory protein Src homology 2 binding domains show distinct regulation of phosphatidylinositol 3-kinase, Bcl-xL, and IL-2 expression in primary human CD4 T lymphocytes. *J. Immunol.* **171,** 166–174.

Salvat, C., Wang, G., Dastur, A., Lyon, N., and Huibregtse, J. M. (2004). The-4 phenylala-
nine is required for substrate ubiquitination catalyzed by HECT ubiquitin ligases. *J. Biol.
Chem.* **279,** 18935–18943.
Salvesen, G. S., and Duckett, C. S. (2002). IAP proteins: Blocking the road to death's door.
Nat. Rev. Mol. Cell Biol. **3,** 401–410.
Suzuki, Y., Nakabayashi, Y., and Takahashi, R. (2001). Ubiquitin-protein ligase activity of
X-linked inhibitor of apoptosis protein promotes proteasomal degradation of caspase-3
and enhances its anti-apoptotic effect in Fas-induced cell death. *Proc. Natl. Acad. Sci. USA*
98, 8662–8667.
Tan, P., Fuchs, S. Y., Chen, A., Wu, K., Gomez, C., Ronai, Z., and Pan, Z. Q. (1999).
Recruitment of a ROC1-CUL1 ubiquitin ligase by Skp1 and HOS to catalyze the
ubiquitination of I kappa B alpha. *Mol. Cell* **3,** 527–533.
Tang, J., Wu, S., Liu, H., Stratt, R., Barak, O. G., Shiekhattar, R., Picketts, D. J., and
Yang, X. (2004). A novel transcription regulatory complex containing death domain-
associated protein and the ATR-X syndrome protein. *J. Biol. Chem.* **279,** 20369–20377.
Vaux, D. L., and Silke, J. (2005). IAPs, RINGs and ubiquitylation. *Nat. Rev. Mol. Cell Biol.*
6, 287–297.
Yang, Y., Fang, S., Jensen, J. P., Weissman, A. M., and Ashwell, J. D. (2000). Ubiquitin
protein ligase activity of IAPs and their degradation in proteasomes in response to
apoptotic stimuli. *Science* **288,** 874–877.

CHAPTER FOURTEEN

PHOSPHATASES AND REGULATION OF CELL DEATH

Chi-Wu Chiang,[†,1] Ling Yan,[*,1] *and* Elizabeth Yang[*,2]

Contents

Abstract

Reversible phosphorylation, catalyzed by kinases and phosphatases, regulates many apoptosis molecules. Phosphorylation inactivates, whereas dephosphorylation activates, the pro-apoptotic functions of select apoptotic regulators, including the BCL2 family, such as BAD, and transcription factors such as FOXOs. The apoptotic function of the BH3 molecule BAD is exquisitely regulated by phosphorylation. Although phosphorylated BAD is sequestered in the cytosol, dephosphorylated BAD translocates to the mitochondria and inactivates BCL-x$_L$

* Departments of Pediatrics, Cancer Biology, Cell and Developmental Biology, Vanderbilt University School of Medicine, Nashville, Tennessee
† Institute of Molecular Medicine and Center for Gene Regulation and Signal Transduction Research, National Cheng Kung University, Tainan, Taiwan
[1] co-first authors
[2] corresponding authors

Methods in Enzymology, Volume 446
ISSN 0076-6879, DOI: 10.1016/S0076-6879(08)01614-5

and BCL2. Analogously, Akt-phosphorylated FOXO1 is cytosolic and inactive as a transcription factor, but dephosphorylated FOXO1 translocates to the nucleus, where it regulates the expression of pro-apoptotic Bim and cell cycle inhibitors. By use of inhibitor experiments and a combination of immunoprecipitations and tagged pull-downs in interaction studies, we identified PP2A enzymes as BAD and FOXO1 phosphatases. PP2A dephosphorylation of BAD is regulated by competitive interaction of 14-3-3, PP2A, and BAD. On survival factor deprivation, PP2A dephosphorylation of pSer112 plays the gatekeeper role for subsequent dephosphorylation at pSer136 and pSer155 by multiple phosphatases. In contrast, PP2A and 14-3-3 can interact with FOXO1 concomitantly, but PP2A dephosphorylates the pThr24 and pSer256 only once 14-3-3 dissociates. Functional assays of cell death, Bim upregulation by FOXO1, and FOXO1 nuclear translocation in the presence of phosphatase inhibitors and phosphatase siRNAs revealed the physiologic significance of PP2A activity on BAD and FOXO1. Demonstrating the role of PP2A in regulating the function of two very different cell death molecules, a BH3 protein and a transcription factor, suggests that activation of pro-apoptotic factors by protein phosphatases may be a general regulatory mechanism in apoptosis.

1. INTRODUCTION

BAD is a prototypical BCL2 family molecule regulated by phosphorylation. Initially, two serine phosphorylation sites were found, Ser112 and Ser136 (Zha *et al.*, 1996). In the phosphorylated state, these sites bind 14-3-3, and BAD is sequestered in the cytosol. In apoptosis, BAD is dissociated from 14-3-3, complexed with BCL-x_L or BCL2, and is mitochondrially localized. Dephosphorylation of pSer136 was noted to be especially important for the apoptotic function of BAD (Datta *et al.*, 1997; Masters *et al.*, 2001; Yaffe *et al.*, 1997). Dephosphorylation of the third phosphorylation site, pSer155, in the BH3 domain, was found to be key in mediating binding to anti-apoptotic counterparts (Datta *et al.*, 2000; Zhou *et al.*, 2000). Other phosphorylation sites have been revealed since, but Ser112, S136, and S155, phosphorylated by Akt and PKA, seem to be the predominant sites regulating survival factor deprivation–induced apoptosis. Other BCL2 family molecules have also been reported to be regulated by reversible phosphorylation of serine/threonine residues, including BCL2, MCL1, BAX, BID, and BIM (Desagher *et al.*, 2001; Gardai *et al.*, 2004; Harada *et al.*, 2004; Opferman, 2006; Ruvolo *et al.*, 2001). The causal role of dephosphorylation in the pro-apoptotic function of BAD has been clearly demonstrated (Chiang *et al.*, 2001; 2003). Although much had been done on the kinases phosphorylating BAD, we reasoned that phosphorylation inactivates, but dephosphorylation *activates* BAD as a cell death molecule; therefore, we embarked on studies of serine/threonine phosphatases that dephosphorylate BAD. We also wished to know whether phosphatases were involved in the activation of pro-apoptotic regulators outside of the BCL2 family and chose to examine the forkhead transcription factor FOXO1.

FOXO transcription factors, which regulate expression of apoptosis and cell cycle arrest genes, such as BIM, FasL, and p27, are also regulated by reversible phosphorylation and involve 14-3-3 binding, analogous to BAD (Arden, 2006; Brunet et al., 1999). FOXOs contain many phosphorylation sites targeted by multiple kinases, but the Akt sites Thr24 and Ser256 are involved in apoptosis induction. We sought to determine the role of serine/threonine phosphatases in the activation of FOXO1 as a cell death transcription factor. FOXO1 represents a very different type of molecule from the BH3 family, and a similar pathway of dephosphorylation exists for this molecule, suggesting that protein phosphatase activation may be a general regulatory mechanism in apoptosis (Yan et al., 2008).

Serine/threonine phosphorylation regulates different functions of apoptosis molecules. Subcellular localization has been shown to be a function of phosphorylation status. BAD phosphorylated at Ser112 and Ser136 and FOXO1 phosphorylated at Thr24 and S256 are retained in the cytosol by 14-3-3, but dephosphorylated BAD is mitochondrial and dephosphorylation of FOXO1 at pSer256 within the NLS is important for nuclear translocation (Zhao et al., 2004). Phosphorylation status also determines protein binding. The apoptotic activity of BAD is exquisitely sensitive to phosphorylation status, and binding to the anti-apoptotic members depends on Ser155 in the BH3 domain being dephosphorylated. For FOXO1, its transcriptional activity requires dephosphorylation at the Akt sites (Huang and Tindall, 2007).

The major classes of serine/threonine phosphatases include the PP1 and PP2A families, Ca^{++}-dependent PP2B, Mg^{++}-dependent PP2C, and the newer PP5 and PP7 subfamilies (Gallego and Virshup, 2005). PP1 and PP2A are two abundant enzymes whose catalytic subunits share homology and have been traditionally distinguished by inhibitor profiles. The PP2A heterotrimeric holoenzymes consist of a catalytic subunit (C), a structural scaffolding subunit (A), and a highly variable regulatory subunit (B). Multiple families of regulatory subunits, each with several members, can complex with the AC core dimer to form a large number of holoenzymes, making studies of PP2A regulation challenging (Janssens et al., 2005). The structure of PP2A was only recently solved and confirmed previous observations that the subunit structure both targets specific substrate binding and inhibits promiscuous interactions (Mumby, 2007). Whereas the regulatory subunits are believed to be rate limiting, the catalytic subunits, used in all holoenzymes, are much more abundant. Our studies thus far focused on identifying the family of serine/threonine phosphatases involved in BAD and FOXO1 dephosphorylation, through catalytic subunit interactions, with much effort in distinguishing between PP1 and PP2A.

Studies of serine/threonine phosphatases in cell death pose many challenges. Among them is that PP1 and PP2A phosphatases are highly abundant and affect a great many signaling pathways, making establishing specificity especially difficult. In particular, not only are BAD and FOXO1 themselves activated by phosphatases, but their upstream kinases are inactivated by the same or similar phosphatases; thus, inhibitors must be used knowledgeably and conclusions drawn judiciously. This chapter will describe the assays we used in studying the biochemistry and functional significance of BAD and FOXO1 dephosphorylation, as well as discuss the technical challenges and pitfalls unique to these studies.

2. IDENTIFICATION OF THE PUTATIVE PHOSPHATASE IN CELLS

2.1. Implicating the phosphatase activity in cells

To study the role of protein phosphatases in apoptosis, it is helpful to first select a suitable cell line system. We chose IL-3–dependent FL5.12 cells to study BAD and FOXO1, because a specific signaling pathway is involved in cell death induced by withdrawal of a single cytokine. A more general death stimulus may activate many phosphatases in multiple signaling pathways and make it difficult to link specific phosphatase activities to the protein of interest.

The study of protein dephosphorylation is greatly aided by phospho-specific antibodies. In the absence such antibodies, a mobility shift associated with phosphorylation is essential. Our initial work on BAD dephosphorylation used the mobility difference between phosphorylated and dephosphorylated species before phospho-specific antibodies were available. In the presence of the survival factor IL-3, multiple phosphorylation–modified forms of BAD are present, including hyperphosphorylated, hypophosphorylated, and nonphosphorylated forms (Zha *et al.*, 1996). The presence of hypophosphorylated and nonphosphorylated forms of BAD implicated protein phosphatases in the regulation of BAD.

Pharmacologic inhibitors are used extensively in the field of serine/threonine phosphatases, particularly because overexpression of active serine/threonine phosphatases is not well tolerated by cells (Honkanen and Golden, 2002). Because the serine/threonine phosphatase holoenzymes within a class are composed of different regulatory subunits, which share a common catalytic subunit, the phosphatase inhibitors inactivate entire classes of phosphatases, rather than specific holoenzymes. As such, they are very useful in determining whether phosphatases are involved at all, because one can simply ask whether a general phosphatase inhibitor changes the phosphorylation status of the protein of interest during cell death. For inhibition to be effective, the inhibitor needs to be in the cells at the time of death stimulation and phosphatase activation. Therefore, cells need to be pretreated with inhibitor before apoptosis is induced. To explore whether phosphatases were involved in regulating the phosphorylation status of BAD, we treated cells with okadaic acid (OA), a general inhibitor of the most abundant serine/threonine phosphatases, PP1 and PP2A. In baseline growth conditions, BAD is both phosphorylated and dephosphorylated, appearing as a doublet. During IL-3 withdrawal, BAD was mostly in the faster-migrating dephosphorylated form. In the presence of okadaic acid, the slower migrating form persisted, suggesting that an OA-sensitive activity was involved in BAD dephosphorylation (Fig. 14.1A). The lack of any OA effect on the serine-to-alanine mutant BAD112A136A confirmed that the observed band shifts were due to phosphorylation changes at the expected sites. Our initial experiment to implicate phosphatases in the regulation of FOXO1 phosphorylation status used a FL5.12 cell line expressing doxycycline-inducible FOXO1, F14 (Plas and Thompson, 2003). The Akt sites on FOXO1, T24, and S256 were phosphorylated in the presence of IL-3 and dephosphorylated after withdrawal of IL-3 but remained phosphorylated in the presence of the PP1 and PP2A inhibitor calyculin A during IL-3 deprivation, indicating a role for PP1 or PP2A in the regulation of FOXO1 phosphorylation (Fig. 14.1B).

To assay for dephosphorylation, it is advantageous for the apoptosis protein to be as phosphorylated as possible to begin with, so that dephosphorylation can be clearly observed. We have found that in steady-state cultures, the relative hyperphosphorylated and hypophosphorylated FOXO1 species varied. To convert as much FOXO1 as possible to the hyperphosphorylated forms, IL-3 was first withdrawn from F14 cells for 3 h, then added back for 30 min to stimulate a burst of kinase activity so that FOXO1 can be maximally phosphorylated, only then was IL-3 washed out from the media to begin the dephosphorylation experiment.

Once a phosphatase activity is implicated, inhibitors with differential activities against different classes of phosphatases can be used to narrow down the putative phosphatase. To identify which phosphatase was acting on BAD in FL5.12 cells, we treated cells with a panel of inhibitors targeting different classes of serine/threonine phosphatases. Both calyculin A and OA inhibit PP1, PP1-like, PP2A, and PP2A-like phosphatases; however, OA shows preference for PP2A (IC50 for PP2A: 0.1 to 0.3 nM, IC50 for PP1: 20 to 50 nM), whereas calyculin A inhibits PP1 and PP2A with similar potency (IC50 for PP2A: 0.25 nM, PP1: 0.4 nM). Fostriecin (FST) is a highly selective inhibitor of PP2A enzymes and inhibits PP2A at 10,000 to 40,000 times lower concentration than that required for PP1 inhibition (IC$_{50}$ for PP2A :1.5 to 5.5 nM, PP1: 45 to 58 μM) (Honkanen and Golden, 2002). Both cyclosporine A and FK506 are inhibitors for the calcium-dependent PP2B (calcineurin). Tautomycin is more selective for PP1 than PP2A (Oikawa, 2002). OA, calyculin A, and fostriecin caused

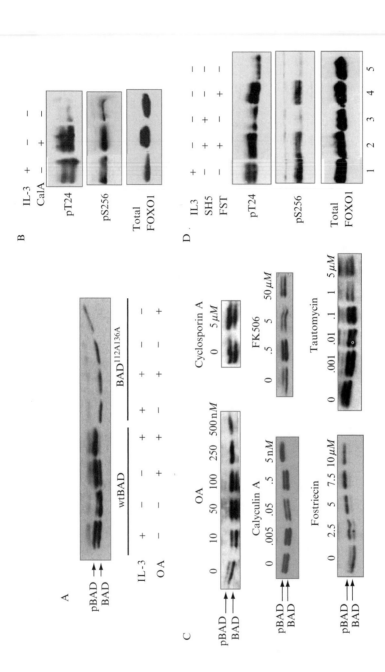

Figure 14.1 (A) FL5.12BCL-x_L/BADwt or FL5.12BCL-x_L/BADI12A136A phosphorylation defective cells were treated with 0.5 μM OA or vehicle in the presence (+) or absence (−) of IL-3 for 3 hours, and western blotted with anti-BAD C20 Ab (Santa Cruz). (B) Fl4 cells induced by doxycycline were treated with 5nM calyculin A in the absence of IL-3 for 3 hours, anti-pS256 Ab (Cell Signaling), and total FOXO1 Ab (Santa Cruz). (C) FL5.12BCL-x_L/BADwt cells were treated with inhibitors in the presence of IL-3 for 3 hours. (D) Induced Fl4 cells were treated with 5 μM SH5 (Akt inhibitor) and 10 μM FST (PP2A selective) in the absence of IL-3 for 3 hours.

dose-dependent band shift of BAD from the lower non/hypophosphory-lated form to the higher phosphorylated form, whereas no band shift was detected with tautomycin, cyclosporine, or FK506 (Fig. 14.1C). These inhibitor studies pointed to the PP2A family as the BAD phosphatase. Similar inhibitor profiles were obtained for FOXO1, which also pointed to PP2A as the phosphatase for the pT24 and pS256 sites of FOXO1.

2.1.1. Methods

1. Culture FL5.12 BCL-x_L/BAD cells in Iscove modified Dulbecco medium supplemented with 10% fetal calf serum and IL-3. For FL5.12 cells expressing inducible FOXO1wt (F14) or AAA mutant (A3), induce with 1 μg/ml doxycycline for 14 h. For IL-3 deprivation, wash cells three times with PBS and reculture in media without IL-3. To ensure that FOXO1 is well phosphorylated to start with, IL-3 was deprived for 3 h, added back for 30 min, then washed out to begin dephosphorylation experiments.
2. To inhibit phosphatases during IL-3 withdrawal, pretreat cells for 1 h with inhibitor or vehicle, wash cells three times with PBS, reculture cells in media without IL-3 but containing inhibitor or vehicle. Final inhibitor concentrations in the media are as follows. Okadaic acid: 250 to 500 nM, calyculin A: 5 nM, fostriecin: 3 to 10 μM.

2.2. Determination of effective phosphatase inhibition in cells

Phosphatase inhibitors have differing cell permeability, and the intracellular concentration of inhibitors could not be predicted from the amount added to the media. The concentrations of inhibitors used in cultures were empirically determined to achieve the desired effect with tolerable toxicity. Because conclusions about phosphatase are drawn on the basis of responses to different inhibitors, we devised a way to check whether the cognate phosphatases were, indeed, inactivated by the inhibitor treatments.

To confirm that PP2A activity within cells was, indeed, selectively inhibited by treatment of OA or fostriecin at the doses used, we assayed PP2A activity biochemically, by use of ^{32}P-labeled substrates. Extracts from cells treated with vehicle, 0.5 μM OA, or 5 μM FST were used to dephos-phorylate the general phosphatase substrate rabbit muscle phosphorylase a, or the more selective PP2A substrate histone H1, both ^{32}P-labeled. Phos-phatase activity, as measured by the release of ^{32}P, was much lower in cells treated with OA or fostriecin than vehicle-treated cells, approximately 50%, 30%, and 50% of untreated, respectively (Fig. 14.2, filled bars). Next, to check how much phosphatase activity remained in the treated cells, extracts were further incubated with vehicle, 5 nM OA, or1 μM OA to inhibit the remaining PP2A or PP1 activity, respectively. We found that addition of 5nM OA to lysates of cells previously treated with OA or fostriecin further

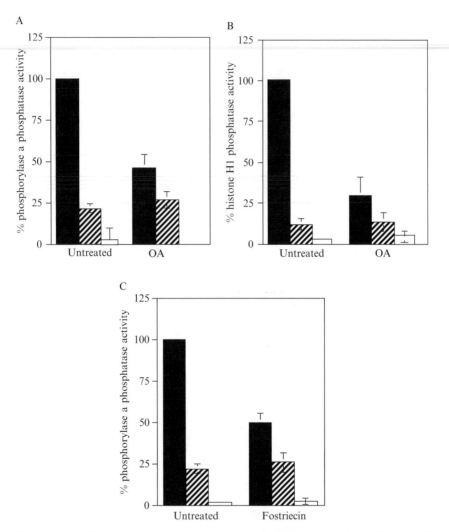

Figure 14.2   (■) Phosphatase activity in cells treated with 0.5 μM OA (A and B) or 5 μM fostriecin (C), using of ^{32}P-phosphorylase a (A and C) or ^{32}P-histone H1 (B) as substrate. Phosphatase activity in cell lysates further treated with 5 nM OA (▨) 1μM OA (□). Activities are expressed as a percentage of ^{32}P release in untreated lysates. Mean ± SD of triplicate assays from at least 2 experiments are graphed.

decreased the release of ^{32}P only modestly, from 50% to 30%, 30% to 15%, and 50% to 25% suggesting that most of the PP2A activity was already inhibited (Fig. 14.2, compare the difference between the filled and hatched bars of the OA and Fostriecin samples). However, addition of 1 μM OA further decreased ^{32}P release, from 30% to 0%, 15% to 7.5%, and 25% to 5%, indicating that PP1 activity had not been effectively inhibited by previous

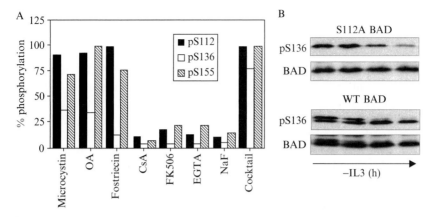

Figure 14.3 (A) Average of 2 experiments of the ratio of densitometry measurements of phosphorylated BAD to total BAD in *in vitro* inhibitor assays in the presence of R18. Ratio at time 0 is set to 100%. (B) Lysates of FL5.12BCL-x$_L$/BADwt or FL5.12BCL-x$_L$/BAD112A cells deprived of IL-3 for 3 hours were immunoblotted for pS136.

OA or fostriecin treatment of cells (Fig. 14.2, compare open and filled bars of OA and Fostriecin samples). By use of this protocol, we demonstrated that the doses of OA and fostriecin we used to treat cells effectively inhibited most of the cellular PP2A activity, while not significantly inhibiting PP1 activity. Thus, we were justified in attributing the OA- and fostriecin-inhibitable BAD and FOXO1 phosphatase activities to PP2A or a PP2A-like enzyme.

2.2.1. Methods
Phosphatase assay

1. Treat cells by adding vehicle (DMSO for OA and methanol for fostriecin), 0.5 μM OA, or 5 μM FST to the media for 3 h.
2. Collect cells, wash, and sonicate in phosphatase assay buffer B (20 mM imidazole, 150 mM NaCl, 14.4 mM β-mercaptoethanol, 1 mM phenyl-methylsulfonyl fluoride, 10 μg/ml aprotinin, and 10 μg/ml leupeptin).
3. Incubate cell extract with either vehicle, 5 nM OA, or 1 μM OA for 15 min on ice to inhibit remaining PP2A activity or PP1 activity, respectively.
4. Add 2.5 μg or 0.5 μg cell extract to a reaction mix containing ^{32}P-labeled phosphorylase a or histone H1 to a final volume of 100 μl.
5. Incubate the phosphatase reaction at 30 °C for 15 min, terminate by adding 1 μg BSA and 25 μl 100% TCA and incubating on ice for 20 min.
6. Centrifuge the reaction for 15 min at 13,000 rpm at 4 °C. Collect the supernatant containing the released ^{32}P and measure by scintillation counting. Limit the amount of cell extracts in the dephosphorylation assay to a maximum of 20% release of total ^{32}P to ensure linearity of the phosphatase reaction.

Preparation of ^{32}P-phosphorylase a

1. Add 200 μl of 10 mg/ml phosphorylase a (Sigma) to 200 μl of phosphorylase kinase buffer (50 mM Tris-Cl, pH7.5, 10 mM MgCl$_2$, 2.5 mM DTT, 2 mM CaCl$_2$, 0.05 mM ATP, 1 μCi/μl γ-^{32}P-ATP).
2. Add 10 μl of 10 mg/ml phosphorylase kinase. Incubate at 30 °C for 1 h.
3. Add 400 μl of saturated ammonium sulfate, mix well, and incubate on ice for at least 30 min.
4. Pellet the ^{32}P-phosphorylase a by centrifugation at 13,000g for 15 min. Wash the pellet twice with the buffer A (50 mM Tris-Cl, pH7.5, 1 mM EDTA, 1 mM DTT, 40% ammonium sulfate). Resuspend the pellet in 200 μl of buffer A.
5. Separate free γ-^{32}P-ATP from the labeled phosphorylase a by filtering through Millipore Ultrafre-MC filter with 30,000 MW cutoff (it takes ~20 min).
6. Wash the labeled phosphorylase a trapped on the filter with buffer A twice.
7. Recover the trapped labeled phosphorylase a by adding 200 μl of phosphatase assay buffer onto the filter, rinse, and collect.
8. Use 2 μl for each phosphatase assay.

Preparation of ^{32}P-histone HI

1. Add 50 μl of 20 mg/ml histone (type III-S, calf thymus, Sigma) into 350 μl PKC kinase assay buffer (50 mM Tris-Cl, pH 7.5, 10 mM MgCl$_2$, 2.5 mM DTT, 2 mM CaCl$_2$, 0.05 mM ATP, 1 μCi/μl γ-^{32}P-ATP, 0.1 mM EGTA, 1× PKC activators [0.05 mg/ml phosphatidylserine and 0.005 mg/ml 1,2-diacyl-sn-glycerol] and 0.1 μM PMA). Incubate at 30 °C for 1 h.
2. Add 100 μl of 100% TCA and incubate on ice for at least 30 min.
3. Pellet the ^{32}P-histone by centrifugation at 13,000g for 15 min. Wash the pellet twice with 1 ml washing buffer (ether/ethanol/HCl, 4/1/0.1 N).
4. Dry the pellet briefly. Resuspend pellet in 800 μl water.
5. Add 200 μl of 100% TCA and incubate on ice for 30 min.
6. Pellet the ^{32}P-histone by centrifugation at 13,000g for 15 min. Wash twice as earlier.
7. Resuspend the pellet in 400 μl phosphatase assay buffer.
8. Use 2 μl for each phosphatase assay.

2.3. Distinguishing between inhibition of phosphatases versus activation of kinases

The phosphorylation status of a molecule is the result of the action of kinases and phosphatases. Many kinases are themselves regulated by phosphatases; therefore, treating cells with phosphatase inhibitors can result in activation of certain kinases. Increased phosphorylation of the protein of interest in the presence of phosphatase inhibitors can be the result of inhibiting direct dephosphorylation by a phosphatase, or the result of increased kinase activity of an upstream kinase. These possibilities need to be distinguished before concluding that the protein of interest is actively dephosphorylated by a phosphatase. To address the possibility that increased BAD phosphorylation seen in cells treated with OA was due to activation of Akt or PKA, IL-3 was withdrawn to induce BAD dephosphorylation, then OA was added to cells. OA treatment did not cause phosphorylation of the already dephosphorylated BAD, indicating that Akt was not being activated by OA in this context, supporting the likelihood that OA was inhibiting a phosphatase activity against BAD.

An alternate approach to rule out kinase activation as the reason for the observed increase in phosphorylation of the protein of interest is to inhibit the kinase. We monitored FOXO1 dephosphorylation in cells simultaneously treated with the PP2A inhibitor fostriecin and the Akt inhibitor SH5 and found that fostriecin treatment of IL-3–deprived cells maintained FOXO1 phosphorylation even when Akt was inhibited by SH5 (Fig. 14.1D, compare lanes 2 and 4). This experiment and similar experiments that used inhibitors of other kinases in conjunction with phosphatase inhibitors enabled us to conclude that the observed effect of fostriecin was not on upstream kinases but more likely on a phosphatase regulating FOXO1.

2.4. Site-specific dephosphorylation

Phospho-specific antibodies greatly facilitate the study of dephosphorylation, because proteins can be phosphorylated at multiple sites, regulated by different phosphatases. For FOXO1, antibodies to pT24 and pS256 suggested that phosphatase regulation for these two sites were similar (Fig. 14.1B and 1D). In the case of BAD, by use of antibodies specific to pS112, pS136, and pS155 in a panel of inhibitor assays, we found that pS112 was selectively dephosphorylated by PP2A, but multiple phosphatases dephosphorylated pS136 (Fig. 14.3A).

Dephosphorylation of specific sites could be further delineated by use of site-specific mutations. Single-site alanine mutants and mutants containing multiple alanine substitutions can reveal not only the phosphatase activity for each site but also how dephosphorylation at one site may be influenced by dephosphorylation of neighboring sites. By use of anti-pS136 antibody in Western blotting of cells expressing wild-type BAD or S112ABAD, we found that dephosphorylation of pS136 was accelerated by the S112A mutation (Fig. 14.3B). This led us to hypothesize that dephosphorylation of pS136 required prior dephosphorylation of pS112, which was confirmed on further experimentation (Chiang et al., 2003). By use of BAD double mutants 136A155A, 112A155A, and 112A136A and phospho-specific anti-bodies to examine dephosphorylation of pS112, pS136, and pS155, we found that fostriecin preferentially inhibited pS112 dephosphorylation.

3. *IN VITRO* PHOSPHATASE ASSAYS

3.1. Deprotecting the phosphorylation sites

Phosphatase activity for individual phosphorylation sites can be further investigated by *in vitro* dephosphorylation assays, in which kinases are not active. Phosphatase assays are simple, because dephosphorylation will occur in almost any buffer incubated 30 °C. However, because serine/threonine phosphatases abound in cells, phosphorylated sites are protected if they are to remain phosphorylated. Serine/threonine phosphorylated proteins are frequently bound by 14-3-3 isoforms, and the association can be very strong (Yaffe, 2004). Although mechanisms to dissociate chaperones would be activated in cells when proteins need to be dephosphorylated, as in apoptosis induction, such processes would not automatically occur *in vitro*. Therefore, when conducting *in vitro* assays, measures need to be taken to ensure that the phosphorylated protein substrate is accessible to phosphatases. To disso-ciate 14-3-3, one can use either the zwitterionic detergent Empigen BB, which dissociates 14-3-3 from keratins, or a synthetic peptide R18 (PHCVPRDLSWLDLEANMCLP), which targets the substrate binding groove of 14-3-3 and displaces 14-3-3 from its partners (Liao and Omary, 1996; Wang *et al.*, 1999). Although no change in the mobility of BAD was observed after 30 mins of lysate incubation, addition of R18 to the lysate clearly resulted in collapse of the BAD doublet into the lower band, demonstrating that 14-3-3 must be dissociated from BAD for phosphatases to gain access to BAD (Fig. 14.4A). An array of inhibitors was added to *in vitro* assays to profile inhibitor sensitivities of individual phosphorylation sites, which revealed that although pS112 and pS155 were mainly depho-sphorylated by PP2A, pS136 was susceptible to dephosphorylation by all classes of phosphatases tested (Fig. 14.3A).

3.1.1. Methods

1. Collect FL5.12 BCL-x_L/BAD cells and lyse in phosphatase buffer A (50 mM Tris-Cl, pH 7.5, 150 mM NaCl, 1 mM EDTA, 0.5 mM EGTA, 0.25% Nonidet P-40, 1 mM phenylmethylsulfonyl fluoride, 10 μg/ml aprotinin, and 10 μg/ml leupeptin) with or without 1% Empigen BB or 25 μM R18 (synthesized), and with or without inhibitors.
2. Incubate cell extracts at 30 °C for 30 min for dephosphorylation.

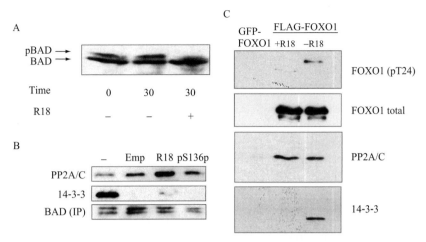

Figure 14.4 (A) FL5.12BCL-x$_L$/BADwt lysed in phosphatase buffer A with or without 25 μM R18 and incubated for 30 minutes at 4°C and western blotted by use of anti-BAD C20 (Santa Cruz). (B) Immunoprecipitation of BAD in the presence or absence of 1% Empigen BB, R18 peptide, of pS136 peptide. (C) FLAG–agarose pull-down of NIH3T3 cells transfected with GFP-FOXO1 or FLAG-FOXO1 in the presence or absence of R18 and western blotted for FOXO1 phosphorylated at T24 or total FOXO1, Pp2A/C, or 14-3-3.

3. For inhibitor profiling of BAD phosphatases, add inhibitors to the lysis buffer: 1 μM microcystin to inhibit PP1 and PP2A, 500 nM OA to inhibit PP2A > PP1, 1 μM FST to inhibit PP2A, 2.5 μM cyclosporine A or 500 ng/ml FK506 to inhibit PP2B, 5 mM EGTA to inhibit PP2C, 10 mM NaF for nonselective inhibition of phosphatases. The inhibitor cocktail contained 5 mM EGTA, 5 mM EDTA, 10 mM NaF, and 1 μM microcystin.
4. Terminate the reaction by adding 4× SDS sample loading buffer and boiling 5 min.

4. INTERACTION OF THE PHOSPHATASE AND APOPTOTIC PROTEIN

Studies that used phosphatase inhibitors suggest, but do not prove, the role of a phosphatase. Demonstration of physical interaction between the protein of interest and a phosphatase will leave little doubt that the phosphatase acts on the protein. Complexes can be isolated by use of immunoprecipitation or pull–down of tagged proteins and probed for the presence of the putative phosphatase. Because inhibitors cannot unequivocally distinguish between PP1 and PP2A, antibodies to both phosphatases should be used to establish definitively the presence of either PP1 or PP2A in the immunocomplex. For both BAD and FOXO1, we repeatedly found PP2A, but never detected PP1, in any of the isolated complexes. Comparing wild-type constructs and mutants with alanine mutations at the phosphorylation sites of interest, such as BAD112A136A155A and FOXO AAA (T24→A, S256→A. S319→A), addresses whether phosphatase recognition of the cell death protein depends on the phosphorylation state. In such interaction studies, one should be leery that phosphatase binding could be inhibited by the presence of phospho-protective proteins, such as 14-3-3, with affinity to the substrate that is so strong as to survive RIPA buffer (such as the case with BAD). To circumvent this problem, complexes can be isolated in the presence or absence of either R18 peptide or Empigen in the lysis buffer, or in survival versus apoptotic conditions, relying on cellular mechanisms to dissociate 14-3-3 after induction of cell death. In proving the BAD and PP2A interaction, we found PP2A/C in the BAD immunocomplex, identifying PP2A itself as the BAD phosphatase. When 14-3-3 was displaced by Empigen BB, R18, or pSer136 peptide, PP2A/C subunit association with BAD was significantly enhanced, indicating that PP2A competes with 14-3-3 for binding to BAD (Fig. 14.4B). In contrast, PP2A/C bound to FOXO1, regardless of whether 14-3-3 is present in the FOXO1 complex, because no difference in PP2A/C signal was observed in the presence or absence of R18 peptide (Fig. 14.4C). However, in this case, binding of PP2A does not mean that it can function, because pT24 remained phosphorylated when 14-3-3 was present. Only when 14-3-3 was displaced from FOXO1 by R18 was the bound PP2A able to dephosphorylate pT24.

For PP1 and PP2A, the regulatory subunit is believed to target the holoenzyme to the substrate, but there are many possible different regulatory subunits (~15), and one usually does not know for which subunit to probe. The catalytic subunits PP2A/C or PP1/C are common to phosphatase holoenzymes, highly abundant, and very easily detectable with existing antibodies. However, PP1 and PP2A have a large number of substrates, only a minute fraction would be bound to any one protein. In addition, the catalytic subunit of PP2A is notoriously sticky and will often nonspecifically bind to reagents in immunoprecipitations or pull-downs. The presence of a faint PP2A/C signal may not give one sufficient confidence that PP2A truly binds in a physiologic manner. We encountered this problem initially with FOXO1 immunoprecipitations and pull-downs. Therefore, we used a two-step process to enrich for bound phosphatases. FLAG-FOXO1 complex was pulled down by anti-FLAG beads, eluted with excess FLAG peptide, then incubated with microcystin-agarose beads. When conjugated to beads, the phosphatase inhibitor microcystin is highly effective in pulling down PP1 and PP2A. The phosphatase that was present in the FLAG-FOXO1 complex would have been eluted along with FOXO1 by FLAG peptide and should be captured by microcystin agarose. Indeed, Western blotting of the microcystin pull-down of the FLAG eluate revealed only PP2A/C, but not PP1, indicating that PP2A was the phosphatase that bound to FLAG-FOXO1 (Fig. 14.5). Interestingly, 14-3-3 was also present in the microcystin pull-down, presumably because it was bound to FLAG-FOXO1, which was bound to PP2A/C. After confirming that there was no direct interaction between PP2A/C and 14-3-3 that we could find, we concluded that, unlike the competitive binding seen for BAD, 14-3-3 and PP2A could bind to FOXO1 at the same time. Thus, phosphatase regulation is not the same for all cell death proteins.

In regard to a potentially weak phosphatase band, crosslinking may enhance the signal and at the same time permit more stringent washing of the immunoprecipitate or pull-down to remove nonspecific associations. In FOXO1 immunoprecipitations, we found that crosslinking the lysates significantly enhanced the signal-to-noise ratio of the PP2A/C band.

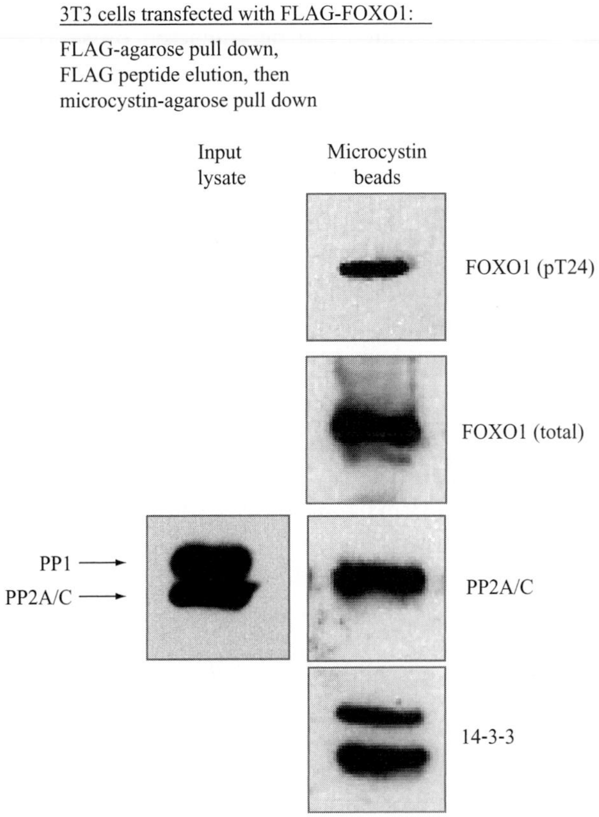

Figure 14.5 Two-step pull-down of FLAG-FOXO1 and PP2A/C. FLAG-FOXO1 complex eluted from FLAG-agarose was subjected to microcystin-agarose pull-down and western blotted.

The scaffolding A subunit does not have the "sticky" problem and usually coexists with the catalytic subunit as a dimer. Detection of the PP2A/A subunit in the immunocomplex corroborates the specificity of PP2A interaction.

Detecting interaction in cells supports that the association is physiologic, but it does not tell us whether the binding is direct. To test for interaction of the phosphatase and the cell death protein in the absence of other cellular factors, recombinant or purified proteins can be mixed *in vitro*. Because phosphatase binding may depend on the phosphorylation state of the substrate, when bacterially produced proteins are used, unphosphorylated and *in vitro* phosphorylated substrates may be compared. We used recombinant GST-BAD *in vitro* phosphorylated by PKA or not, or GST-FOXO1 *in vitro* phosphorylated by Akt or not, and purified PP2A (commercially available) to test for direct interaction. Reciprocally, recombinant GST-PP2A/A and GST-PP2A/C fusion proteins were used to pull-down BAD or FOXO1. Abundant production of GST fusions of pro-apoptotic proteins can be toxic to bacteria, which can be alleviated by short inductions at temperatures lower than 37 °C.

4.1. Methods

4.1.1. Two-step pull-down

1. Transfect NIH3T3 cells with FLAG-FOXO1 by use of lipofectimine 2000. Collect cells after 30 h and lyse in IP buffer (142.5 mM KCl, 5 mM MgCl$_2$, 10 mM HEPES, 0.1% Nonidet P-40, 1 mM PMSF, 10 μg/ml aprotinin, and 10 μg/ml leupeptin). Preclear lysate with agarose for 1 h.
2. Wash FLAG-agarose (Sigma, Cat:A2220, 50% matrix slurry used at 10% of the lysate volume) with IP buffer three times and resuspend in IP buffer. Add FLAG-agarose to lysate and shake at 4 °C for 2 h. Spin down agarose beads at 14,000 rpm for 10 min and wash with IP buffer three times, elute FLAG-FOXO1 from beads with 2 mg/ml FLAG peptide.
3. Add microcystin-agarose (Upstate Cell Signaling solutions, Cat: 16-147, 50% matrix slurry used at 10% of the lysate volume, washed as earlier) to eluted FLAG-FOXO1 mixture and incubate at 4 °C for 2 h. Wash beads with IP buffer three times. Resuspend in loading dye and run on SDS-PAGE for Western blot. The antibodies are listed as following: Anti-Flag M2 (Sigma, Cat: F3165, dilution: 1:1000), anti-phospho-T24-FOXO1, anti-phospho-S256-FOXO1 and anti-PP2A A subunit (Cell signaling Technology, Cat: 9464, Cat: 9461 and Cat: 2041, dilution: 1:500), anti-FOXO1 (Santa Cruz Biotechnology, Cat: sc-11350, dilution: 1:1000), anti-PP2A catalytic subunit (BD Biosciences, Cat: 610555, dilution: 1: 2000), anti-PP1 and anti-14-3-3 (Santa Cruz Biotechnology, Cat: sc-7482 and sc-1657, dilution: 1:1000).

4.1.2. Crosslinking lysates

1. Make lysate as usual. Weigh appropriate amount of water–soluble cross-linker DTSSP (Pierce Biotechnology, Cat: 21578) for a final concentration of 2 mM and add to lysate directly. Pipet several times to dissolve crosslinker. Add protein A + G sepharose (for preclearing) and mix for 2 h at 4 °C.
2. Stop the crosslinking reaction by adding Tris-Cl, pH 7.5, to a final concentration of 20 mM, quenching for 15 min at room temperature.
3. Immunoprecipitate as usual.

5. LINKING PHOSPHATASE ACTIVITY TO CELL DEATH PROTEIN FUNCTION

5.1. Cell death assays

For the specifically identified phosphatase to have physiologic significance, a functional effect of inhibiting the phosphatase must be demonstrated. The most obvious way is to assay for improved survival of cells during apoptosis induction when the particular phosphatase activity is inhibited, either by pharmacologic inhibitors or by siRNA knockdown. The presence of an inhibitor in the medium uniformly affects cells in culture, but the effects of inhibitors are not specific. Knockdowns are specific to a particular phosphatase, but the siRNA may not be taken up by all cells. This can be circumvented by assaying apoptosis of only the cells with siRNA, such as gating on cells with a cotransfected fluorescent indicator. If the effect of phosphatase inhibition is transient or subtle, it can be easily obscured by untransfected cells, which would continue to die as usual. A technical limitation we found is that an efficient PP2A knockdown itself can cause cell death after a few days, probably because of PP2A's involvement in so many signaling pathways, making it nearly impossible to measure PP2A activity on one specific cell death protein in a prolonged time course experiment. In choosing an inhibitor for survival studies, cell permeability and off-target toxicity are considered. For our cell death studies, we chose the phosphatase inhibitor calyculin A for its high cell permeability and low toxicity and fostriecin for its PP2A selectivity. By pretreating FL5.12 cells with 5 nM calyculin A or 3 μM fostriecin in the presence of IL-3, then withdrawing IL-3 from the media but continuing inhibitor treatment, we were able to demonstrate that PP2A inhibition rescued cells expressing wild-type BAD or FOXO1 from IL-3 starvation but not cells expressing phosphorylation-defective mutants. Cell death studies such as this allow one to conclude that the putative phosphatase has an effect on survival that is mediated through the expected phosphorylation sites. However, survival is a final outcome, which can result from any number of factors and does not prove that PP2A action on BAD or FOXO directly causes cell death.

5.2. Measuring intermediate pro-apoptotic functions

Because phosphatase inhibitors can affect many signaling pathways that influence cell death, assaying a specific cell death protein function addresses the significance of phosphatase activity more directly. For example, we examined Bim upregulation as a specific apoptotic function of the transcription factor FOXO1 (Dijkers *et al.*, 2000). IL-3 deprivation causes dephosphorylation of FOXO1, which activates transcription of pro-apoptotic Bim, leading to significantly increased Bim protein levels after IL-3 withdrawal. Demonstrating that Bim level was not upregulated when PP2A was inhibited by treating

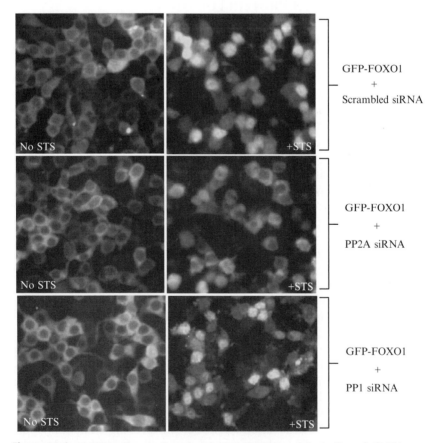

Figure 14.6 BOSC cells transfected with GFP-FOXO1 and the indicated siRNA were treated with 2 μM staurosporine (STS) or not for 1 hour.

IL-3–deprived cells with calyculin A allowed us to conclude that PP2A activity is necessary for FOXO1's transcriptional function in apoptosis.

The effect of phosphatase knockdown can also be assessed directly by a cell death protein function that can be measured shortly after delivery of phosphatase siRNA. Phosphorylation at T24 and S256 localizes FOXO1 to the cytoplasm, where it is transcriptionally inactive. FOXO1 dephosphorylated at these sites translocates into the nucleus, where it can transcribe cell death genes. We specifically tested the effect of PP2A knockdown in FOXO1 nuclear translocation during rapid apoptosis induction by use of GFP-FOXO1. Within 30 min to 1 h of staurosporine treatment ($2\ \mu M$), cytoplasmic GFP-FOXO1 translocates into the nucleus in nearly all cells. In the presence of PP2A siRNA (Dharmacon, Cat: M-003598-00), but not PP1 siRNA (Santa Cruz Biotechnology, Cat: sc-43545), GFP-FOXO1 is still cytoplasmic in most of the cells at 1 h (Fig. 14.6). Measuring an intermediate activity of FOXO1 dephosphorylation by use of PP2A siRNA unequivocally demonstrated a functional role of PP2A in the activation of FOXO1 as a pro-apoptotic transcription factor.

6. CONCLUSION

Our studies of dephosphorylation of the cell death proteins BAD and FOXO1 drew from an extensive field of serine/threonine phosphatase research. We presented how we adapted phosphatase inhibitor experiments, interaction assays, phosphatase knockdowns, as well as functional studies to address issues particular to cell death, and how the results were interpreted. These approaches are by no means exhaustive, other techniques have been used in phosphatase work on other cell death proteins. Data from all of these experiments support the emerging role of phosphatases in the regulation of cell death.

ACKNOWLEDGMENT

Work described in this chapter was supported by a fellowship from the American Heart Association to CWC, a grant to EY from Hope Street Kids Foundation, and NIH RO1CA92498.

REFERENCES

Arden, K. C. (2006). Multiple roles of FOXO transcription factors in mammalian cells point to multiple roles in cancer. *Exp. Gerontol.* **41,** 709–717.
Brunet, A., Bonni, A., Zigmond, M. J., Lin, M. Z., Juo, P., Hu, L. S., Anderson, M. J., Arden, K. C., Blenis, J., and Greenberg, M. E. (1999). Akt promotes cell survival by phosphorylating and inhibiting a Forkhead transcription factor. *Cell* **96,** 857–868.

Chiang, C. W., Harris, G., Ellig, C., Masters, S. C., Subramanian, R., Shenolikar, S., Wadzinski, B. E., and Yang, E. (2001). Protein phosphatase 2A activates the proapoptotic function of BAD in interleukin- 3–dependent lymphoid cells by a mechanism requiring 14–3-3 dissociation. *Blood* **97,** 1289–1297.

Chiang, C. W., Kanies, C., Kim, K. W., Fang, W. B., Parkhurst, C., Xie, M., Henry, T., and Yang, E. (2003). Protein phosphatase 2A dephosphorylation of phosphoserine 112 plays the gatekeeper role for BAD-mediated apoptosis. *Mol. Cell Biol.* **23,** 6350–6362.

Datta, S. R., Dudek, H., Tao, X., Masters, S., Fu, H., Gotoh, Y., and Greenberg, M. E. (1997). Akt phosphorylation of BAD couples survival signals to the cell-intrinsic death machinery. *Cell* **91,** 231–241.

Datta, S. R., Katsov, A., Hu, L., Petros, A., Fesik, S. W., Yaffe, M. B., and Greenberg, M. E. (2000). 14–3-3 proteins and survival kinases cooperate to inactivate BAD by BH3 domain phosphorylation. *Mol. Cell* **6,** 41–51.

Desagher, S., Osen-Sand, A., Montessuit, S., Magnenat, E., Vilbois, F., Hochmann, A., Journot, L., Antonsson, B., and Martinou, J. C. (2001). Phosphorylation of bid by casein kinases I and II regulates its cleavage by caspase 8. *Mol. Cell* **8,** 601–611.

Dijkers, P. F., Medema, R. H., Lammers, J. W., Koenderman, L., and Coffer, P. J. (2000). Expression of the pro-apoptotic Bcl-2 family member Bim is regulated by the forkhead transcription factor FKHR-L1. *Curr. Biol.* **10,** 1201–1204.

Gallego, M., and Virshup, D. M. (2005). Protein serine/threonine phosphatases: Life, death, and sleeping. *Curr Opin Cell Biol.* **17,** 197–202.

Gardai, S. J., Hildeman, D. A., Frankel, S. K., Whitlock, B. B., Frasch, S. C., Borregaard, N., Marrack, P., Bratton, D. L., and Henson, P. M. (2004). Phosphorylation of Bax Ser184 by Akt regulates its activity and apoptosis in neutrophils. *J. Biol. Chem.* **279,** 21085–21095.

Harada, H., Quearry, B., Ruiz-Vela, A., and Korsmeyer, S. J. (2004). Survival factor-induced extracellular signal-regulated kinase phosphorylates BIM, inhibiting its association with BAX and proapoptotic activity. *Proc. Natl. Acad. Sci. USA* **101,** 15313–15317.

Honkanen, R. E., and Golden, T. (2002). Regulators of serine/threonine protein phosphatases at the dawn of a clinical era? *Curr. Med. Chem.* **9,** 2055–2075.

Huang, H., and Tindall, D. J. (2007). Dynamic FoxO transcription factors. *J. Cell Sci.* **120,** 2479–2487.

Janssens, V., Goris, J., and Van Hoof, C. (2005). PP2A: The expected tumor suppressor. *Curr. Opin. Genet. Dev.* **15,** 34–41.

Liao, J., and Omary, M. B. (1996). 14–3-3 proteins associate with phosphorylated simple epithelial keratins during cell cycle progression and act as a solubility cofactor. *J. Cell Biol.* **133,** 345–357.

Masters, S. C., Yang, H., Datta, S. R., Greenberg, M. E., and Fu, H. (2001). 14–3-3 inhibits Bad-induced cell death through interaction with serine-136. *Mol. Pharmacol.* **60,** 1325–1331.

Mumby, M. (2007). The 3D structure of protein phosphatase 2A: New insights into a ubiquitous regulator of cell signaling. *ACS Chem. Biol.* **2,** 99–103.

Oikawa, H. (2002). Synthesis of specific protein phosphatase inhibitors, tautomycin and tautomycetin toward structure-activity relationship study. *Curr. Med. Chem.* **9,** 2033–2053.

Opferman, J. T. (2006). Unraveling MCL-1 degradation. *Cell Death Differ.* **13,** 1260–1262.

Plas, D. R., and Thompson, C. B. (2003). Akt activation promotes degradation of tuberin and FOXO3a via the proteasome. *J. Biol. Chem.* **278,** 12361–12366.

Ruvolo, P. P., Deng, X., and May, W. S. (2001). Phosphorylation of Bcl2 and regulation of apoptosis. *Leukemia* **15,** 515–522.

Wang, B., Yang, H., Liu, Y. C., Jelinek, T., Zhang, L., Ruoslahti, E., and Fu, H. (1999). Isolation of high-affinity peptide antagonists of 14–3-3 proteins by phage display. *Biochemistry* **38,** 12499–12504.

Yaffe, M. B. (2004). Master of all things phosphorylated. *Biochem. J.* **379,** e1–e2.

Yaffe, M. B., Rittinger, K., Volinia, S., Caron, P. R., Aitken, A., Leffers, H., Gamblin, S. J., Smerdon, S. J., and Cantley, L. C. (1997). The structural basis for 14–3-3:phosphopeptide binding specificity. *Cell* **91,** 961–971.

Yan, L., Lavin, V. A., Moser, L. R., Cui, Q., Kanies, C., and Yang, E. (2008). PP2A regulates the pro-apoptotic activity of FOXO1. *J. Biol. Chem.* **283,** 7411–7420.

Zha, J., Harada, H., Yang, E., Jockel, J., and Korsmeyer, S. J. (1996). Serine phosphorylation of death agonist BAD in response to survival factor results in binding to 14–3-3 not BCL-X(L). *Cell* **87,** 619–628.

Zhao, X., Gan, L., Pan, H., Kan, D., Majeski, M., Adam, S. A., and Unterman, T. G. (2004). Multiple elements regulate nuclear/cytoplasmic shuttling of FOXO1: characterization of phosphorylation- and 14–3-3–dependent and –independent mechanisms. *Biochem. J.* **378,** 839–849.

Zhou, X. M., Liu, Y., Payne, G., Lutz, R. J., and Chittenden, T. (2000). Growth factors inactivate the cell death promoter BAD by phosphorylation of its BH3 domain on Ser155. *J. Biol. Chem.* **275,** 25046–25051.

CHAPTER FIFTEEN

ANALYSIS OF NEURONAL CELL DEATH IN MAMMALS

Marcello D'Amelio,* Virve Cavallucci,* Adamo Diamantini,[†] and Francesco Cecconi*

Contents

Abstract

Apoptosis, often defined as programmed cell death, plays a very important role in many physiologic and pathologic conditions. Therefore, detecting apoptotic cells or monitoring the cells progressing to apoptosis is an essential step in basic and/or applied research.

* Laboratory of Molecular Neuroembryology, IRCCS Fondazione Santa Lucia, Rome, Italy and Dulbecco Telethon Institute at the Department of Biology, University of Rome "Tor Vergata", 00133 Rome, Italy
[†] Laboratory of Neuroimmunology, IRCCS Fondazione Santa Lucia, Rome

Methods in Enzymology, Volume 446
ISSN 0076-6879, DOI: 10.1016/S0076-6879(08)01615-7

Apoptosis is characterized by many biologic and morphologic changes of cells, for example, cytochrome *c* release from mitochondria, activation of caspases, DNA fragmentation, membrane blebbing, and formation of apoptotic bodies. On the basis of these changes, various assays have been designed to detect or quantify apoptotic cells.

The goal of this chapter is to provide readers with a scientific guide to proven methods that highlight the current strategies for detecting apoptosis in the nervous system.

1. INTRODUCTION

Programmed cell death (PCD) has been recognized as an important event in the normal development of the mammalian nervous system and during the course of illness, as has been clearly demonstrated in the case of a number of human neurodegenerative disorders and animal models of human disease.

Current knowledge about the genetic regulation of PCD is mainly based on studies of the nematode *Caenorhabditis elegans*. This model system affords a powerful combination of genetic and molecular tools for studying PCD; in fact, *ced-3*, *ced-4*, *ced-9,* and *egl-1* genes have been identified as essential for PCD in worms (Horvitz, 1999). Also, mammalian homologs were discovered; some homologs were found to be part of the apoptotic machinery downstream of cytochrome *c* release, such as caspases (ced-3 homologs) (Nicholson, 1999) and the apoptotic proteases activating factor, Apaf1 (ced-4 homolog) (Zou *et al.*, 1997), whereas others were identified as the mitochondrial anti-apoptotic factors, such as Bcl2 and Bcl-X_L (ced-9 homologs) or pro–apoptotic factors, such as Bax, Bad, and Bid (egl-1 homologs) (Korsmeyer, 1999). The genetic disruption of these genes has demonstrated that they have functions similar to those of *C. elegans* homologs, thus opening a window for analysis of molecular mechanisms of mammalian neuronal death.

In this chapter, we describe the methods that have been successfully used to detect apoptosis in basic research. First, we describe the assay for studying cell death by use of two cell systems: (1) primary neuronal cultures derived from perinatal individuals and (2) cell lines of embryonic origin (ETNA cells). Second, we describe techniques to detect apoptosis during brain development. Finally, we describe a method to visualize the early apoptosis event(s) in synapses of mouse adult brain.

2. APOPTOSIS DETECTION *IN VITRO*

In this section we describe the methods to detect neuronal apoptosis in two cell systems: primary neurons and cell lines of embryonic origin (ETNA cells). The examples we report here are provided with the appropriate controls (i.e., embryos expressing the human amyloid precursor protein harboring the Swedish familial Alzheimer's disease mutation, APPswe, and embryos expressing APPswe transgene but apoptosome-deficient caused by Apaf1 inactivation are compared). The ETNA cells, as well, are $+/+$ (wild type) or $-/-$ (*Apaf1*-deficient).

We observe apoptosis induction on two different apoptotic stimuli (APPswe transgene expression for primary neurons and staurosporine for ETNA cells) and, in addition, we show that these stimuli induce cell death by means of the mitochondrion-apoptosome pathway.

2.1. Reagents for primary neuronal cell culture and ETNA cell culture

- *Dissection solution*: 0.025% trypsin (Trypsin, liquid, GIBCO 15090-046) diluted in $1\times$ PBS.
- MEM#1 medium composition: 25 mM glucose (D-(+)-glucose, Sigma-Aldrich G5400); 1% FBS (fetal bovine serum, heat inactivated, GIBCO 10500); 2 mM glutamine (L-glutamine solution, Sigma-Aldrich G7513); 0.1 mg/ml gentamicin (Gentamicin, liquid, GIBCO 15750); MEM medium (minimum essential medium, liquid, GIBCO 21090). NOTE: Sterilize immediately by filtration by use of a 0.22-μm filter, and place the medium in the incubator for 30 min to allow it to equilibrate to 37 °C and 5% CO_2.
- MEM#2 medium composition: 25 mM glucose; 5% FBS; 5% HS (horse serum, heat inactivated, GIBCO 26050); 2 mM glutamine; 0.1 mg/ml gentamicin; MEM medium. NOTE: Sterilize immediately by filtration with a 0.22-μm filter and place the medium in the incubator for 30 min to allow it to equilibrate to 37 °C and 5% CO_2.
- Neurobasal medium composition: 2% B-27 minus AO (B-27 Supplement Minus AO, liquid, GIBCO 10889-038); 2 mM glutamine; 0.1 mg/ml gentamicin; neurobasal medium (Neurobasal^{TM}-A Medium, liquid, GIBCO 10888-022). NOTE: Place the medium in the incubator for 30 min to allow it to equilibrate to 37 °C and 5% CO_2.
- DMEM medium composition: 10% FBS; 1:100 P/S (penicillin-streptomycin solution, liquid, GIBCO 15070); DMEM medium (D-MEM, liquid, GIBCO 31966).

2.2. Primary neuronal cell culture preparation

1. Sacrifice pregnant mouse by cervical dislocation at embryonic day (E) 14.5. In our case we use *Apaf1*$^{+/-}$ female crossed with a double transgenic male (APPswe^{+} and *Apaf1*$^{+/-}$).

2. Take out the uterus and place it in cold $1\times$ PBS. Remove the embryos from the uterus and separate them from placenta and yolk sac; place the embryos in a fresh dish of cold $1\times$ PBS.

3. Decapitate the embryos and place the heads in a dissection dish, then remove the brains and place them in a fresh dissection dish.

4. Remove the cerebella and use them for DNA extraction and successive genotyping (in our case, APPswe$^+$|*Apaf1*$^{+/+}$ and APPswe$^+$|*Apaf1*$^{-/-}$ embryos).

5. Collect the cortices and incubate them in 1.5 ml dissection solution for 7 min at 37 °C.

6. Add 150 μl FBS (100 μl for 1 ml of dissection solution) to inhibit trypsin activity; centrifuge 3 min at 220g and discard the supernatant.

7. Add 2 ml MEM#1 medium and triturate the tissue by pipetting it seven times through a 1000-μl pipette; pipet slowly to avoid formation of bubbles. Centrifuge 10 min at 220g and discard the supernatant.

8. Resuspend the pellet in 2 ml MEM#2 medium by pipetting it once through a 1000-μl pipette and count the cells by use of a hemocytometer.

9. Plate the cells in the well of a 6-well plate (polylysine coated): plate 10^6 cells in 2 ml MEM#2 medium and place the multiwell plate in a humidified 37 °C, 5% CO_2 incubator.

10. One hour after plating, replace the medium with 2 ml neurobasal medium and continue to incubate plates for up 1 to 2 weeks.

2.3. ETNA cell culture

ETNA cells were derived from striatum of E14.5 *Apaf1*$^{-/-}$ or wild-type embryos, as described previously (Cozzolino *et al.*, 2004).

ETNA cells are routinely grown in DMEM medium, at 33 °C in an atmosphere of 5% CO_2 in air.

3. ASSESSMENT OF CELL DEATH

Under specific apoptotic stimuli, a number of pores are created in the mitochondrial membrane and bring about its permeabilization. The result of this permeabilization is the release, into the cytoplasm, of cytochrome *c*, which, in the presence of dATP, induces the formation of the Apaf1-containing macromolecular complex called the apoptosome. This complex, in turn, binds and activates procaspase-9. Mature caspase-9 remains bound to the apoptosome, recruiting and activating executioner caspase-3 and/or caspase-7. The activation of caspase-3 and the cleavage of its nuclear substrates kill the cells (Green and Kroemer, 2004).

Consequently, cytochrome c release, caspase-3,7 activation, the detection of substrates cleaved by caspase-3, and, finally, the fragmentation of DNA are hallmarks of apoptosis.

3.1. Cytochrome c Release

Cytochrome c, in physiologic conditions, localizes into the internal mitochondrial membrane and, therefore, colocalizes with mitochondrial markers (we use MnSOD). On apoptotic stimuli, cytochrome c is released into the cytoplasm; in these experimental conditions, the cytochrome c and MnSOD have a different pattern that we can document by means of double-labeling confocal immunofluorescence microscopy (see Fig. 15.1).

3.1.1. Reagents for the detection of cytochrome c release

- Staurosporine, STS (Sigma–Aldrich S4400).
- Blocking solution composition: 2% horse serum in 1× PBS.
- Primary antibodies
  Rabbit anti-manganese superoxide dismutase polyclonal antibody, Anti-MnSOD (Stressgene SOD-110); dilute 1:100.
  Purified mouse anti-cytochrome c monoclonal antibody, clone 6H2.B4 (BD Biosciences Pharmingen 556432); dilute 1:100.
- Secondary antibodies
  Goat anti-mouse IgG, Alexa Fluor® 488 (Molecular Probes A-11017); dilute 1:300.

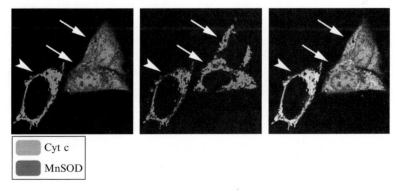

Figure 15.1 Cytochrome c detection in ETNA$^{+/+}$ cells is abolished after STS treatment. Double-labeling immunofluorescence of ETNA$^{+/+}$ cells treated for 12 h with 5 μM staurosporine (STS); images were taken by a confocal microscope. Cytochrome c (green), MnSOD (red), and merged patterns are shown. The white arrowhead points to a cell with cytochrome c localized into the internal mitochondrial membrane, and the white arrows indicate cells with released cytochrome c. (See color insert.)

Donkey anti-rabbit IgG, Cy^{TM}3-conjugated (Jackson ImmunoResearch Laboratories 711-166-152); dilute 1:300.

DAPI solution: 1 μg/ml DAPI (Sigma–Aldrich D9542) diluted in 1 × PBS.

3.1.2. Procedure for the detection of cytochrome *c* release

1. Plate the cells in DMEM medium and, at 80% confluence, add apoptotic stimulus: 5 μM staurosporine.
2. 12 h after apoptosis induction, wash the cells in PBS and fix with 4% paraformaldehyde in 1× PBS for 15 min.
3. Permeabilize with 0.2% Triton X-100 in 1× PBS for 5 min and incubate for 30 to 60 min in blocking solution.
4. Incubate for 1 h at 37 °C with primary antibody diluted in blocking solution.
5. Wash the cells with blocking solution and incubate for 1 h with labeled secondary antibody.
6. After incubation, wash twice with PBS and treat for 5 min in the dark with DAPI solution.
7. Wash twice with 1× PBS and apply a coverslip with aqueous mounting medium suitable for fluorescence (Gel/Mount, Biomeda M01) and analyze the cells by use of confocal microscopy (see Fig. 15.1).

3.2. Detection of caspase-3 activity

The caspases are synthesized as inactive pro-enzymes (or pro-caspases) and, on apoptotic stimuli, they are processed by proteolytic cleavage to form active enzymes. Caspase-3 exists in cells as an inactive 32-KDa pro-enzyme; in apoptotic conditions, it is cleaved into active 17- and 19-KDa subunits. Active caspase-3 plays a critical role in the execution phase of apoptosis. Important targets of caspase-3 include nuclear proteins, such as poly ADP ribose polymerase (PARP).

We can monitor the caspase-3 activation by use of three complementary techniques:

1. Detection of cleaved caspase-3 active form (Fig. 15.2A)
2. Examination of cleavage of candidate substrates (Fig. 15.2B)
3. Assessment of cleavage of small fluorogenic substrates (Fig. 15.3).

3.2.1. Techniques I and II: Immunoblotting analysis to detect cleaved caspase-3 and cleaved PARP
3.2.1.1. Reagents for immunoblotting analysis

- TBS-T composition: 20 mM Tris-HCl, pH 7.5; 160 mM NaCl; 0.1% Tween 20.
  Blocking solution: 5% skim milk in TBS-T

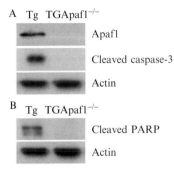

Figure 15.2 Active caspase-3 detection in APPswe⁺ transgenic neurons. (A) Total lysate prepared from APPswe⁺ transgenic neurons at 12 days *in vitro* (DIV) were subjected to SDS-PAGE and analyzed for active caspase-3. (B) Total lysate was prepared from APPswe⁺ transgenic neurons (12 DIV), it was subjected to SDS-PAGE, and analyzed for PARP processing. Note that in total lysate of neurons lacking Apaf1, caspase-3 is not cleaved and that PARP, a substrate of caspase-3, is not processed.

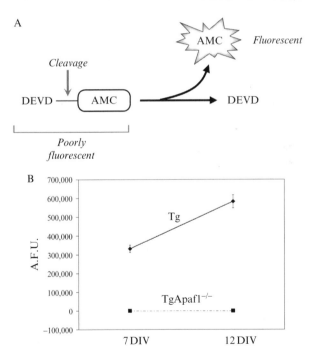

Figure 15.3 Detection of caspase activity with fluorogenic substrate. (A) Scheme showing release of AMC from a tetrapeptide-coupled substrate as a consequence of caspase activation. (B) Caspase-3 activity in APPswe⁺ primary neurons at 7 and 12 days *in vitro* (DIV). Lack of caspase-3 activity in transgenic neurons lacking Apaf1. Caspase-3 activity is expressed as arbitrary fluorescence units (AFU).

- Primary antibodies

    Cleaved caspase-3 (Asp175) antibody (Cell Signaling 9661); dilute 1:1000.

    Cleaved PARP (Asp214) antibody (mouse specific) (Cell Signaling 9544); dilute 1:1000.

    Monoclonal antibody to Apaf1 (18H2) (Alexis Biochemicals ALX-804-350); dilute 1:1000.

    Anti-actin (Sigma-Aldrich A5060); dilute 1:10000.

- Secondary antibody

    Goat anti-rabbit IgG (H + L) horseradish peroxidase conjugate (BIO-RAD 170-6515); dilute 1:3000.

    Anti-rat IgG (whole molecule) peroxidase conjugate (Sigma-Aldrich A5795); dilute 1:3000.

- ECL Plus^{TM} Western blotting detection reagents (Amersham RPN2132).

3.2.1.2. Immunoblotting procedure

1. Load cell lysates (4 to 10 μg of proteins) onto SDS-PAGE gel and blot gel onto PVDF membrane (Immobilon-P, Millipore IPVH20200).
2. Wash membrane with TBS-T and incubate 1 h at room temperature with blocking solution.
3. Incubate overnight at 4 °C with primary antibody, dilute in 2% skim milk in TBS-T.
4. Wash three times with TBS-T (10 min each) and incubate 1 h at room temperature with secondary antibody diluted in 1% skim milk in TBS-T.
5. Wash three times with TBS-T (10 min each) and reveal target proteins with the ECL (enhanced chemoluminescence) system (see Fig. 15.2).

3.2.2. Technique III: Cleavage of peptide substrate

In this procedure, the activity of caspase-3 in cultured primary neurons is measured in an indirect manner. The fluorogenic substrate, Ac-DEVD-AMC, is cleaved by caspase-3, and the cleavage can be monitored by excitation at 380 nm and emission at 460 nm.

3.2.2.1. Reagents for caspase-3 activity assay

- Lysis buffer for caspase-3 assay composition: 100 mM HEPES, pH 7.4; 0.1% Chaps; 1 mM EDTA; 1 mM PMSF; 10 mM DTT. NOTE: This solution can be stored at 4 °C, but PMSF and DTT have to be added immediately before use.
- Fluorogenic substrate: Ac-DEVD-AMC (caspase-3 substrate II, fluorogenic, Calbiochem 235425).

3.2.2.2. Caspase-3 activity assay protocol

1. At a given time point, stop the incubation of cell culture, place the plate in ice, remove the medium, and wash twice with $1\times$ PBS.
2. Add 20 μl lysis buffer for caspase-3 assay (this volume is recommended for a well of 6-well plate or 35-mm plate), scrape the cells, and collect them in Eppendorf tubes.
3. Freeze the samples in nitrogen liquid and thaw at 37 °C three times.
4. Centrifuge 5 min at $11,500g$ in a refrigerated centrifuge at 4 °C.
5. Recover the supernatant and determine concentration of solubilized protein by Bradford method (BIO-RAD protein assay).
6. Use 10 to 20 μg of protein, adjust to 50 μl with lysis buffer for caspase-3 assay and add 50 μl of buffer containing 100 μM fluorogenic substrate (final 50 μM).
7. Aliquot the samples in the wells of 96-well plate and incubate in the dark for 30 min at 37 °C.
8. Measure the fluorescence with 380 nm excitation wavelength and 460 nm emission wavelength (see Fig. 15.3).

3.3. Flow cytometry in the study of apoptosis

In the 30 years since it was developed, flow cytometry has evolved into a technology capable of rapidly analyzing and separating cells on the basis of physical, biochemical, immunologic, or functional properties.

In this section we describe an application of flow cytometry in identifying distinct cellular populations in different phases of the apoptotic pathway.

As outlined previously, one of the earliest features of apoptosis is a morphologic change in the plasma membrane. Viable cells maintain an asymmetric distribution of different phospholipids between the inner and outer layer of the plasma membrane. The distribution of phospholipids in the plasma membrane changes during early steps of apoptosis, and this involves the translocation of the membrane phospholipid phosphatidylserine (PS) from the internal layer to the external layer of the cell membrane. In presence of Ca^{2+}, Annexin V has a high specificity and affinity for PS. Thus, the binding of Annexin V to cells with exposed PS provides a highly sensitive method for detecting early cellular apoptosis (van Engeland et al., 1998; Vermes et al., 1995). In late apoptosis, membrane permeabilization allows Annexin V to enter the cell and bind to PS on the membrane's cytoplasmic face and the fluorescent DNA-binding dye propidium iodide (PI) to enter the nucleus. Consequently, by use of dual staining (Annexin V and propidium iodide) three populations of cells are distinguishable in two-color flow cytometry (Fig. 15.4).

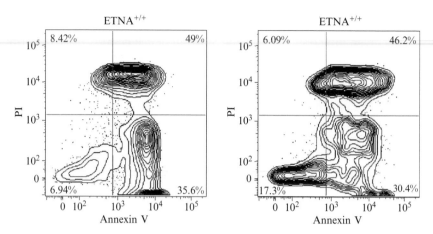

Figure 15.4 Quadrant analysis from flow cytometry apoptosis assay for ETNA cells. A slight increase (49 vs 46.2%) in the PI-positive and Annexin V–positive cells, quadrant III, indicates late apoptosis/necrosis of ETNA$^{+/+}$ cells exposed to 5 μM STS for 24 h. In comparison, in ETNA$^{-/-}$ cells the same treatment produces a higher percentage (17.3 vs 6.94%) in quadrant I, representing cells in which the plasma membrane was intact (PI-negative) and annexinV-negative (cells neither apoptotic nor necrotic).

3.3.1. Flow cytometry reagents

Staurosporine, STS (Sigma–Aldrich S4400).
Trypsin-EDTA (GIBCO 25300).
Binding buffer composition: 10 mM HEPES-NaOH, 150 mM NaCl, 5 mM KCl, 1mM MgCl$_2$, 1.8 mM CaCl$_2$. NOTE: Prepare it fresh each time.
FITC Annexin V (BD Pharmingen^{TM} 556419).
Propidium iodide (Sigma–Aldrich P4170).

3.3.2. Flow cytometry procedure

1. Maintain ETNA cultures in the same conditions as described previously. Before treatment, cells were seeded at $4 \times 10^4/cm^2$.
2. After overnight culture, add to medium culture 5 μM staurosporine (STS) dissolved in DMSO.
3. After 24 h of exposure to STS, collect the cells by trypsinization, and centrifuge at 180g for 5 min, remove the medium and wash twice with 1× PBS.
4. Resuspend in 100 μl binding buffer, add 1 mg/ml FITC Annexin V and incubate 10 min in the dark at room temperature.
5. Add to the cell suspension 1 mg/ml propidium iodide and analyze by flow cytometry with appropriate software (FACS Canto and FACS Diva; Becton Dickson, San Jose, CA).

6. Plot scatter profiles for fluorescence intensity in FL-1 (Annexin V) and FL-2 (PI).
7. Determine the percentage of cells in each quadrant with Flowjo software, version 8.4 (see Fig. 15.4).

4. Apoptosis Detection in Brain Development

In the context of neural development, programmed cell death refers to the physiologically occurring cell death seen at various stages of development in different neural populations.

The potential significance of this cell death for normal nervous system morphogenesis was proposed approximately 70 years ago (reviewed in Cowann *et al.*, 1984; Hamburger, 1992).

The term "programmed cell death" is not entirely synonymous with "apoptosis," even though most naturally occurring cell death in the developing brain fulfills "apoptotic" criteria.

In this section we describe two histochemical methods to detect apoptotic cells in the wild-type mouse developing nervous system and in the homozygous mutant mouse ($Apaf1^{-/-}$) lacking a pivotal apoptosis gene (Cecconi *et al.*, 1998).

4.1. Terminal deoxynucleotidyltransferase (TdT)–mediated dUTP nick end labeling (TUNEL)

The TUNEL assay was developed for apoptotic cells detection in histologic sections. The labeling technique in the TUNEL assay is functionally based on the activity of the TdT enzyme, which transfers dUTP to the $3'$-OH groups of fragmented DNA.

Normal nuclei have very low levels of $3'$-OH breaks and, in contrast to apoptotic nuclei, do not produce an appreciable signal (Fig. 15.5A).

4.1.1. TUNEL reagents

- Proteinase K (20 mg/ml) (Boehringer Mannheim 745723).
- Equilibration buffer composition: 100 mM sodium cacodylate, 0.1 M dithiothreitol, 5 mM cobalt chloride, 30 mM Tris-NaOH, pH 7.2.
- Reaction buffer composition: 1 μM digoxigenin-dUTP (Boehringer Mannheim 1093088) in equilibration buffer.
- TdT enzyme (Boehringer Mannheim 220582).
- Stop solution composition: 300 mM NaCl, 30 mM sodium citrate.
- Peroxidase conjugated anti-digoxigenin antibody: anti-digoxigenin-POD, Fab fragments (Roche 11207733910).

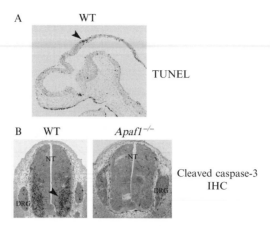

Figure 15.5 Apoptosis detection in mouse brain development. (A) TUNEL staining on sagittal sections from E9.5 mouse embryo. Paraffin-embedded sections of mouse embryo were deparaffinized, rehydrated, and treated with proteinase K before TUNEL staining. The dUTP-labeled fragmented DNA was then visualized by use of a fluorescein-labeled antidigoxigenin antibody with green fluorescence. Image was taken with a fluorescence microscope, converted into grey scale, and inverted; the TUNEL-positive cells in the neural plate of midbrain are shown as black dots (black arrowhead). (B) Immunohistochemical staining of cleaved caspase-3 on paraffin-embedded coronal sections from E10.5 wild type and *Apaf1*$^{-/-}$ litter mates. Images were taken with a fluorescence microscope, converted into grey scale, and inverted; the immunopositive cells in the wild-type embryo are shown as black dots (black arrowhead); note that in *Apaf1*$^{-/-}$ active caspase-3 is not detectable.

4.1.2. TUNEL protocol

1. Excise the uterus of pregnant mouse (approximate gestational day, 15.5) and place it in cold 1× PBS. Remove the embryos from the uterus and separate from the placenta and yolk sac. The yolk sac is used for genomic DNA extraction and genotyping.
2. Fix the embryos with 4% (w/v) paraformaldehyde (overnight, 4 °C).
3. Rinse the embryos with 1× PBS, dehydrate, and include them in paraffin.
4. Prepare 10-μm sections by use of a microtome and store slices at room temperature.
5. Deparaffinize and rehydrate tissue sections.
6. Tissue pretreatment
   - Apply fresh proteinase K (20 μg/ml) directly on the tissue section and incubate for 15 min at room temperature.
   - Wash with 1× PBS for 2 min, twice.
   - Gently aspirate around the section and immediately apply 100 μl of equilibration buffer and incubate at room temperature for at least 30 min.

7. Gently aspirate around the section and immediately pipette onto the section 100 μl of TdT solution (20 U/100 μl) prepared in reaction buffer. Incubate in a humid chamber at 37 °C for 1 h.
8. Stop labeling reaction by incubating the section in stop buffer for 15 min at room temperature. Wash three times with 1× PBS, 5 min each.
9. Gently tap off excess PBS, apply anti-digoxigenin conjugate antibody, and incubate in the dark at room temperature for 30 min.
10. Wash the sections four times with distilled water (5 min each) and mount under a glass coverslip with aqueous mounting medium suitable for fluorescence (Gel/Mount, Biomeda M01).
11. Analyze by fluorescence microscopy with appropriate excitation and emission filters (Fig. 15.5A).

4.2. Cleaved caspase-3 immunostaining

Caspase-3 is a crucial mediator of apoptosis, and it is essential for normal brain development (Kuida et al., 1996). Here we describe an immunohisto-chemical analysis of paraffin-embedded mouse embryo by use of cleaved caspase-3 (Asp175) antibody to detect neuronal cells undergoing apoptosis.

4.2.1. Reagents for caspase-3 immunostaining

- PBS/Triton solution composition: 0.3% Triton X-100 in 1× PBS.
- Sodium citrate buffer composition: 0.1 M sodium citrate trisodium salt dihydrate ($C_6H_5Na_3O_7 \cdot 2H_2O$). Adjust pH 6.0.
- Blocking solution composition: 5% normal goat serum in PBS/Triton solution.
- Primary antibody
  Cleaved caspase-3 (Asp175) antibody (Cell Signaling 9661); dilute 1:100.
- Secondary antibody
  Donkey anti-rabbit IgG Alexa Fluor®555 (Molecular Probes A-21425)

4.2.2. Caspase-3 immunostaining procedure

1. Deparaffinize, rehydrate, and wash tissue sections in dH_2O for 5 min.
2. Antigen unmasking
   - Place sections in 10 mM sodium citrate buffer, pH 6.0, at room temperature.
   - Bring sections to the boiling point in sodium citrate buffer by use of microwave, then maintain at 95 to 99 °C for 10 min.
   - Cool sections for 30 min on bench top.
   - Rinse sections in dH_2O three times (5 min each) and aspirate around the section dH_2O.

3. Block each section with 100 μl blocking solution for 1 h at room temperature.
4. Remove blocking solution and add 50 μl primary antibody to each section (dilute antibody in blocking solution).
5. Gently aspirate around the section the antibody solution and rinse sections in dH_2O three times (5 min each).
6. Add 50 μl fluorochrome-conjugated secondary antibody diluted in PBS/Triton and incubate in the dark at room temperature for 1 h.
7. Remove secondary antibody solution and wash sections three times with wash buffer (5 min each).
8. Apply a coverslip with aqueous mounting medium suitable for fluorescence (Gel/Mount, Biomeda M01) and analyze the sections by confocal microscopy with appropriate excitation wavelength, depending on fluorochrome (see Fig. 15.5B).

5. DETECTION OF SYNAPTIC APOPTOSIS (SYNAPTOSIS)

Emerging data from neurologic studies indicate that, in the neuron, sublethal apoptotic activity can lead to a limited form of apoptosis in synapses, thus causing synaptic failure without neuronal cell death. For example, in Alzheimer's disease, a synaptic apoptosis (also termed synaptosis) is thought to cause an early loss of synapses leading to the initial cognitive decline.

With this preamble, we here describe another application of the flow cytometry that makes it possible to identify "apoptotic synapses" with a mouse brain crude synaptosomal fraction (CSF) preparation.

Several excellent references provide detailed information on this experimental system, and recent articles from the laboratory of Gregory M. Cole are the best in this regard (Gylys *et al.*, 2004a,b).

As mentioned previously, we use the apoptosis marker Annexin V to measure phosphatidylserine (PS) exposure and the viability marker calcein AM to analyze only "viable synaptosomes."

5.1 Reagents synaptic apoptosis detection

- Homogenization buffer composition: 4 mM HEPES, pH 7.4; 320 mM sucrose; 1 mM EGTA; 1 mM PMSF; 1× protease inhibitor cocktail (Sigma-Aldrich P8340). NOTE: This solution can be stored at 4 °C, but PMSF and DTT have to be added immediately before to use.
- TBS-T composition: 20 mM Tris-HCl, pH 7.5; 160 mM NaCl; 0.1% Tween 20.

- Primary antibodies

    Mouse anti-postsynaptic density protein 95 (PSD-95) monoclonal antibody (Chemicon MAB1598), dilute 1:1000.

    Monoclonal anti-ß-tubulin (Sigma-Aldrich T4026), dilute 1:5000.

- Secondary antibody

    Goat anti-mouse IgG (H + L) horseradish peroxidase conjugate (BIO-RAD 170-6516).

- ECL Plus^{TM} Western blotting detection reagents (Amersham RPN2132).

- Krebs–Ringer phosphate buffer (KRP buffer) composition: 118 mM NaCl; 5 mM KCl; 4 mM MgSO$_4$; 1 mM CaCl$_2$; 1 mM KH$_2$PO$_4$; 16 mM sodium phosphate buffer, pH 7.4; 10 mM glucose.

- Staurosporine (Sigma-Aldrich S4400).

- PE Annexin V (BD Pharmingen^{TM} 556421).

- Calcein acetoxymethyl (AM) ester (Molecular Probes C3100MP).

5.2. Synaptic apoptosis detection procedure

1. Sacrifice the mouse by cervical dislocation, dissect the brain and place it in a dissection dish.
2. Rapidly place the brain, without cerebellum, in cold homogenization buffer.
3. Homogenize the tissues in 3.6 ml cold homogenization buffer with 10 strokes of a tight-fitting glass Douce tissue grinder (7 ml; Wheaton).
4. Transfer the homogenate in Eppendorf tubes (store an aliquot for total protein extraction).
5. Centrifuge 10 min at 1000g in a refrigerated centrifuge at 4 °C.
6. Recover the supernatant and centrifuge it 15 min at 12,000g in a refrigerated centrifuge at 4 °C.
7. Discard the supernatant, resuspend the pellet in 2.5 ml homogenization buffer and centrifuge 15 min at 13,000g in a refrigerated centrifuge at 4 °C.
8. For CSF characterization, resuspend the final pellet in homogenization buffer and determine concentration of solubilized protein by Bradford method (BIO-RAD Protein Assay).
9. The CSF characterization is performed by immunoblotting assay loading 2 μg of proteins onto SDS-PAGE gel and with a primary antibody against post-synaptic density protein 95 (PSD-95). This protein is used to evaluate the enrichment of synaptosomes in the CSF preparation and the ß-tubulin is used as loading control (see Fig. 15.6A).
10. For FACS analysis of CSF, wash the final pellet with 2 ml KRP by centrifuging it 4 min at 2000g in a refrigerated centrifuge at 4 °C.

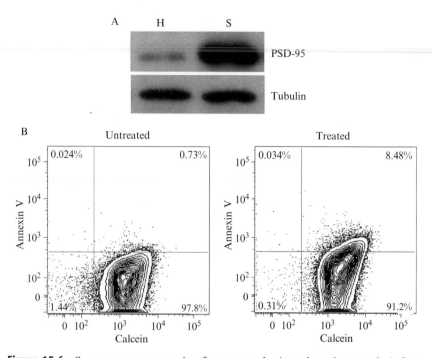

Figure 15.6 Synaptosome preparation from mouse brain and quadrant analysis from flow cytometry apoptosis assay for synaptosomes. (A) Representative Western blot analysis of brain homogenate, H, and synaptosomes, S. Equal amounts of protein were loaded and PSD-95 protein was used as a marker of synapse enrichment. (B) The plot shows calcein AM fluorescence vs Annexin V fluorescence for synaptosomes untreated or treated with STS. The percentage of total particles is shown for each quadrant. An increase (8.48 vs 0.73%) in the Annexin V–positive particles (quadrant III) indicates early apoptosis of synaptosomes.

11. Discard the supernatant and resuspend the pellet in 1 ml KRP.
12. Incubate with 0.1 μM staurosporine for 30 min at 32 °C.
13. Centrifuge 4 min at 2000g, discard the supernatant, and resuspend the pellet in 500 μl KRP.
14. Add to 20 μl of this resuspension the dye (100 nM calcein AM or 1 mg/ml PE Annexin V) diluted in KRP buffer (final volume 100 μL).
15. Incubate for 10 min at room temperature, then dilute in 1× PBS (final volume 500 μl) for immediate flow cytometry analysis.
16. Run on a FACS Canto flow cytometer polystyrene beads (1.4 and 4.5 μm) to standardize forward scatter measurements then acquire the samples.
17. Analyze particles whose size is 1.4 and 4.5 μm, according to previous studies (Gylys *et al.*, 2004b).
18. Determine the percentage of particles in each quadrant by use of Flowjo software, version 8.4 (see Fig. 15.6B).

6. Conclusions

Considering that neuronal apoptosis is an integral part of life, it is obvious that it could play a role in the pathogenesis of neurologic disorders. Diseases such as neuroblastomas, gliomas, and glioblastomas are linked to suppression of apoptosis. Other diseases are linked to increased apoptosis such as bacterial infections (e.g., *Neisseria meningitidis*), autoimmune disorders (e.g., multiple sclerosis), stroke, and neurodegenerative disorders (e.g., Alzheimer's disease). Consequently, great interest has arisen in devising therapeutic strategies for regulating apoptotic processes (D'Amelio *et al.*, 2008; Fischer and Schulze-Osthoff, 2005).

The methods previously described for studying neuronal mammalian cell death offer opportunities not only to observe, "ante mortem" and "in real time", the progression of apoptotic processes *in vivo* or *ex vivo* (bioptic specimen) but they also offer a way to evaluate the efficacy of novel therapeutic agents designed to prevent or cure pathologic conditions in which the equilibrium of life-and-death is dysregulated.

ACKNOWLEDGMENTS

We are grateful to Elisabetta Ferraro for the images of Fig. 15.1 and to Martin Wilmot Bennett for excellent editorial work. Our research is supported by Fondazione Telethon, Compagnia di San Paolo, MUR, and AIRC. FC is an Associate Telethon Scientist.

REFERENCES

Cecconi, F., Alvarez-Bolado, G., Meyer, B. I., Roth, K. A., and Gruss, P. (1998). Apaf1 (CED-4 homolog) regulates programmed cell death in mammalian development. *Cell* **94,** 727–737.

Cowan, W. M., Fawcett, J. W., O'Leary, D. D. M., and Stanfield, B. B. (1984). Regressive event in neurogenesis. *Science* **225,** 1258–1265.

Cozzolino, M., Ferraro, E., Ferri, A., Rigamonti, D., Quondamatteo, F., Ding, H., Xu, Z. S., Ferrari, F., Angelici, D. F., Rotilio, G., Cattaneo, E., Crrì, M. T., *et al.* (2004). Apoptosome iactivation rescues proneural anneural cells from neurodegeneration. *Cell Death Differ.* **11,** 1179–1191.

D'Amelio, M., Tino, E., and Cecconi, F. (2008). The apoptosome: Emerging insights and new potential targets for drug design. *Pharm. Res.* **25,** 740–751.

Fischer, U., and Schulze-Osthoff, K. (2005). Apoptosis-based therapies and drug targets. *Cell Death Differ.* **12,** 942–961.

Green, D. R., and Kroemer, G. (2004). The pathophysiology of mitochondrial cell death. *Science* **305,** 626–629.

Gylys, K. H., Fein, J. A., Wiley, D. J., and Cole, G. M. (2004a). Rapid annexin-V labeling in synaptosomes. *Neurochem. Int.* **44,** 125–131.

Gylys, K. H., Fein, J. A., Yang, F., and Cole, G. M. (2004b). Enrichment of presynaptic and postsynaptic markers by size-based gating analysis of synaptosome preparations from rat and human cortex. *Cytometry A.* **60,** 90–96.

Hamburger, V. (1992). History of the discovery of neuronal death in embryos. *J. Neurobiol.* **23,** 1116–1123.

Horvitz, H. R. (1999). Genetic control of programmed cell death in the nematode. *Caenorhabditis elegans. Cancer Res.* **59,** 1071s–1706s.

Korsmeyer, S. J. (1999). BCL2 gene family and the regulation of programmed cell death. *Cancer Res.* **59,** 1693s–1700s.

Kuida, K., Zheng, T. S., Na, S., Kuan, C., Yang, D., Karasuyama, H., Rakic, P., and Flavell, R. A. (1996). Decreased apoptosis in the brain and premature lethality in CPP32-deficient mice. *Nature* **384,** 368–372.

Nicholson, D. W. (1999). Caspase structure, proteolytic substrates, and function during apoptotic cell death. *Cell Death Differ.* **6,** 1028–1042.

van Engeland, M., Nieland, L. J. W., Ramaekers, F. C. S., Schutte, B., and Reutelingsperger, C. P. M. (1998). Annexin V-affinity assay: A review of an apoptosis detection system based on phosphatidylserine exposure. *Cytometry* **31,** 1–9.

Vermes, I., Haanen, C., Steffens-Nakken, H., and Reutelingsperger, C. P. M. (1995). A novel assay for apoptosis flow cytometric detection of phosphatidylserine expression on early apoptotic cells using fluorescein labeled annexin V. *J. Immunol. Methods* **184,** 39–51.

Zou, H., Henzel, W. J., Liu, X., Lutschg, A., and Wang, X. (1997). Apaf-1, a human protein homologous to *C. elegans* CED-4, participates in cytochrome cdependent activation of caspase-3. *Cell* **90,** 405–413.

CHAPTER SIXTEEN

DISSECTING APOPTOSIS AND INTRINSIC DEATH PATHWAYS IN THE HEART

Danielle Weidman,* James Shaw,[†] Joseph Bednarczyk,*
Kelly M. Regula,* Natalia Yurkova,* Tong Zhang,*
Floribeth Aguilar,* *and* Lorrie A. Kirshenbaum*,[†],[‡]

Contents

Abstract

The limited regenerative capacity of postnatal ventricular myocytes coupled with their meager ability for genetic manipulation has presented a major technical obstacle for deciphering apoptosis initiation and execution signals in the heart. In this report, we describe the technical approaches used to study the intrinsic death pathways in postnatal ventricular myocytes during acute hypoxic injury. Discussed are methods for hypoxia, recombinant adenovirus–mediated gene transfer, cellular viability assays using the vital dyes calcein acetomethoxyester and ethidium homodimer-1, analysis of nuclear morphology by use of Hoechst dye 33258, and assessment of the state of the mitochondrial permeability transition pore. Our work has established that hypoxia triggers perturbations to mitochondria consistent with loss of mitochondrial membrane potential, permeability transition pore opening, and apoptotic cell death by the intrinsic pathway.

* Department of Physiology, Faculty of Medicine University of Manitoba, Winnipeg, Manitoba R2H2A6
[†] Department of Pharmacology and Therapeutics, Faculty of Medicine University of Manitoba, Winnipeg, Manitoba R2H2A6
[‡] Institute of Cardiovascular Sciences, St. Boniface Gen Hosp Res Centre, Faculty of Medicine, University of Manitoba, Winnipeg, Manitoba R2H2A6

Methods in Enzymology, Volume 446
ISSN 0076-6879, DOI: 10.1016/S0076-6879(08)01616-9

1. INTRODUCTION

From a historical perspective, the postnatal heart has been viewed as a nonproliferative organ with a limited capacity for regeneration after injury. This has largely been attributed to the irreversible exit from the cell cycle that occurs shortly after birth, with the subsequent growth of the postnatal heart occurring by hypertrophic rather than hyperplastic processes. This is in contrast to other cell types of the body, including liver, skin, and gut that retain a capacity to readily exit and reenter the cell cycle, perhaps explaining why these organs have a greater propensity for proliferative disorders such as cancer.

The functional loss of force-generating cardiac cells by an apoptotic process is purported to underlie the ventricular remodeling and decline in ventricular performance in individuals with coronary artery disease or acute myocardial infarction (Anversa *et al.*, 1997; Foo *et al.*, 2005; Narula *et al.*, 1997). Indeed, several reports document apoptosis during hypoxia, ischemia, and after myocardial infarction (Anversa *et al.*, 1997; Gottlieb *et al.*, 1994; Kajstura *et al.*, 1996; Moissac *et al.*, 2000; Olivetti *et al.*, 1997; Regula *et al.*, 2002). Interestingly, the acknowledged ability of adult myocytes to reenter the cell cycle and synthesize DNA is limited and inadequate to restore ventricular function in patients with heart failure (Beltrami *et al.*, 2003). Therefore, one therapeutic approach in reducing morbidity and mortality in patients with heart failure would be to preserve the number of existing myocytes by suppressing apoptosis (Mercier *et al.*, 2005; Nam *et al.*, 2004; Yussman *et al.*, 2002).

To begin to understand how apoptosis is regulated, seminal studies have been carried out in the worm *C. elegans* and in immortalized cell lines derived from a multitude of lineages. From this body of work, many fundamental apoptotic and cytoprotective pathways have been characterized. However, the possibility exists that the signaling pathways and molecular targets that underlie apoptosis in the heart may not be universally conserved or equivalent to those identified in other cell types. Unfortunately, the molecular dissection of apoptotic signaling pathways in the heart has been hampered by several technical impediments, such as the limited ability of postnatal cardiomyocytes to be maintained in long-term culture, coupled with their meager capacity for genetic manipulation by conventional transfection methods (Kirshenbaum *et al.*, 1993).

In attempt to address the physiologic role of apoptosis in the context of the heart, we have embarked on *in vitro* studies that used neonatal and adult ventricular myocytes. Previously we reported that oxygen deprivation of neonatal and adult ventricular myocytes triggers mitochondrial perturbations consistent with the loss of mitochondrial membrane potential ($\Delta\psi_{m}$), permeability transition pore opening, cytochrome *c* release, and apoptosis

(Baetz *et al.*, 2005; Gurevich *et al.*, 2001; Regula *et al.*, 2002). Interventions that suppressed mitochondrial perturbations during hypoxia also suppressed cell death (Baetz *et al.*, 2005; Gurevich *et al.*, 2001). Herein we describe our methods for the establishment and application of primary cultured postnatal cardiomyocytes for the study of hypoxia, adenoviral gene transfer, and the analysis of the extent of cell death and mitochondrial perturbations.

2. PROTOCOLS

2.1. Hypoxia

As a result of the limited regeneration capacity of cardiac cells, irreversible damage occurs to the myocardium under hypoxic conditions. To study the effects of hypoxia on the heart, we first cultured 2-day-old Sprague-Dawley rat postnatal ventricular myocytes as previously described (Kirshenbaum and Schneider, 1995). Rat pups were euthanized by cervical dislocation, and hearts were removed by midline sternotomy. Hearts were washed with PBS^{-2} containing 10 g/L of glucose (hereafter referred to as PBS), minced with scissors, and repeatedly enzymatically digested with trypsin (740 U/digestion), collagenase (370 U/digestion), and DNase (2880 U/digestion) together at 35 °C for a total of six 10-min digestions. Cardiac myocytes were isolated by Percoll gradient centrifugation (Pharmacia, GE Health Care) at Percoll densities of 1.05, 1.06, and 1.082 g/ml. Cells were preplated on noncoated 150-mm culture plates for 45 min in a tissue culture incubator (37 °C, 5% CO_2) to allow any remaining fibroblasts to be plated. Isolated cardiac myocytes were subsequently plated on sterile Primaria^{TM} 6-well tissue culture plates (Becton Dickinson, Inc.) at a plating density of 1×10^6 cells per 35-mm well and incubated overnight in Dulbecco's Modified Eagle Medium/Ham's nutrient mixture F-12 (1:1) (Gibco, Cat. No. 12500–039), 17 mM HEPES, 3 mM NaHCO$_3$, 2 mM glutamine, 10 μg/ml gentamicin, and 10% fetal bovine serum. The following day, serum-free media (Dulbecco's Modified Eagle Medium/nutrient mixture F-12 (1:1) supplemented with 17 mM HEPES, 3 mM NaHCO$_3$, 2 mM glutamine, 1 nM Na$_2$SeO$_4$, 5 μg/ml transferrin, 1 μg/ml insulin, 1 nM LiCl, 25 μg/ml ascorbic acid, 1 nM thyroxine, and 10 μg/ml gentamicin, hereafter referred to as DFSF (DMEM/F12 serum free)) was deoxygenated by bubbling 95% N$_2$ and 5% CO_2 gas through the media for 1 h at room temperature (Gurevich *et al.*, 2001; Regula *et al.*, 2002). After washing the cells once with PBS to remove residual serum, the hypoxic media was applied to the myocytes. The cells were incubated for 24 h in a sealed hypoxic chamber that was continuously gassed with 95% N$_2$ and 5% CO_2 to mimic hypoxic conditions. The hypoxic chamber was designed to be housed within a 37 °C water jacketed tissue culture incubator to maintain the temperature and otherwise similar culture conditions as the normoxic control cells. During the incubation period, the environment inside the hypoxic chamber was maintained at $P_{O_2} \leq 5$ to 10 mmHg by continuous gas flow, while the nitrogen content of the chamber was controlled by a Proox 110 gas monitor (Reming Bioinstruments). Normoxic myocytes were maintained under standard conditions with 95% O_2 and 5% CO_2 at 37 °C for comparison.

2.2. Adenoviral gene transfer

Although several potential methods are available for delivering genes to cells, such as DEAE- dextran sulfate, calcium phosphate precipitation, lipid-coated particles, and electroporation, ventricular myocytes have a technical disadvantage that makes them readily impervious to conventional plasmid-based gene transfer protocols. One proven method to circumvent this technical impediment is viral gene delivery systems (Bett *et al.*, 1994; Graham and Prevec, 1991; Kirshenbaum, 1997; Kirshenbaum *et al.*, 1993). In this regard, we established that adenovirus particles, when packaged to encode exogenous DNA, could deliver foreign genes into cardiomyocytes with an efficiency that approximates 90%. Therefore, the development of adenovirus, lentivirus, and adeno-associated virus techniques has provided a means to genetically modify postnatal cardiac cells with putative regulators of the cell death machinery. For these experiments, we generated recombinant adenoviruses harboring deletions in the E1 region that have been repackaged encoding our gene of interest. Cells were infected by simply incubating the recombinant virus particles with the cells in DFSF media for a minimum of 4 h with a multiplicity of infection of 10 to 20, which achieves an efficiency of gene delivery of $\geq$90% (Kirshenbaum *et al.*, 1993).

2.3. Analysis of cellular viability

Traditional techniques for assessing cell viability have included the use of trypan blue dye exclusion technique and MTT assays. However, given the relative nonspecificity and poor uptake of the trypan blue and the inability to readily use MTT assay on cultured cells, we adapted a cell viability procedure that uses fluorescent vital dyes to simultaneously assess cardiac gene expression and viability by epifluorescence microscopy. Myocytes plated on glass coverslips were assessed for viability by staining with the vital dyes calcein acetoxymethyl ester (AM) and ethidium homodimer-1 (Invitrogen, Inc.) to determine the number of live and dead cells, respectively (Gurevich *et al.*, 2001; Kirshenbaum *et al.*, 1996). The intracellular hydrolysis of calcein AM results in green fluorescence, thereby indicating live cells. Dead cells appear red as a result of ethidium homodimer-1 (Invitrogen), which is a DNA stain excluded by the intact membranes of

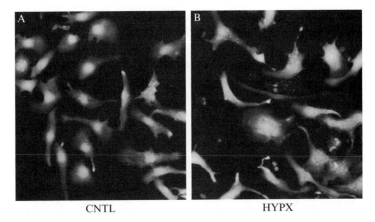

Figure 16.1 Viability staining of postnatal ventricular myocytes by epifluorescence microscopy under normal (CNTL, panel A) and hypoxic (HYPX, panel B) conditions. Original magnification 100×. (See color insert.)

living cells. The dyes were prepared in 10 ml of 37 °C DFSF media containing 2 μl of calcein-AM (4mM stock) and 3 μl of ethidium homodimer-1 (2mM stock). This type of analysis is generally performed after the cells have been genetically manipulated or exposed to hypoxia. Thus, myocytes on coverslips were treated according to the experiment and subsequently washed once with PBS before a 30-min incubation at 37 °C in the dark with DFSF containing the two dyes. After the incubation period, the media was removed and the cells were washed with PBS and treated with fresh DFSF. The PBS washes were sometimes omitted if it was found that dead or nearly dead cells were washing away and not being counted. To visualize the stained cells, the glass coverslips were inverted on glass slides and immediately observed by epifluorescence microscopy using an Olympus AX70 Research microscope. As shown in Fig. 16.1, the microscope has the capacity to visualize both living (green) and dead (red) cells simultaneously, facilitating data collection by eliminating the need to look at each color individually. To quantify the results, the numbers of living and dead cells are counted manually for each experimental condition and reported as the percentage of dead cells relative to the total number of cells counted.

2.4. Nucleosomal DNA fragmentation

To visualize the nuclear morphology of cardiac myocytes, cells were cultured on glass coverslips in 24-well tissue culture plates and subjected to experimental treatment such as hypoxia or adenoviral-mediated gene transfer. Myocytes were subsequently washed with PBS and fixed with 400 μl per well of ice-cold 70% ethanol for 15 min at 4 °C. The cells were washed three times with PBS and nuclei were stained with Hoechst 33258, a dye that fluoresces blue when bound to DNA. The 10 mg/ml stock solution of Hoechst 33258 (Invitrogen, Inc.) was diluted by adding 2 μl into 10 ml of PBS. Each glass coverslip with myocytes was incubated with 300 μl of this staining solution for 3 min in the dark at room temperature (Baetz *et al.*, 2005; Kirshenbaum *et al.*, 1996). After staining, the coverslips were washed three times with PBS and mounted on glass slides with FluorSave Reagent (Invitrogen, Inc.). Analysis was conducted on an Olympus AX70 Research microscope equipped with excitation and emission filters, enabling the detection of Hoechst 33258 stains. The number of condensed, abnormally shaped, or hyperchromatic nuclei was determined and reported as percent apoptosis relative to the total number of cells counted. This technique allows for hyperchromatic nuclear DNA fragmentation to be readily visualized in the presence of counter staining with cell specific markers or with no background fluorescence, as shown in Fig. 16.2.

2.5. Mitochondrial permeability transition (PTP)

This technique, initially described by Palo Bernardi (Bernardi and Petronilli, 1996; Petronilli *et al.*, 1998; 1999), highlights the ability to monitor mitochondrial PTP opening by fluorescence microscopy and has been adapted by our laboratory for the study of the PTP in cardiac myocytes. For these experiments, ventricular myocytes were cultured on glass coverslips and subjected to experimental treatment. The cells were incubated with the esterified form of calcein acetoxymethyl ester (calcein-AM, Invitrogen, Inc.), which enters both the cytoplasm and the mitochondria. The presence

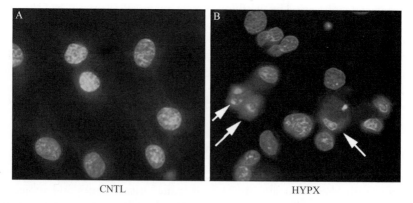

CNTL   HYPX

Figure 16.2 Epifluorescence microscopy of postnatal ventricular myocytes stained with Hoechst 33258 dye to assess nucleosomal DNA fragmentation in normoxic control cells. Labels are the same as for Fig. 16.1. Original magnification 600×.

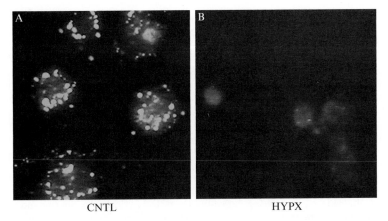

Figure 16.3 Mitochondrial permeability transition pore assay of postnatal ventricular myocytes under normoxic and hypoxic conditions. Labels are the same as for Fig. 16.1. Original magnification 600×.

of cobalt chloride ($CoCl_2$) in the media exclusively quenches the cytoplasmic signal that would otherwise be detected (Gurevich *et al.*, 2001). In the event that the permeability transition occurs after an apoptotic signal, the calcein leaks out of the mitochondria, resulting in a loss of green fluorescence because of the $CoCl_2$. Therefore, the opening of the PTP is marked by a reduction in green fluorescence as shown in Fig. 16.3. For assessing PTP opening in myocytes, the calcein AM dye was prepared in 10 ml of DFSF containing 1 μl of a 0.4 μM calcein-AM stock and 50 μl of a 1 M $CoCl_2$ stock (5 mM). After a wash with PBS, the cells were incubated with this media for 30 min in the incubator (37 °C, 5% CO_2). After incubation, the myocytes were immediately washed with PBS and treated with fresh DFSF. The coverslips were then inverted on glass slides and immediately visualized using an Olympus AX70 Research microscope. Fluorescence signals were assessed quantitatively by measuring average integrated optical density of fluorescent signal. All fluorescent images were captured using a high-speed Sensys (Photometrics, Inc., Waterloo, Ontario, Canada) or CoolSNAP-Pro (Media Cybernetics) digital camera and Image Pro-plus software (Media Cybernetics).

ACKNOWLEDGMENTS

We are grateful to Dr. Harvey Weisman for critical comments on the manuscript and Hongying Gang, for expert technical assistance. This work was supported by grants to L.A.K. from the Canadian Institute for Health Research. J. S. holds a studentship from the Manitoba Health Research Council, and L.A.K. is a Canada Research Chair in Molecular Cardiology.

REFERENCES

Anversa, P., Olivetti, G., Leri, A., Liu, Y., and Kajstura, J. (1997). Myocyte cell death and ventricular remodeling [see comments]. *Curr. Opin. Nephrol. Hypertens.* **6,** 169–176.

Baetz, D., Regula, K. M., Ens, K., Shaw, J., Kothari, S., Yurkova, N., and Kirshenbaum, L. A. (2005). Nuclear factor-kappaB–mediated cell survival involves transcriptional silencing of the mitochondrial death gene BNIP3 in ventricular myocytes. *Circulation* **112,** 3777–3785.

Beltrami, A. P., Barlucchi, L., Torella, D., Baker, M., Limana, F., Chimenti, S., Kasahara, H., Rota, M., Musso, E., Urbanek, K., Leri, A., Kajstura, J., Nadal-Ginard, B., and Anversa, P. (2003). Adult cardiac stem cells are multipotent and support myocardial regeneration. *Cell* **114,** 763–776.

Bernardi, P., and Petronilli, V. (1996). The permeability transition pore as a mitochondrial calcium release channel: a critical appraisal. *J. Bioenerg. Biomembr.* **28,** 131–138.

Bett, A. J., Haddara, W., Prevec, L., and Graham, F. L. (1994). An efficient and flexible system for construction of adenovirus vectors with insertions or deletions in early regions 1 and 3. *Proc. Natl. Acad. Sci. USA* **91,** 8802–8806.

Foo, R. S., Mani, K., and Kitsis, R. N. (2005). Death begets failure in the heart. *J. Clin. Invest.* **115,** 565–571.

Gottlieb, R. A., Burleson, K. O., Kloner, R. A., Babior, B. M., and Engler, R. L. (1994). Reperfusion injury induces apoptosis in rabbit cardiomyocytes. *J. Clin. Invest.* **94,** 1621–1628.

Graham, F., and Prevec, L. (1991). Manipulation of adenovirus vectors. *In* "Gene Transfer and Expression Protocols." (E. J. Murray, Ed.), pp. 109–128. Clifton, NJ: The Human Press Inc.

Gurevich, R. M., Regula, K. M., and Kirshenbaum, L. A. (2001). Serpin protein CrmA suppresses hypoxia-mediated apoptosis of ventricular myocytes. *Circulation* **103,** 1984–1991.

Kajstura, J., Cheng, W., Reiss, K., Clark, W. A. R., Sonnenblick, E. H., Krajewski, S., Reed, J. C., and Anversa, P. (1996). Apoptotic and necrotic myocyte cell deaths are independent contributing variables of infarct size in rats. *Lab. Invest.* **74,** 86–107.

Kirshenbaum, L. A. (1997). Adenovirus mediated–gene transfer into cardiomyocytes. *Mol. Cell Biochem.* **172,** 13–21.

Kirshenbaum, L. A., Abdellatif, M., Chakraborty, S., and Schneider, M. D. (1996). Human E2F-1 reactivates cell-cycle progression in ventricular myocytes and represses cardiac gene transcription. *Dev. Biol.* **179,** 402–411.

Kirshenbaum, L. A., MacLellan, W. R., Mazur, W., French, B. A., and Schneider, M. D. (1993). Highly efficient gene transfer into adult ventricular myocytes by recombinant adenovirus. *J. Clin. Invest.* **92,** 381–387.

Kirshenbaum, L. A., and Schneider, M. D. (1995). Adenovirus E1A represses cardiac gene transcription and reactivates DNA synthesis in ventricular myocytes, via alternative pocket protein- and p300-binding domains. *J. Biol. Chem.* **270,** 7791–7794.

Mercier, I., Vuolo, M., Madan, R., Xue, X., Levalley, A. J., Ashton, A. W., Jasmin, J. F., Czaja, M. T., Lin, E. Y., Armstrong, R. C., Pollard, J. W., and Kitsis, R. N. (2005). ARC, an apoptosis suppressor limited to terminally differentiated cells, is induced in human breast cancer and confers chemo- and radiation-resistance. *Cell Death Differ.* **12,** 682–686.

Moissac, D., Gurevich, R. M., Zheng, H., Singal, P. K., and Kirshenbaum, L. A. (2000). Caspase activation and mitochondrial cytochrome *c* release during hypoxia-mediated apoptosis of adult ventricular myocytes. *J. Mol. Cell Cardiol.* **32,** 53–63.

Nam, Y. J., Mani, K., Ashton, A. W., Peng, C. F., Krishnamurthy, B., Hayakawa, Y., Lee, P., Korsmeyer, S. J., and Kitsis, R. N. (2004). Inhibition of both the extrinsic and intrinsic death pathways through nonhomotypic death-fold interactions. *Mol. Cell* **15,** 901–912.

Narula, J., Kharbanda, S., and Khaw, B. A. (1997). Apoptosis and the heart. *Chest* **112,** 1358–1362.

Olivetti, G., Abbi, R., Quaini, F., Kajstura, J., Cheng, W., Nitahara, J. A., Quaini, E., Di Loreto, C., Beltrami, C. A., Krajewski, S., Reed, J. C., and Anversa, P. (1997). Apoptosis in the failing human heart. *N. Engl. J. Med.* **336,** 1131–1141.

Petronilli, V., Miotto, G., Canton, M., Brini, M., Colonna, R., Bernardi, P., and Di Lisa, F. (1999). Transient and long-lasting openings of the mitochondrial permeability transition pore can be monitored directly in intact cells by changes in mitochondrial calcein fluorescence. *Biophys. J.* **76,** 725–734.

Petronilli, V., Miotto, G., Canton, M., Colonna, R., Bernardi, P., and Di Lisa, F. (1998). Imaging the mitochondrial permeability transition pore in intact cells. *Biofactors* **8,** 263–272.

Regula, K. M., Ens, K., and Kirshenbaum, L. A. (2002). Inducible expression of BNIP3 provokes mitochondrial defects and hypoxia-mediated cell death of ventricular myocytes. *Circ. Res.* **91,** 226–231.

Yussman, M. G., Toyokawa, T., Odley, A., Lynch, R. A., Wu, G., Colbert, M. C., Aronow, B. J., Lorenz, J. N., and Dorn, G. W. (2002). Mitochondrial death protein Nix is induced in cardiac hypertrophy and triggers apoptotic cardiomyopathy. *Nat. Med.* **8,** 725–730.

CHAPTER SEVENTEEN

QUANTIFICATION OF VASCULAR ENDOTHELIAL CELL APOPTOSIS *IN VIVO*

Xuefeng Zhang* *and* Sareh Parangi*,†

Contents

Abstract

Regulation of vascular endothelial cell survival and apoptosis plays a crucial role during development and numerous physiologic and pathologic processes. Analyzing endothelial apoptosis *in vivo* is necessary not only for the understanding of many physiologic and pathologic processes but also for evaluating treatments that induce endothelial cell apoptosis. This chapter describes one of the more widely used protocols for detecting and quantifying vascular endothelial cell apoptosis by double-staining vascular endothelial cells with immunofluorescence and terminal deoxynucleotidyl transferase dUTP nick-end labeling (TUNEL).

1. INTRODUCTION

Regulation of vascular endothelial cell survival and apoptosis plays a crucial role during development and numerous physiologic and pathologic processes. Formation of new blood vessels, defined as angiogenesis, constitutes a crucial step for many diseases, including malignant tumors (Folkman, 1995).

* Department of Surgery, Beth Israel Deaconess Medical Center and Harvard Medical School, Boston, Massachusetts
† Department of Surgery, Massachusetts General Hospital and Harvard Medical School, Boston, Massachusetts

Methods in Enzymology, Volume 446
ISSN 0076-6879, DOI: 10.1016/S0076-6879(08)01617-0

With better understanding of angiogenesis, the vascular endothelial cell has become a novel target for antiangiogenic treatment, with induction of endothelial cell apoptosis as an important part of the new therapeutic intervention (Folkman, 2003).

Given the obvious importance of studying vascular endothelial cell apoptosis, many methods have been developed for analyzing vascular endothelial cell apoptosis both *in vitro* and *in vivo*. For *in vivo* studies, one feasible method is *in situ* labeling of endothelial cells and apoptotic cells in the same sample section with two different fluorochromes; the apoptotic endothelial cells can, therefore, be recognized by colocalization of both labeling signals.

Before starting, it is necessary to study the fluorescence microscope or confocal microscope that will be used to choose three fluorochromes for apoptotic cells, endothelial cells, and nuclei counterstaining, respectively. For example, fluorescein for apoptotic cells, Texas Red for endothelial cells, and DAPI for nuclei. However, many other combinations can be chosen as long as the equipment allows.

2. Protocols

2.1. Fluorescence immunohistochemistry for vascular endothelial cells

Several membrane molecules have been used as markers to label vascular endothelial cells, among which CD31 is one of the most widely used. CD31 immunohistochemistry is routinely performed in many medical centers on paraffin-embedded human tissue. However, according to our experience, frozen sections have to be used for CD31 immunohistochemistry on mouse tissue. Both snap-frozen tissue (unfixed) and prefixed frozen tissue can be used. Tissue can be fixed in 4% paraformaldehyde in phosphate-buffered saline (PBS) at 4 °C for 2 to 4 h, incubated in 30% sucrose at 4 °C overnight, and embedded in OCT compound, and then frozen in liquid nitrogen, and stored at −80 °C. To get frozen sections, tissue needs to be sectioned in a cryostat. When sectioning the samples, it is better to keep adjacent sections for CD31/TUNEL double-fluorescent staining, as well as CD31 and H&E staining. The CD31 and H&E sections are very useful reference for fluorescence microscopy. Five-micrometer-thick sections are ideal for most immunohistochemistry, but sections up to 10 μm in can also be used. The tissue sections need to be mounted on positively charged Superfrost slides (Fisher Scientific, Co.). If snap-frozen tissue (without fixation) is used, the sections need to be washed in PBS and fixed in 4% paraformaldehyde at 4 °C for 20 min. However, if samples had been fixed before frozen, the sections can be washed in PBS and then be blocked directly. Before blocking, the sample should be circled with a PAP pen, which provides a thin filmlike hydrophobic barrier around a specimen on a slide. Blocking reagent is usually serum from the species that the secondary antibody is made from, diluted to 5% in PBS. After adding the blocking reagent to the PAP pen circle, the sections are incubated in a humidified chamber at room temperature for 1 h. A humidified chamber must be used for all the incubations to prevent the evaporation of the reagent. After blocking, the sections can be briefly rinsed in PBS and followed by primary antibody incubation. An anti-human CD31, anti-mouse CD31 antibody, or CD31 antibody for other species needs to be chosen according to the sample being used. We use anti-mouse CD31 antibody (1:200 diluted in blocking reagent) from BD Pharmingen regularly, which produces consistent results (Zhang *et al.*, 2005). The primary antibody incubation can be either 2 h at room temperature or overnight at 4 °C. The sections are then washed three times in PBS. The secondary antibody is chosen according to the primary antibody that has been used. If a rat anti-mouse CD31 antibody has been used, the secondary antibody would be an anti-rat IG. Because most commercially available TUNEL kits use fluorescein (excitation wavelength 494 nm, emission wave length 521 nm) as the fluorescent tracer, a secondary antibody labeled with another fluorochrome with a different excitation wavelength will need to be chosen. Of course, this depends on the fluorescent or confocal microscope that will be used. Because most fluorescent microscopes and confocal microscopes are equipped with filters for both green and red fluorescence, a secondary antibody conjugated with Texas red (excitation wavelength 595 nm, emission wave length 615 nm), or Alexa Fluor 594, can be chosen. The application concentration of antibodies from different manufacturers may vary, and the concentration needs to be determined in a preliminary experiment. Because the fluorochromes are sensitive to light, the sections need to be protected from light after addition of the fluorescence conjugated secondary antibody. Normally, 1 h at room temperature is sufficient for secondary antibody incubation. The sections are then washed three times in PBS (5 min each). Before proceeding to the TUNEL, it is better to check the sample under a fluorescent microscope to make sure the CD31 staining worked. If possible, an extra section can be used for this purpose; this section can also serve as a negative control for the TUNEL.

2.2. Labeling of apoptotic cells

Many ways exist to detect apoptosis at different stages on histologic sections. TUNEL (terminal deoxynucleotidyl transferase biotin-dUTP nick end labeling) method is one of the most commonly used (Gavrieli *et al.*, 1992) and is often used in conjunction with other methods such as immunohistochemistry. One of the characteristics of apoptosis is the degradation of DNA after the activation of Ca/Mg-dependent endonucleases. This DNA cleavage leads to strand breaks within the DNA. The TUNEL method identifies apoptotic cells *in situ* by use of terminal deoxynucleotidyl transferase (TdT) to transfer fluorescence-dUTP to these strand breaks of cleaved DNA. The fluorescence-labeled cleavage sites are then visualized directly by fluorescence microscopy.

Here we describe the use of one of many commercially available kits for TUNEL (Promega Corp). The sections stained with CD31 first need to be permeabilized in 0.2% Triton X-100 in PBS for 15 min at room temperature. Remember to protect the sections from light. The sections are then washed twice in PBS (5 min each), and each sample needs to be circled again with a PAP pen (because Triton may have washed the old circle away). Then 100 μl equilibration buffer is added to each section, and the sections are incubated at room temperature for 10 min. If a positive control is preferred to make sure that the assay works, it can be prepared it in advance while keeping other sections in PBS. For a positive control, incubate a section in DNase I, grade I (20 U/ml) for 10 min at room temperature, wash twice in PBS, and then proceed to equilibration. Be careful not to contaminate other sections with DNase, which will cause false-positive readings. During the equilibration, the TUNEL reaction mixture can be prepared by mixing equilibration buffer, nucleotide mix, and TdT enzyme. Remember to include a negative control without adding TdT to determine the background reactivity. Note that the TUNEL reaction mixture should be prepared immediately before use and should not be stored, and the TUNEL reaction mixture must be kept on ice until use. After removing the equilibration buffer from the sample, add 50 μl TUNEL reaction mixture on each sample and cover the section with a plastic coverslip (included in the kit) and incubate for 60 min at 37 °C in a humidified chamber in the dark. On finishing the reaction, incubate the sections in 2× SSC for 15 min to stop the reaction, then wash the sections in PBS. To reduce the background, the sections can be washed in 0.1% Triton X-100, 0.5% BSA in PBS three times.

2.3. Counterstaining for nuclei

To quantify apoptotic endothelial cells, it is necessary to counterstain the nuclei with a third fluorochrome. Counterstaining not only enables you to quantify the total number of endothelial cells in each visual field, but also verifies that the TUNEL signals are localized in the nuclei, thus identifying the false-positive labeling. Choice of the fluorescent dye depends on the fluorescence microscope or confocal microscope that will be used. For example, if the microscope is equipped with a UV filter, DAPI will be a good choice. Other choices include propidium iodide, which can be visualize under similar wavelength as that for Texas Red, if the microscope allows you to choose another fluorescence dye for CD31 staining. Because

DAPI or propidium iodide–containing mounting solutions are widely available, counterstaining and coversliping can be done in one simple step. Many mounting media, such as Vectashield mounting medium from Vector Laboratories, or antifade reagent from Molecular Probes, can also prevent rapid loss and photobleaching of fluorescence during microscopic examination and storage. However, when choosing a mounting medium, it is important to make sure that the medium is compatible with all fluorescent proteins that have been used in the staining.

2.4. Microscopy and quantification of apoptotic endothelial cells

Once the staining is completed, images should be captured under a fluorescence/confocal microscope as soon as possible, although the section cans be kept at 4 °C in the dark for up to weeks if "antifading" mounting media is used. While capturing pictures under a fluorescence/confocal microscope,

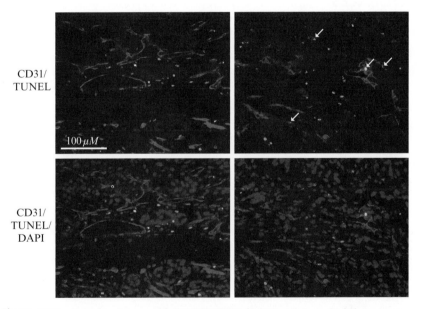

Figure 17.1 Sample pictures of CD31/TUNEL/DAPI staining. Staining was performed on xenograft tumors grown in mouse. Left panel shows tumor from control mouse, and right panel shows tumor from mouse that received treatment that induced endothelial cell apoptosis. Apoptotic endothelial cells can be determined by colocalization of TUNEL signal within the nuclei of CD31-positive endothelial cells, and total endothelial cell number can be determined by counting nuclei localized in endothelial cells, which are labeled by CD31 staining. Please note that to be true positive, TUNEL-positive signals must be localized within nuclei. Figure is adapted from Zhang, X., et al. (2005). *Clin. Cancer Res.* **11**, 2337–2344. (See color insert.)

it is better to use adjacent sections stained with H&E and CD31 as reference. This is especially necessary if tumor sections are analyzed, because necrosis is very common in large tumors, whereas necrotic areas need to be avoided when capturing pictures for apoptotic studies.

Once you have sufficient pictures of the samples (the number depends on the statistic features of the apoptotic endothelial cells in the samples and the study you are doing), the number of endothelial cells and apoptotic endothelial cells can be counted in each picture. Figure 17.1 shows sample pictures of CD31/TUNEL/DAPI staining. Total endothelial cell number can be determined by counting nuclei localized in endothelial cells, which are labeled by CD31 staining, whereas apoptotic endothelial cells can be determined by colocalization of TUNEL signal within the nuclei of CD31 positive endothelial cells. Apoptotic endothelial cells can be quantified as the ratio of apoptotic endothelial cells to the total number of endothelial cells in each visual field.

REFERENCES

Folkman, J. (1995). Angiogenesis in cancer, vascular, rheumatoid and other disease. *Nat. Med.* **1**, 27–31.

Folkman, J. (2003). Angiogenesis and apoptosis. *Semin. Cancer Biol.* **13**, 159–167.

Gavrieli, Y., Sherman, Y., and Ben-Sasson, S. A. (1992). Identification of programmed cell death *in situ* via specific labeling of nuclear DNA fragmentation. *J. Cell Biol.* **119**, 493–501.

Zhang, X., Galardi, E., Duquette, M., Delic, M., Lawler, J., and Parangi, S. (2005). Antiangiogenic treatment with the three thrombospondin-1 type 1 repeats recombinant protein in an orthotopic human pancreatic cancer model. *Clin. Cancer Res.* **11**, 2337–2344.

CHAPTER EIGHTEEN

RECOMBINANT TRAIL AND TRAIL RECEPTOR ANALYSIS

Nicholas Harper* *and* Marion MacFarlane*

Contents

Abstract

Death receptors are a subgroup of the tumor necrosis factor receptor superfamily (TNFRSF) and mediate activation of what is widely known as the "extrinsic" apoptosis pathway. TRAIL (tumor necrosis factor–related apoptosis-inducing ligand) is one of the most recent death receptor ligands identified. The TRAIL receptor family consists of four distinct membrane-bound receptors, named TRAIL-R1 to -R4. TRAIL-R1 and TRAIL-R2 belong to the "death receptor" subfamily of the TNFRSF. Unlike other death receptor ligands, such as FasL/CD95L and TNF, TRAIL seems to display selective toxicity by killing tumor and transformed,

* MRC Toxicology Unit, University of Leicester, Leicester, LE1 9HN, UK

Methods in Enzymology, Volume 446
ISSN 0076-6879, DOI: 10.1016/S0076-6879(08)01618-2

but not normal, cells. Importantly, there also seems to be a complete lack of apparent toxicity, specifically hepatotoxicity, when TRAIL is used *in vivo*. Taken together, these observations led to TRAIL being proposed as a potential anti-tumor therapeutic, thus explaining the intense activity surrounding TRAIL and TRAIL receptor research over the past few years.

This chapter describes a number of methods for the production of recombinant TRAIL in *E. coli* followed by labeling of recombinant TRAIL with either biotin or fluorochromes. These recombinant TRAIL preparations are then used to study various aspects of TRAIL signaling from cell surface receptor levels to the composition of death receptor complexes. This combination of direct binding and functional analysis provides a very powerful approach to aid in further characterization of TRAIL/TRAIL–Receptor regulation and signaling.

1. INTRODUCTION

The TRAIL receptor family consists of four distinct membrane-bound receptors (reviewed in LeBlanc and Ashkenazi, 2003). TRAIL-R1 and TRAIL-R2 belong to the death receptor subfamily of the tumor necrosis factor receptor superfamily (TNFRSF), which also includes TNF-R1 and Fas/CD95.

Like all death receptors, TRAIL-R1/-R2 contain an 80-amino acid motif within their cytoplasmic domain termed the death domain (DD) that is critically required for induction of cell death (Tartaglia *et al.*, 1993). In contrast, TRAIL-R3 (DcR1) and TRAIL-R4 (DcR2) either completely lack, or contain, a truncated intracellular domain and, as a consequence, are unable to initiate apoptotic signaling. These TRAIL receptors have, therefore, been proposed to act as "decoy" receptors (Pan *et al.*, 1997; Sheridan *et al.*, 1997). Like other members of the TNFRSF, all four TRAIL receptors are also characterized by a series of cysteine-rich repeats within their extracellular domains.

Although several reports demonstrate that TRAIL can activate other signaling pathways such as the transcription factors NFkB and AP-1, its main function seems to be the induction of apoptosis (Harper *et al.*, 2001; MacFarlane, 2003; MacFarlane *et al.*, 2000; Varfolomeev *et al.*, 2005).

TRAIL induces apoptosis by binding to either TRAIL-R1 or TRAIL-R2, which then leads to receptor aggregation and recruitment of the adaptor protein FADD to the cytoplasmic DD motif of the receptors. FADD, in turn, recruits the initiator procaspase-8 to form the death-inducing signaling complex (DISC) (reviewed in Peter and Krammer, 2003). The "enforced proximity" of procaspase-8 molecules within the DISC leads to autoactivation, resulting in its subsequent maturation and release into the cytoplasm. Activated caspase-8 is then able to process the effector caspases-3 and -7, as well as activating the proapoptotic Bcl-2 homolog, Bid. Bid, in turn, acts as a mitochondrial amplification signal, necessary in some cell types for induction of cell death (Li *et al.*, 1998).

TRAIL is of particular interest as a potential therapeutic agent because it seems to display selective toxicity, inducing apoptosis in tumor cells but not most normal cells (Pitti *et al.*, 1996; Wiley *et al.*, 1995). This has led to a number of clinical trials that have used either TRAIL ligand (Apo2L/TRAIL—Genentech, South San Francisco, CA) or agonistic antibodies directed against either TRAIL-R1 or -R2 (HGS-ETR1 and HGS-ETR2—Human Genome Sciences, Rockville, MD) for induction of cell death. Of those primary malignancies that have been demonstrated to be resistant to TRAIL and TRAIL receptor agonistic antibodies (MacFarlane *et al.*, 2002), a number can be sensitized when some of these reagents are used in combination with other potential chemotherapeutics such as proteasome inhibitors or histone deacetylase inhibitors (Dyer *et al.*, 2007; Inoue *et al.*, 2004; Koschny *et al.*, 2007).

A fuller understanding of TRAIL signaling, in a particular cell type, relies on a combination of both functional assays in intact cells and robust TRAIL/TRAIL–Receptor binding assays. This chapter describes complementary methods for the study of TRAIL signaling, all of which rely on the successful generation of recombinant TRAIL. The application of these various methods to evaluate TRAIL-induced apoptosis, TRAIL receptor expression levels, TRAIL DISC composition, and TRAIL/TRAIL–Receptor internalization is discussed.

2. Generation of Recombinant TRAIL

TRAIL, unlike a number of other TNFSF members, seems to function as a "soluble" protein and as a consequence can be produced relatively easily in large quantities by expression in *E. coli*. CD95L, by comparison, is insoluble when expressed in bacteria and seems to require posttranslational modification, specifically glycosylation, to aid solubility (Peter *et al.*, 1995, and unpublished results). In addition, "soluble" CD95L requires artificial oligomerization for activity. As a result, commercial preparations of CD95L are synthesized as soluble, secretable fusion proteins expressed in mammalian cells. Crosslinking is provided either by an internal oligomerization tag, such as a leucine zipper, or through crosslinking of the ligand with antibodies directed against an additional common peptide tags such as the Flag peptide.

TRAIL, like other TNFSF members, is a type I transmembrane protein; therefore, production in *E. coli* requires removal of both the transmembrane and intracellular domains of the protein to aid solubility. Truncated TRAIL (MacFarlane *et al.*, 1997) or receptor-selective TRAIL mutants (MacFarlane *et al.*, 2005) (residues 95 to 281), cloned N-terminally in the plasmid expression vector, pet28 (Novagen, Nottingham UK), are fused "in-frame" with a hexa-histidine tag to allow for a relatively simple purification protocol on nickel or cobalt-charged affinity resins.

2.1. Protocol 1: Induction of His-tagged TRAIL in *E. coli* with IPTG

1. *E. coli* BL-21 DE3 (pLys) are transformed with pet28-TRAIL by use of standard laboratory protocols.
2. For induction of recombinant TRAIL protein, a 10-ml culture in LB-medium, containing kanamycin (25 μg/ml), is inoculated and then grown overnight at 37 °C with constant shaking.
3. The following day, this culture is used to initiate a larger (400 ml) culture required for protein induction.
4. The culture is grown to an OD of approximately ~0.6 (log phase), then protein is induced with isopropylthiogalactoside (IPTG) (1 mM) for 3 h at 27 °C. A sample is taken before the addition of IPTG (un-induced) to assess the efficiency of protein induction compared with induced cultures.
5. Bacterial pellets are then collected by centrifugation at 5000 rpm for 15 min and washed one time with ice-cold PBS. Pellets can then be further processed or stored at −80 °C until required.[1]

2.2. Protocol 2: Purification of His-TRAIL with a nickel affinity resin

1. Previously frozen or freshly prepared, induced bacterial pellets (from 400 ml of culture) are first thawed, then lysed in a Triton lysis buffer (20 mM Tris/HCl (pH 7.5), 150 mM NaCl, 10% glycerol, 1% Triton X-100) containing Complete[TM] protease inhibitors without EDTA [2] (Roche, Sussex UK) for 30 min on ice. To ensure full lysis, the bacteria can be sonicated on ice.
2. Lysates are then cleared by centrifugation at 13,000 rpm for 30 min at 4 °C then supplemented with 500 mM imidazole to a final concentration of 20 mM.
3. 6×His-tagged TRAIL is then purified by batch purification[3] with Ni^{2+}-NTA agarose beads (Qiagen, Sussex UK) for 1.5 h at 4 °C.
4. Beads are washed with several bed volumes of lysis buffer.
5. Bound TRAIL is then eluted from the beads by use of either 150 mM EDTA or 100 mM imidazole.

---

[1] IPTG-induced bacterial pellets have successfully been stored at −80 °C for many years without any apparent effect on the stability/activity of the subsequently isolated recombinant TRAIL.
[2] EDTA will interfere with binding to the nickel-containing affinity resin.
[3] TRAIL purification can be automated by use of any of the available prepacked IMAC columns.

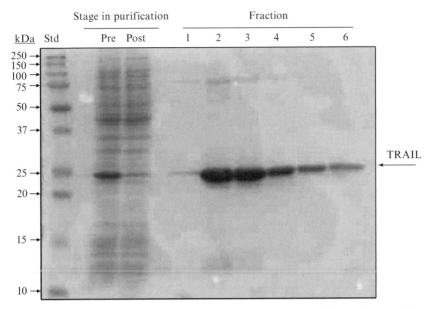

Figure 18.1 Purification of recombinant His$_6$-TRAIL. Recombinant human His$_6$-TRAIL was purified as described in Protocol 2. Samples of *E. coli* bacterial supernatant (both pre- and post- purification) and eluted fractions (1–6) were assessed for purity by SDS-PAGE. Fractions 2 and 3 contain the highest levels of TRAIL protein. In this purification, Fractions 2–5 were pooled and protein concentration and apoptosis activity assessed as described in Protocol 3. The postpurification supernatant confirms efficient purification of TRAIL. Molecular Weight standards (Std) are shown in Lane 1 (Precision Plus Protein Standards; BioRad).

A typical TRAIL purification is shown in Fig. 18.1. After elution, a small sample of each elution fraction is removed for analysis and individual fractions frozen at −80 °C. After analysis of protein content (using a suitable protein assay reagent) and purity (by SDS-PAGE), those fractions containing the highest purified yield are pooled and then aliquoted. Any precipitate that may have formed after freeze–thawing of the individual fractions can be removed by centrifugation before pooling and aliquoting.

The protein content of the final aliquots is again checked, as is the apoptosis-inducing activity of purified TRAIL on a suitable TRAIL-sensitive, tumor cell line (see Protocol 3). Recombinant TRAIL can then be successfully stored at −80 °C for a number of years with no noticeable loss in activity.

2.2.1. Assessment of the apoptosis-inducing activity of recombinant TRAIL with Annexin V

The apoptosis-inducing activity of recombinant TRAIL can be conveniently assessed by treatment of a TRAIL-sensitive cell line, such as BJAB or Jurkat, with a range of concentrations of recombinant TRAIL and incubation times of up to 6 h at 37 °C. The extent of apoptosis can then be quantified by Annexin V labeling of apoptotic cells followed by FACs analysis. Annexin V is a vascular protein that binds in a Ca^+-dependent manner to the phospholipid, phosphatidylserine, which is present on the external cell membrane of apoptotic cells (Martin *et al.*, 1995). What follows is a method for Annexin V-FITC labeling of a hematopoietic cell line, grown in suspension, using a commercially available form of Annexin V-FITC routinely used within our laboratory.

2.3. Protocol 3: Labeling of apoptotic cells with Annexin V-FITC

1. Cells (1×10^6 cells), either left untreated, or exposed to TRAIL at 37 °C for varying times are resuspended in 1 ml Annexin buffer (10 mM HEPES/NaOH, pH 7.4, 150 mM NaCl, 5 mM KCl, 1 mM MgCl$_2$, 1.8 mM CaCl$_2$) using tubes suitable for flow cytometry (BD Falcon, Cat # 353052).
2. 1.5 μl Annexin V-FITC (e.g., Bender Medsystems, Cat # BMS306FI) is added to the cell suspension and cells are then incubated for 8 min at room temperature (~22 °C).
3. At the end of the incubation time, cells are placed on ice; 2 μl propidium iodide (PI; 50 μg/ml PBS) is added, and the cell suspension then incubated on ice for 2 min.
4. Samples are stored on ice and the extent of apoptosis assessed by FACs analysis using a Becton Dickinson FACsCalibur (or equivalent), with excitation/emission wavelengths of 488 nm/519 nm for Annexin V-FITC (FL1) and 488 nm/575 nm for PI (FL2). A minimum of 10,000 events are acquired (flow rate ~200 events/sec), and cells are analyzed with CellQuest Pro software (Fig. 18.2).
5. Cells are initially analyzed by dot-plot analysis of FSC (forward scatter) versus SSC (side scatter) and the main cell population (Region 1 = R1) then further analyzed by dot-plot analysis of propidium iodide (FL2) versus Annexin V-FITC (FL1). A quadrant marker is then used to quantify the percentage of cells in the population that are either: nonapoptotic (Annexin V^-/PI$^-$: lower left quadrant), apoptotic (Annexin V^+/PI$^-$: upper left quadrant), or "secondary" necrotic (Annexin V^+/PI$^+$: upper right quadrant).

The preceding method should be carefully optimized for each individual cell line in terms of the amount of Annexin V-FITC used and the optimal incubation time required to obtain differential labeling of Annexin V^+ versus Annexin V^- cells. For example, differential labeling of Jurkat T cells requires only 1.5 μl Annexin-V-FITC and an incubation time of

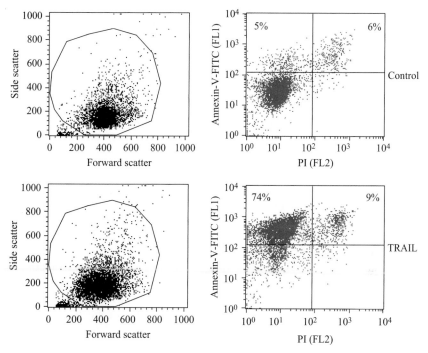

Figure 18.2   Annexin V staining of BJAB cells treated with recombinant His$_6$-TRAIL. BJAB cells (1×10^6 cells/ml) were either left untreated (Control) or treated with recombinant human TRAIL (500 ng/ml) for 4 h at 37 °C and the extent of apoptosis assessed by staining with Annexin V-FITC/PI as described in Protocol 3. Samples were then analysed by flow cytometry (FACSCalibur; Becton Dickinson) and CellQuest Pro software. The right panels show dot-plots of Annexin V-FITC (FL1) versus PI (FL2) -stained cells. Annexin V-FITC$^+$/PI$^-$ cells are scored as apoptotic (upper left quadrant) and Annexin V-FITC$^+$/PI$^+$ cells (upper right quadrant) are scored as "secondary" necrotic. This assay confirms the apoptosis-inducing activity of the purified recombinant human TRAIL preparation from Figure 1. [a] Data kindly provided by Laura Hall.

8 min at room temperature, whereas the cell line, BJAB, requires up to 15 μl Annexin-V-FITC and an incubation time of 30 min.

The preceding Annexin-V–labeling method can also be adapted for use on adherent cells (such as MCF-7 cells) exposed to TRAIL (or other apoptotic stimuli). In this case, the adherent cell culture supernatant (containing apoptotic cells that have lost adherence) is carefully retained and the remaining adherent cells collected by gentle trypsinization. These two cell populations are then combined and immediately followed by a recovery period of at least 20 min at 37 °C in complete medium (to allow resealing of the outer cell membrane). Cells are then gently pelleted by centrifugation and resuspended in Annexin buffer for staining of apoptotic cells with Annexin V-FITC/PI as outlined previously (Protocol 3; Step 1 onward).

2.3.1. Labeling of recombinant TRAIL

One of the benefits of generating ligands, such as TRAIL, "in-house" is that they can be produced in sufficient quantities to allow subsequent modification of the purified ligand depending on the specific application required.

Described in the following sections are two such recombinant TRAIL modifications, together with methods that rely on the use of a suitably labeled ligand. Many modifying reagents that are used to label proteins, such as TRAIL, do so non-specifically (e.g., N-hydroxysuccinimide esters which label primary amino groups lysine and N-terminal). It should be noted that we have rarely seen decreased activity after TRAIL modification with these reagents. Care should, however, be taken to titrate those labeling reagents used to ensure that the activity and receptor binding properties of the subsequent TRAIL-conjugates is retained.

To retain the biologic activity of the subsequent labeled ligand, we have essentially modified the ligand purification protocol to allow the labeling reactions to be performed with recombinant TRAIL still immobilized on agarose beads (see Protocol 2). Although this may not be ideal to accurately control the extent of labeling, its benefit is in obviating the need for extensive dialysis to remove any unincorporated label, thereby retaining the biologic activity of the labeled ligand. Subsequent to purification and labeling, methods are also available to assess the extent of label-incorporation per milligram of purified protein, thus helping to retain consistency between batches of labeled TRAIL protein.

2.4. Protocol 4: Preparation of biotin-labeled TRAIL

Biotin binds with very high affinity to both avidin and streptavidin. This makes it an ideal tag for either capture of biotinylated proteins on avidin/streptavidin conjugated to insoluble supports (agarose or Sepharose^{TM} beads), or for detection/labeling of biotinylated proteins with the many avidin/streptavidin conjugates that are commercially available.

The most common biotin labeling reagents use amine-reactive chemistry to label primary amino groups in proteins (N-terminus and lysines). Importantly, because of its relatively small size (~250 Da), incorporation of biotin, in general, does not significantly affect the biological activity of the labeled protein.

1. As mentioned previously, we label TRAIL as part of the standard purification protocol, primarily to avoid the need for extensive dialysis protocols and to retain the maximum amount of biological activity.
2. TRAIL is initially purified as outlined in Protocol 2 (Steps 1 to 4, *only*), except that the ligand is then retained immobilized on Ni-NTA-agarose beads in lysis buffer. Because of the presence of primary amines in this buffer (which will interfere with labeling), the agarose beads are then extensively washed with PBS or a bicarbonate buffer at 4 °C (pH 7.5 to 9.0). The beads are then resuspended in 1 ml of PBS (or bicarbonate buffer).

3. Immediately before labeling, a solution of biotinyl aminocaproic acid–*N*-hydroxysuccinimide ester (Roche, Sussex UK) is dissolved in DMSO (20 mg/ml). 5 to 20 μl of this solution is added to the beads that are then labeled for 1 h at 4 °C on an end-to-end shaker[4]. *Note*: any excess labeling solution should be discarded because of the labile nature of the active group.

4. After biotin labeling, excess labeling reagent is quenched by the addition of Tris (pH 8.0) to a final concentration of 1 *M* for an additional 15 min. Beads are then washed with lysis buffer and recombinant biotin-TRAIL eluted as before (Protocol 2, Step 5), by use of either EDTA or imidazole.

Biotinylated TRAIL is then aliquoted and stored as described previously (Protocol 2). Before use, biotin-TRAIL must be checked to confirm both its apoptosis-inducing activity (Protocol 3) and the incorporation of biotin. Successful labeling with biotin can be determined by Western blotting. Essentially, a small amount of biotin-labeled TRAIL is subjected to SDS-PAGE followed by Western blotting. After blocking of the transfer membrane, biotinylated TRAIL can be detected by a short incubation with streptavidin conjugated to horseradish peroxidise, followed by detection by chemiluminescence.

If necessary, the extent of biotin incorporation can also be assessed with HABA (4'-hydroxyazobenzene-2-carboxylic acid). Free biotin displaces HABA from avidin, causing a decrease in absorbance at 500 nm that is measured and compared with a standard curve to determine molar incorporation of biotin after labeling (Janolino, 1996).

Many other biotinylation reagents are available from companies such as Roche and Perbio Science UK Ltd, Northumberland UK. Reagents with longer/shorter linkages, thiol/photo-cleavable linkages or containing different reactive moieties (e.g., carboxyl or carbohydrate) are available depending on the end-user's requirements (Savage *et al.*, 1992).

2.5. Protocol 5: Labeling TRAIL with fluorochromes

The protocol we have developed for labeling recombinant TRAIL with fluorochromes is essentially a further modification of the standard TRAIL purification protocol (Protocol 2). In our laboratory, we have used the Alexa Fluor® 647 protein labeling kit (Molecular Probes, Paisley, UK), which is primarily designed for antibody labeling, but in practice will label any protein. The reactive dye has a succinimidyl ester moiety that reacts with primary amines (essentially using the same chemistry as the biotin-labeling reagent used previously in Protocol 4). Again, to maintain biological activity, TRAIL is labeled while still immobilized on the Ni-NTA–agarose beads during the purification procedure.

---

[4] Although many reagents suggest that the labeling reaction should be carried out at room temperature, we have observed that sufficient labeling occurs at 4 °C.

1. Recombinant TRAIL is purified as per Protocol 2 (Steps 1 to 4 *only*).
2. As before, beads are washed several times with lysis buffer followed by extensive washing with PBS (or bicarbonate buffer) and then resuspended in 1 ml of PBS.
3. The Alexa Dye is warmed to room temperature and dissolved in a small volume of PBS. Beads are then resuspended in 1 ml of PBS and a range of different volumes of dye (e.g., 5 to 20 μl) added to each tube.
4. TRAIL is then labeled for 1 h at 4 °C on an end-to-end shaker and beads then washed in PBS. Beads are then eluted with either EDTA or imidazole (Protocol 2, Step 5). Because of the lability of the Alexa-647-TRAIL, Alex-labeled TRAIL should be stored in the dark in smaller aliquots than the unlabeled or biotin-labeled TRAIL variants. It should be noted, however, that we have not found any stability issues when Alexa-647-TRAIL is stored aliquoted in black-walled tubes at −80 °C.

Before use, Alexa-Fluor–TRAIL must first be checked to confirm biological activity. Because of the 650-nm emission wavelength of the Alexa-Fluor-647 dye (FL4), there is no direct interference with either FITC-labeled Annexin-V (FL1) or propidium iodide (FL2); therefore, the Annexin-V–staining protocol and FACS analysis (Protocol 3) can again be used.

By use of this particular fluorochrome-labeling reagent, we have never observed any reduction in biological activity because of overlabeling. However, significant underlabeling with Alexa-647 will result in either a weak, or no, fluorescent signal. To maintain consistency from batch-to-batch, the degree of labeling can be determined by measuring the absorbance of the TRAIL conjugate at both 280 nm (protein) and 650 nm (fluorescent dye). Then, by correcting for A_{280} of the Alexa dye (A_{280} − $(0.03 \times A_{650})$, and assuming an approximate extinction coefficient for the His-tagged TRAIL of 25,580 cm^{-1}M^{-1}, the protein concentration of Alexa-647-TRAIL can be determined. The degree of labeling can then be calculated with the following formula: moles dye/moles of protein = $A_{650}/239,00 \times$ protein concentration (239,00 cm^{-1}M^{-1} is the approximate molar coefficient of Alexa 647 at 650 nm).

3. ANALYSIS OF TRAIL RECEPTORS

3.1. Protocol 6A: The use of TRAIL receptor–specific antibodies to assess cell surface expression of TRAIL receptors

What follows is a series of protocols for staining of TRAIL receptors with a number of reagents obtained from eBiosciences (San Diego, CA) that we have found work well in our hands. The protocols can, of course, be modified for use with alternative receptor-labeling reagents. The following reagents: phycoerythrin (PE)-conjugated anti-TRAIL-R1 (12-6644), PE-conjugated anti-TRAIL-R2 (12-9908), and a PE-conjugated isotype-matched control antibody (12-4714) were all obtained from eBiosciences.

At least 1×10^6 cells/cell line, or specific treatment, is required. For staining, this cell population is then divided equally into four reactions: Control, no antibody; Isotype control, isotype-matched antibody; anti-TRAIL-R1; and anti-TRAIL-R2. All subsequent steps are performed with the cells on ice and/or solutions at 4 °C, unless indicated otherwise.

1. Cells are collected by centrifugation and resuspended in fresh medium at a concentration of 1×10^6/ml.
2. 250 μl of cell suspension is aliquoted into four individual flow cytometry tubes. Cells are then pelleted by centrifugation, resuspended in the indicated volume of normal goat serum (Control, 50 μl; Isotype control, 40 μl; anti-TRAIL-R1, 40 μl; and anti-TRAIL-R2, 45 μl), and left to recover on ice for 5 min.
3. The following volume of each fluorochrome-conjugated antibody is then added to the appropriate tube: Isotype control, 10 μl; anti-TRAIL-R1, 10 μl; -anti-TRAIL-R2, 5 μl. Cells are incubated on ice for 1 h in the dark.
4. Cell are then washed three times with PBS and analyzed by flow cytometry (FACSCalibur, Becton Dickinson) using excitation and emission wavelengths of 488 nm and 575 nm, respectively (FL2). A typical analysis of cell surface expression of TRAIL-R1 and TRAIL-R2 is shown in Fig. 19.3A.

3.2. Protocol 6B: The use of receptor-specific antibodies in adherent cells

For determination of "cell surface" expression of TRAIL-R1 or TRAIL-R2 in adherent cells, a modification of the preceding protocol is required.

1. Adherent cells are first removed from the plastic matrix by gentle agitation in a mild solution of trypsin and EDTA (essentially following conventional protocols).
2. As before, (Protocol 6A), at least 1×10^6 cells/cell line, or specific treatment, is required.
3. Cells are then centrifuged, resuspended in fresh "complete" medium, and then left to recover at 37 °C for 30 min (to allow for recovery of the cells after gentle trypsinization).
4. 250 μl of cell suspension is then aliquoted into four individual flow cytometry tubes. After centrifugation, cells are resuspended in normal goat serum, incubated on ice for 5 min, and then stained as described previously in Protocol 6A (Step 3, onward).

When initial studies were performed on the subcellular localization of TRAIL receptors after TRAIL receptor stimulation, it was found that most receptors exhibited an intracellular, specifically perinuclear, localization with relocalization occurring after TRAIL stimulation (Zhang *et al.*, 2000). In view of these findings, it is therefore advisable to also assess total cellular TRAIL receptor expression, as well as cell surface levels. By fixing and permeabilizing cells before receptor antibody staining, the preceding protocol can be modified to measure "total" TRAIL receptor expression (i.e., both cell surface and the intracellular pool). Essentially, the difference between the mean fluorescence intensities of the "cell surface" and "total" TRAIL receptor expression levels can then provide some indication as to the contribution of the "intracellular" TRAIL receptor pool. To success-fully label the "total" TRAIL receptor pool, cells are first fixed in formalin (10%) for 20 min at room temperature. After centrifugation, cells are washed once in PBS and then permeabilized by incubating for 5 min at room temperature in 0.5 ml PBS, containing 0.1% saponin and 1% BSA. The permeabilized cells are then equally divided between four flow cyto-metry tubes (125 μl/tube), centrifuged, then resuspended in normal goat serum and stained for TRAIL-R1 and TRAIL-R2 expression as described in Protocol 6A (Step 2 onward).

It is important to note that, by use of receptor-specific antibodies, the relative expression levels of an individual TRAIL receptor can only be reliably compared across various treatments within the same cell line or between different cell lines. However, because of the potentially different binding affinities of each TRAIL-Receptor–specific antibody, care should be taken when comparing the relative levels/ratio of the different TRAIL receptors (even within the same cell line).

The cell surface expression of TRAIL-R1 and TRAIL-R2 in BJAB cells, assessed by flow cytometry using receptor-specific antibodies, is shown in Fig. 18.3A. These data demonstrate that BJAB cells express significant levels of both TRAIL-R1 and TRAIL-R2 at the cell surface. On the basis of these data alone, one might be tempted to predict that this particular cell line is sensitive to TRAIL. However, it is now clear that the cell surface expression level of TRAIL receptors does *not* always correlate with, or, indeed, reliably predict, sensitivity to TRAIL. Consequently, several other measures, such as DISC analysis and a functional assessment of apoptosis are required to fully characterize each individual cell line in terms of TRAIL sensitivity.

3.3. Protocol 7: The use of TRAIL conjugated to fluorochromes

Many pharmacologic agents have been shown to sensitize cells to TRAIL by upregulating cell surface TRAIL receptor expression (Ganten *et al.*, 2004). By use of Alexa Fluor©-647–conjugated TRAIL (TRAIL-AF647),

A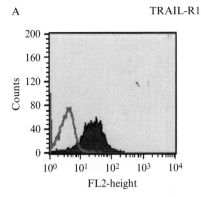

TRAIL-R1

Gate: G1

Marker	% Gated	Mean
All	100.00	28.56

TRAIL-R2

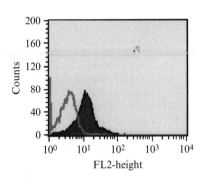

Gate: G1

Marker	% Gated	Mean
All	100.00	13.54

B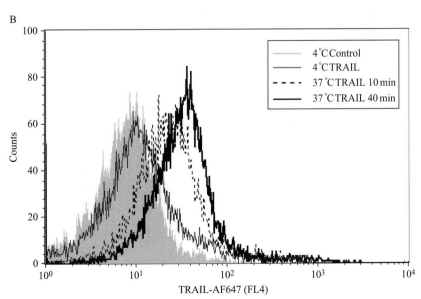

Figure 18.3 (*Continued*)

C   0 min   5 min

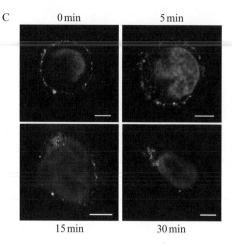

15 min   30 min

Figure 18.3   Cell surface labelling of TRAIL Receptors in BJAB cells. (A) Cell surface expression of TRAIL-R1 and TRAIL-R2 as assessed by flow cytometry using TRAIL receptor-specific antibodies. BJAB cells (1×10^6 cells/ml) were labelled with PE-labelled IgG1 (grey line, both panels), PE-TRAIL-R1 (solid line; upper panel) or PE-TRAIL-R2 (solid line; lower panel) as described in Protocol 6A. Surface expression of TRAIL-R1/TRAIL-R2 and IgG1 was measured by flow cytometry (FACSCalibur; Becton Dickinson) using excitation and emission wavelengths of 488 and 575 nm (FL2), respectively. Data was then analysed using CellQuest Pro software. The mean-fluorescence intensity (MFI) of TRAIL-R1–stained cells was 28.6 and the MFI of TRAIL-R2–stained cells was 13.5, compared with a MFI value of 3.0 for IgG1-stained cells (Isotype control). (B) Uptake and Internalization of TRAIL assessed by Flow Cytometry using TRAIL-AF647. BJAB cells (1×10^6 cells/ml) were incubated for the indicated times at 4 °C or 37 °C, in the presence or absence of recombinant TRAIL-AF647 (1 µg/ml), to assess uptake and rapid internalization of TRAIL-AF647 as described in Protocol 7. Following acid washes, TRAIL-AF647 fluorescence was detected by flow cytometry (FACSCalibur; Becton Dickinson) using excitation and emission wavelengths of 635 and 668 nm (FL4), respectively and data analysed using CellQuest Pro software. The histogram shows selected times during a kinetic analysis of TRAIL-AF647 internalization in BJAB cells. [b]modified with full copyright permission (Kohlhaas *et al.*, 2007). (C) Internalization of TRAIL assessed by Confocal Microscopy using Biotinylated-TRAIL and Streptavidin-labelled Alexa-568. Receptor-mediated internalization of TRAIL was assessed using Biotin-TRAIL as described in Protocol 8. BJAB cells (1×10^6 per sample) were chilled to 4 °C for 1 h followed by treatment with biotinylated-TRAIL (500 ng/ml) for 45 min at 4 °C. Cells were then washed three times with ice-cold PBS (to remove excess TRAIL) and treated with Streptavidin labelled Alexa-568 (*red*) for 1 h at 4 °C. Cells were washed and either fixed after 4 °C treatment (0 min) or released up to 37 °C for 5, 15 or 30 min (to allow TRAIL to internalize) and then fixed in 4% paraformaldehyde for 10 min at room temperature. After fixation, cells were counterstained with the DNA dye Hoechst-33342 (*blue*) for up 5 min to stain nuclei and then visualised using a Zeiss LSM510 with Axiovert 200 confocal microscope. Results shown are of one representative cell from each time point. The *white* bar represents 5 µm. [b]modified with full copyright permission (Kohlhaas *et al.*, 2007). (See color insert.)

as generated in Protocol 5, the TRAIL binding capacity of a cell can be conveniently measured by flow cytometry. To avoid dissociation of the fluorescently labeled ligand from the receptors, or internalization of the ligand-associated receptor complex, cells should first be "loaded" with Alexa-conjugated TRAIL on ice.

1. To obtain both a convenient and direct measure of the TRAIL-binding capacity of BJAB cells, cells (1 × 10^6 cells) are "loaded" with TRAIL-AF647 (1 μg/ml), or medium alone, for 1 h on ice.
2. Cells are then washed twice with ice-cold PBS and the extent of TRAIL-AF647 binding (to cell surface receptors only), is then immediately assessed by flow cytometry (see Step 7).

Alternately, fluorescently labeled TRAIL, in conjunction with FACS analysis, can instead be used to directly assess cellular uptake of recombinant TRAIL, thus providing a quantitative measure of ligand internalization:

3. In this case, BJAB cells (1 × 10^6 cells) are incubated either at 4 °C, in the presence or absence of TRAIL-AF647 (to block endocytosis), or at 37 °C in the presence of ligand (TRAIL-AF647; 1 μg/ml) for the indicated times to assess TRAIL internalization.
4. Samples are then rapidly chilled on ice (to inhibit endocytosis) and pelleted by brief centrifugation at 4 °C.
5. After two washes with prechilled "wash buffer" (20 mM HEPES, pH 7.4, 150 mM NaCl, 5 mM KCl, 1 mM CaCl$_2$, 1 mM MgCl$_2$), cell surface–associated ligand is then efficiently removed by resuspension in prechilled "acid wash" solution (0.2 M NaCl, 0.2 M acetic acid) for 5 min on ice.
6. After three washes in "wash buffer," cells are resuspended in ice-cold PBS containing 2% (w/v) fetal bovine serum.
7. Either TRAIL-AF647 binding (Steps 1 and 2 *only*) or TRAIL-AF647 uptake/internalization (Steps 3 to 6 *only*) should then be analyzed immediately by flow cytometry by use of excitation and emission wavelengths of 635 nm and 668 nm (FL4), respectively.

A typical analysis of TRAIL receptor–mediated internalization of recombinant TRAIL in BJAB cells, assessed by flow cytometry using Alexa-Fluor$^©$ 647–conjugated TRAIL, is shown in Fig. 18.3B. Cells were incubated with TRAIL-AF647, either on ice to block endocytosis or at 37 °C for increasing times to permit internalization. Increasing the temperature to 37 °C results in a rapid internalization of TRAIL, with uptake detected at 10 min and proceeding for up to 40 min. By use of this method, concentrations of TRAIL as low as 50 ng/ml resulted in uptake of detectable TRAIL (Kohlhaas *et al.*, 2007).

3.4. Protocol 8: Analysis receptor-mediated TRAIL internalization using biotinylated TRAIL

The role of receptor internalization in death receptor signaling is controversial with some studies showing that it is required for both TNF-R1 and CD95-mediated apoptosis, but not for TRAIL-mediated apoptosis (Kohlhaas et al., 2007; Lee et al., 2006; Schneider-Brachert et al., 2004). Internalization of TRAIL can be conveniently assessed with a combination of biotinylated-TRAIL together with an appropriate Alexa-Fluor© streptavidin conjugate (Molecular Probes).

To prevent "inappropriate/premature" internalization, it is imperative that cells are always kept on ice during the initial ligand "loading" phase. Once "loaded", cells can then be washed (in prewarmed medium) and brought up to 37 °C to permit internalization and start the experiment. After the required time periods at 37 °C, internalization is then "stopped" by returning cells to ice.

1. BJAB cells (1×10^6 cells) are first chilled on ice for 1 h before "loading" up with ligand.
2. Cells are then "loaded" with biotinylated TRAIL (500 ng/ml) or medium alone (Control) for 45 min on ice.
3. After three washes with ice-cold PBS (to remove excess TRAIL), cells are stained with Alexa Fluor© 568–conjugated streptavidin (AF568) for 1 h on ice.
4. Cells are then either immediately adhered to poly-L-lysine–coated slides and fixed (see Step 5) after treatment at 4 °C (0 min, control), or biotin-TRAIL is allowed to internalize by rapidly switching the cells to 37 °C (usually by resuspending the cells in prewarmed medium). Internalization is then "stopped" at the indicated time points by returning the cells to ice, and excess biotin-TRAIL is removed by washing at 4 °C. If required, cell surface–bound ligand can be removed by use of a mild acid wash (0.2 M acetic acid in PBS) before analysis.
5. Cells are then adhered to polylysine-coated slides and fixed in 4% paraformaldehyde for 10 min at room temperature. After three washes in PBS, nuclei are counterstained with the DNA dye, Hoechst 33342 (1 μg/ml) for 5 min.
6. Biotin-TRAIL internalization is then analyzed with a Zeiss LSM510, with Axiovert 200, confocal microscope.

A typical analysis of receptor-mediated internalization of biotinylated TRAIL in BJAB cells, visualized by confocal microscopy, is shown in Fig. 18.3C. Cells loaded at 4 °C show clear labeling of TRAIL on the cell surface, and no biotin-TRAIL seems to be internalized (Fig. 18.3C, 0 min). On warming the cells to 37 °C, internalization of TRAIL is observed as rapidly as 5 min and is further increased by 15 min, with marked staining of small vesicles (Fig. 18.3C). More complete internalization has occurred by

30 min, with increased vesicular staining concentrated in larger intracellular compartments.

It should be noted that, in cells that express significant amounts of cell surface TRAIL receptors, there is also the possibility that Alexa Fluor-conjugated–TRAIL (as produced in Protocol 5) could be used, in conjunction with confocal microscopy, to directly probe TRAIL/TRAIL receptor internalization.

3.5. Protocol 9: Isolation of TRAIL receptor complexes–TRAIL DISC analysis

On ligand binding, the TRAIL receptors, TRAIL-R1/-R2, bind adaptor molecules such as FADD through a death domain motif in their intracellular domain. FADD, a bipartite molecule, in turn binds the apical procaspase-8 through its death-effector domain motifs. Caspase-8 is then believed to be activated through a proximity-induced effect that relies on adjacent procaspase-8 molecules (Muzio et al., 1998). This receptor-bound cell surface complex is known as the death-inducing signaling complex or DISC. In our laboratory, we have successfully used biotin-labeled TRAIL to isolate the TRAIL DISC (Harper et al., 2001; 2003).

The number of cells required for DISC analysis depends on cell type. Generally, we use ~50 × 10^6 cells for each DISC isolation.

1. Cells are treated with biotin-labeled TRAIL (500 ng/ml) for the required time periods at 37 °C (to optimally assess TRAIL DISC formation), then cells are immediately washed two times with ice-cold PBS.
2. Cells are then immediately lysed (3 ml lysis buffer/50 × 10^6 cells), in a Triton lysis buffer (20 mM Tris/HCl (pH 7.5), 150 mM NaCl, 10% glycerol, 1% Triton X-100) containing Complete^{TM} protease inhibitors (Roche, Sussex UK), for 30 min on ice.
3. Lysates are then clarified by centrifugation at 13,000 rpm for 30 min and the supernatants collected.
4. Biotinylated TRAIL-bound complexes can then be simply precipitated overnight with streptavidin-Sepharose^{TM} beads (beads should be pre-washed with lysis buffer) at 4 °C.
5. Beads are then washed with several volumes of fresh lysis buffer, again containing Complete^{TM} protease inhibitors, and isolated complexes are then solubilized directly in SDS sample buffer.

A typical TRAIL DISC, isolated from BJAB cells and analyzed by SDS-PAGE and Western blotting, is shown in Fig. 18.4. By use of receptor-specific antibodies and FACs analysis, we previously demonstrated that BJAB cells express significant cell surface levels of both TRAIL-R1 and TRAIL–R2 (Fig. 18.3A). Here we show that exposure of BJAB cells to TRAIL at 37 °C results in rapid formation of a DISC, with binding of TRAIL-R1 and

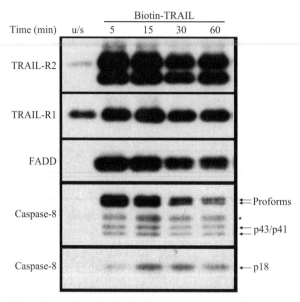

Figure 18.4 TRAIL induces DISC formation in BJAB cells in a time-dependent manner. BJAB cells (2.5×10^7 cells per sample) were treated with biotinylated-TRAIL (500 ng/ml) at 37 °C for the indicated times and then washed and lysed as described in Protocol 9. Analysis of the known DISC proteins, TRAIL-R2/-R2, FADD and Caspase-8, was then carried out by Western blot analysis as described in Protocol 9. Unstimulated cells (u/s), lysed and then treated with 500 ng/ml biotinylated-TRAIL, served as a control and confirmed *in vitro* binding of biotin-TRAIL to TRAIL-R1/-R2. * represents a nonspecific band detected by the Caspase-8 antibody. [b] modified with full copyright permission (Kohlhaas *et al.*, 2007).

TRAIL-R2 together with recruitment of FADD and procaspase-8 evident within 5 min (Fig. 18.4). Caspase-8 is, in turn, partially processed to its p43/p41 fragments, and a significant amount of its catalytically active large subunit is also detectable by 15 min. By 60 min, although significant amounts of TRAIL-R1 and -R2 are still bound to biotin-TRAIL, much less FADD and caspase-8 are present, presumably because these proteins have already dissociated from the TRAIL/TRAIL-Receptor complex.

A number of commonly used cell types express extremely low levels of TRAIL receptors (e.g., Jurkat T cells). In these cells, we have often failed to detect TRAIL receptors directly by Western blotting of whole cell lysates. This is either due to poor affinity of some commercially available TRAIL receptor antibodies on Western blots or, more likely, the low level of expression of TRAIL receptors in these cells. In these particular cell types, some form of initial "concentration" step is therefore required before TRAIL receptor detection by Western blotting.

In our laboratory, we have addressed this issue by routinely using biotin-TRAIL to essentially "concentrate" all TRAIL receptors from unstimulated cell lysates onto streptavidin beads before SDS-PAGE and Western blot analysis. Importantly, fewer cells are required for this than for "conventional" DISC analysis; however, it does allow for the analysis of TRAIL receptors from many more cells than a crude cell lysate would allow. In this respect, it should also be noted that incubation of live cells at 4 °C or 37 °C with biotinylated TRAIL, followed by washing and cell lysis, results in TRAIL binding of *only* cell surface–associated TRAIL receptors. On the other hand, the addition of biotin-TRAIL to precleared, unstimulated, cell lysates as described previously, results in precipitation of the "total" cellular TRAIL receptor pool (both cell surface and intracellular TRAIL receptors).

There seems to be a critical requirement for aggregation of death receptors to induce signaling and caspase-8 activation (Boldin *et al.*, 1995; Martin *et al.*, 1998). After TRAIL treatment, TRAIL receptors have been shown to form SDS–stable aggregates (Kischkel *et al.*, 1995). These receptor complexes can also be isolated by DISC precipitation (Protocol 9) and can then be "crudely" analyzed either by conventional SDS-PAGE under non–reducing conditions or by gel filtration column chromatography.

4. CONCLUDING REMARKS

The ability to generate biologically active recombinant TRAIL, as well as biotin or fluorescently labeled TRAIL variants, has provided our laboratory with tools that have proved to be invaluable in the assessment of TRAIL/TRAIL-Receptor signaling in tumor cell lines and primary tumor cells. As described previously, an assessment of cell surface TRAIL receptor expression alone, in most cases, gives little/no indication of TRAIL sensitivity. However, by use of a combination of the standard protocols listed herein an assessment of cellular sensitivity to TRAIL, TRAIL DISC analysis, and TRAIL-TRAIL-Receptor complex internalization can be routinely assessed, thus providing key information on TRAIL signaling in a particular cell type.

Finally, it is important to note that many of the protocols presented here can also be applied to study the biological function of other TNF family members.

ACKNOWLEDGMENTS

Work described here was supported by the UK Medical Research Council at the MRC Toxicology Unit in Leicester. Many thanks go to past and present members of the Apoptosis Laboratory, 416. In particular, we wish to thank Prof. Gerald M. Cohen for his continued support, Drs. Andrew Craxton, Susan Kohlhass, Michael Pinkoski, and Laura Hall for permission to use previously published or unpublished data, and Davina Twiddy for help with TRAIL receptor antibody staining protocols.

REFERENCES

Boldin, M. P., Mett, I. L., Varfolomeev, E. E., Chumakov, I., Shemer-Avni, Y., Camonis, J. H., and Wallach, D. (1995). Self-association of the "death domains" of the p55 tumor necrosis factor (TNF) receptor and Fas/APO1 prompts signaling for TNF and Fas/APO1 effects. *J. Biol. Chem.* **270**, 387–391.

Dyer, M. J., MacFarlane, M., and Cohen, G. M. (2007). Barriers to effective TRAIL-targeted therapy of malignancy. *J. Clin. Oncol.* **25**, 4505–4506.

Ganten, T. M., Haas, T. L., Sykora, J., Stahl, H., Sprick, M. R., Fas, S. C., Krueger, A., Weigand, M. A., Grosse-Wilde, A., Stremmel, W., Krammer, P. H., and Walczak, H. (2004). Enhanced caspase-8 recruitment to and activation at the DISC is critical for sensitisation of human hepatocellular carcinoma cells to TRAIL-induced apoptosis by chemotherapeutic drugs. *Cell Death Differ.* **11**(Suppl 1), S86–S96.

Harper, N., Farrow, S. N., Kaptein, A., Cohen, G. M., and MacFarlane, M. (2001). Modulation of tumor necrosis factor apoptosis-inducing ligand-induced NF-kappa B activation by inhibition of apical caspases. *J. Biol. Chem.* **276**, 34743–34752.

Harper, N., Hughes, M. A., Farrow, S. N., Cohen, G. M., and MacFarlane, M. (2003). Protein kinase C modulates tumor necrosis factor-related apoptosis-inducing ligand-induced apoptosis by targeting the apical events of death receptor signaling. *J. Biol. Chem.* **278**, 44338–44347.

Inoue, S., MacFarlane, M., Harper, N., Wheat, L. M., Dyer, M. J., and Cohen, G. M. (2004). Histone deacetylase inhibitors potentiate TNF-related apoptosis-inducing ligand (TRAIL)-induced apoptosis in lymphoid malignancies. *Cell Death Differ.* **11**(Suppl 2), S193–S206.

Janolino, V. G., Fontecha, J., and Swaisgood, H. E. (1996). A spectrophotometric assay for biotin-binding sites of immobilized avidin. *Appl. Biochem. Biotech.* **56**, 1–7.

Kischkel, F. C., Hellbardt, S., Behrmann, I., Germer, M., Pawlita, M., Krammer, P. H., and Peter, M. E. (1995). Cytotoxicity-dependent APO-1 (Fas/CD95)associated proteins form a death-inducing signaling complex (DISC) with the receptor. *EMBO J.* **14**, 5579–5588.

Kohlhaas, S. L., Craxton, A., Sun, X. M., Pinkoski, M. J., and Cohen, G. M. (2007). Receptor-mediated endocytosis is not required for tumor necrosis factor-related apoptosis-inducing ligand (TRAIL)induced apoptosis. *J. Biol. Chem.* **282**, 12831–12841.

Koschny, R., Holland, H., Sykora, J., Haas, T. L., Sprick, M. R., Ganten, T. M., Krupp, W., Bauer, M., Ahnert, P., Meixensberger, J., and Walczak, H. (2007). Borte-zomib sensitizes primary human astrocytoma cells of WHO grades I to IV for tumor necrosis factor-related apoptosis-inducing ligand-induced apoptosis. *Clin. Cancer Res.* **13**, 3403–3412.

LeBlanc, H. N., and Ashkenazi, A. (2003). Apo2L/TRAIL and its death and decoy receptors. *Cell Death Differ.* **10**, 66–75.

Lee, K. H., Feig, C., Tchikov, V., Schickel, R., Hallas, C., Schutze, S., Peter, M. E., and Chan, A. C. (2006). The role of receptor internalization in CD95 signaling. *EMBO J.* **25**, 1009–10023.

Li, H., Zhu, H., Xu, C. J., and Yuan, J. (1998). Cleavage of BID by caspase 8 mediates the mitochondrial damage in the Fas pathway of apoptosis. *Cell* **94**, 491–501.

MacFarlane, M. (2003). TRAIL-induced signalling and apoptosis. *Toxicol. Lett.* **139**, 89–97.

MacFarlane, M., Ahmad, M., Srinivasula, S. M., Fernandes-Alnemri, T., Cohen, G. M., and Alnemri, E. S. (1997). Identification and molecular cloning of two novel receptors for the cytotoxic ligand TRAIL. *J. Biol. Chem.* **272**, 25417–35420.

MacFarlane, M., Cohen, G. M., and Dickens, M. (2000). JNK (c-Jun N-terminal kinase) and p38 activation in receptor-mediated and chemically-induced apoptosis of T-cells: Differential requirements for caspase activation. *Biochem. J.* **348**(Pt 1), 93–101.

MacFarlane, M., Harper, N., Snowden, R. T., Dyer, M. J., Barnett, G. A., Pringle, J. H., and Cohen, G. M. (2002). Mechanisms of resistance to TRAIL-induced apoptosis in primary B cell chronic lymphocytic leukaemia. *Oncogene* **21,** 6809–6818.

MacFarlane, M., Kohlhaas, S. L., Sutcliffe, M. J., Dyer, M. J., and Cohen, G. M. (2005). TRAIL receptor-selective mutants signal to apoptosis via TRAIL-R1 in primary lymphoid malignancies. *Cancer Res.* **65,** 11265–11270.

Martin, D. A., Siegel, R. M., Zheng, L., and Lenardo, M. J. (1998). Membrane oligomerization and cleavage activates the caspase-8 (FLICE/MACHalpha1) death signal. *J. Biol. Chem.* **273,** 4345–4349.

Martin, S. J., Reutelingsperger, C. P., McGahon, A. J., Rader, J. A., van Schie, R. C., LaFace, D. M., and Green, D. R. (1995). Early redistribution of plasma membrane phosphatidylserine is a general feature of apoptosis regardless of the initiating stimulus: inhibition by overexpression of Bcl-2 and Abl. *J. Exp. Med.* **182,** 1545–1556.

Muzio, M., Stockwell, B. R., Stennicke, H. R., Salvesen, G. S., and Dixit, V. M. (1998). An induced proximity model for caspase-8 activation. *J. Biol. Chem.* **273,** 2926–2930.

Pan, G., Ni, J., Wei, Y. F., Yu, G., Gentz, R., and Dixit, V. M. (1997). An antagonist decoy receptor and a death domain-containing receptor for TRAIL. *Science* **277,** 815–818.

Peter, M. E., Hellbardt, S., Schwartz-Albiez, R., Westendorp, M. O., Walczak, H., Moldenhauer, G., Grell, M., and Krammer, P. H. (1995). Cell surface sialylation plays a role in modulating sensitivity towards APO-1-mediated apoptotic cell death. *Cell Death Differ.* **2,** 163–171.

Peter, M. E., and Krammer, P. H. (2003). The CD95(APO-1/Fas) DISC and beyond. *Cell Death Differ.* **10,** 26–35.

Pitti, R. M., Marsters, S. A., Ruppert, S., Donahue, C. J., Moore, A., and Ashkenazi, A. (1996). Induction of apoptosis by Apo-2 ligand, a new member of the tumor necrosis factor cytokine family. *J. Biol. Chem.* **271,** 12687–12690.

Savage, M. D., Mattson, G., Desai, S., Nielander, G. W., Morgensen, S., and Conklin, E. J. (1992). "Avidin-Biotin Chemistry: A Handbook." Pierce Chemical Co. (Product #15055).

Schneider-Brachert, W., Tchikov, V., Neumeyer, J., Jakob, M., Winoto-Morbach, S., Held-Feindt, J., Heinrich, M., Merkel, O., Ehrenschwender, M., Adam, D., Mentlein, R., Kabelitz, D., and Schutze, S. (2004). Compartmentalization of TNF receptor 1 signaling: Internalized TNF receptosomes as death signaling vesicles. *Immunity* **21,** 415–428.

Sheridan, J. P., Marsters, S. A., Pitti, R. M., Gurney, A., Skubatch, M., Baldwin, D., Ramakrishnan, L., Gray, C. L., Baker, K., Wood, W. I., Goddard, A. D., Godowski, P., *et al.* (1997). Control of TRAIL-induced apoptosis by a family of signaling and decoy receptors. *Science* **277,** 818–821.

Tartaglia, L. A., Ayres, T. M., Wong, G. H., and Goeddel, D. V. (1993). A novel domain within the 55 kd TNF receptor signals cell death. *Cell* **74,** 845–853.

Varfolomeev, E., Maecker, H., Sharp, D., Lawrence, D., Renz, M., Vucic, D., and Ashkenazi, A. (2005). Molecular determinants of kinase pathway activation by Apo2 ligand/tumor necrosis factor-related apoptosis-inducing ligand. *J. Biol. Chem.* **280,** 40599–40608.

Wiley, S. R., Schooley, K., Smolak, P. J., Din, W. S., Huang, C. P., Nicholl, J. K., Sutherland, G. R., Smith, T. D., Rauch, C., Smith, C. A., and Goodwin, R. G. (1995). Identification and characterization of a new member of the TNF family that induces apoptosis. *Immunity* **3,** 673–682.

Zhang, X. D., Franco, A. V., Nguyen, T., Gray, C. P., and Hersey, P. (2000). Differential localization and regulation of death and decoy receptors for TNF-related apoptosis-inducing ligand (TRAIL) in human melanoma cells. *J. Immunol.* **164,** 3961–3970.

CHAPTER NINETEEN

Analysis of Tnf-Related Apoptosis-Inducing Ligand *In Vivo* Through Bone Marrow Transduction and Transplantation

Keli Song, Raj Mariappan, *and* Roya Khosravi-Far

Contents

Department of Pathology, Beth Israel Deaconess Medical Center and Harvard Medical School, Boston, Massachusetts

Methods in Enzymology, Volume 446

ISSN 0076-6879, DOI: 10.1016/S0076-6879(08)01619-4

Abstract

Tumor necrosis factor (TNF)–related apoptosis-inducing ligand (TRAIL) is a member of the TNF family of cytokines. TRAIL has gained much attention because of its ability to preferentially kill tumor cells with no apparent toxic side effects. Recently, different TRAIL receptor agonists, including TRAIL itself and various agonistic monoclonal antibodies against the two apoptosis-inducing human TRAIL receptors, have been developed as novel cancer thera-peutics and are currently under investigation in clinical trials. However, the mechanisms by which TRAIL mediates its selective antineoplastic activity are still not well understood. In addition to playing a role in cancer immune surveillance and tumor suppression, TRAIL has been associated with immune homeostasis, inflammatory diseases, and autoimmunity. In light of the multi-functional role of TRAIL in mediating various pathologic conditions and the potential benefits of TRAIL-based therapies, the study of the physiologic signif-icance of TRAIL is of great importance. Here, we describe a syngeneic system for the characterization of the *in vivo* function of TRAIL. By use of this model, in which the full-length murine TRAIL protein is overexpressed in the hematopoie-tic cells of wild-type mice, the *in vivo* tumoricidal activity of TRAIL overexpres-sion can be studied on syngeneic murine tumor cell challenge, and the potential toxicity of TRAIL protein to normal tissues can also be analyzed.

1. INTRODUCTION

TRAIL is a type II membrane protein. Although TRAIL mRNA has been found in a broad range of tissues and cells in the body (Wiley *et al.*, 1995), TRAIL protein has mainly been detected in cells of the immune system, such as helper and cytotoxic (CD4$^+$, CD8$^+$) T cells natural killer (NK) cells, B cells, macrophages, monocytes, and dendritic cells. Yet, among those cell types, only murine NK cells constitutively express TRAIL, whereas the other human and murine cell types express TRAIL after activa-tion (Kayagaki *et al.*, 1999; Smyth *et al.*, 2003). Functional TRAIL expres-sion has also been found in some organ tissues, particularly in the eye. Human and mouse ocular tissue cells constitutively express TRAIL, and tumor surveillance by TRAIL in the eye has been shown in experimental mice (Lee *et al.*, 2002).

The initial discoveries of the antitumor properties of TRAIL were made by use of several recombinant soluble forms of human and mouse TRAIL protein in cell culture and in experimental animal studies by different laboratories. The *in vivo* antineoplastic activity of recombinant TRAIL was first demonstrated in xenograft animal models, in which soluble human TRAIL was administered to immune-deficient mice bearing human tumor cells (Ashkenazi *et al.*, 1999; Walczak *et al.*, 1999). Although the human and murine forms of TRAIL protein share 65% amino acid identity and demonstrate cross-species activity on certain cell types (Wiley *et al.*, 1995), the cross-reactivity is not identical between the two species (Walczak *et al.*, 1999; Wiley *et al.*, 1995). Therefore, a xenogenic system may be an inadequate model for the evaluation of the full profile of TRAIL activity. Here we describe a syngeneic system in which bone marrow cells explanted from donor wild-type mice are transduced with a retroviral vector expressing full-length murine TRAIL and subsequently transplanted into recipient mice. This bone marrow transduction and transplantation model results in the expression of TRAIL in the hematopoietic cells of the recipient mice. Because TRAIL protein is mainly expressed in cells from a hematopoietic origin, this model is consistent with certain physiologic responses. In our study, mice that overexpress TRAIL in their hematopoietic cells do not have detectable abnormalities (Song *et al.*, 2006). This model can be used to evaluate the effects of TRAIL overexpression *in vivo*, including the antitumor activity of TRAIL on syngeneic murine tumor cell challenge and the potential toxicity of TRAIL protein to normal tissue cells.

The first step in generating TRAIL-transduced mice involves the transfection of a retroviral packaging cell line with retroviral cDNA to make a retroviral infection medium. Next, bone marrow cells harvested from wild-type donor mice are transduced *in vitro* with the retroviral infection medium. Transduced bone marrow cells are then transplanted into lethally irradiated recipient mice of the same strain, as outlined in Fig. 19.1. The overall procedure entails the following steps.

2. GENERATION OF RETROVIRAL INFECTION MEDIUM

A retroviral infection medium is produced by transient transfection of a retrovirus-packaging cell line, BOSC23 (ATCC, Rockville, MD), with a retroviral cDNA vector and an ecotropic glycoprotein-encoding vector. The transfection can be performed by use of either Lipofectamine (Invitrogen, Carlsbad, CA) according to the manufacturer's instructions or the calcium-phosphate method, as previously described (Song *et al.*, 2006). The pMIG retroviral vector encoding a modified humanized green fluorescent protein (GFP), which is used to generate a negative TRAIL control sample, the pMIG/mTRAIL retroviral vector encoding full-length mouse TRAIL (mTRAIL), as well as humanized GFP and the pCl$_3$-EcotR vector encoding ecotropic glycoprotein, have been previously described (Song *et al.*, 2006). The titer of the retroviral infection medium is generally on the order of 10^6 to 10^7 viral particles/ml.

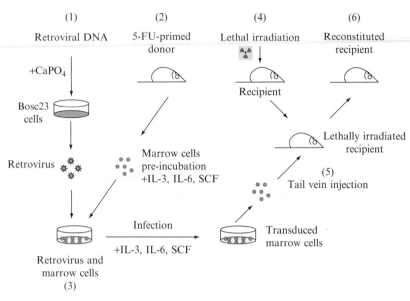

Figure 19.1 Diagrammatic outline of retrovirus-mediated mouse TRAIL gene transfer into mouse bone marrow cells followed by bone marrow transplantation. Order of the steps is indicated by numbers in parentheses. (1) Control or TRAIL-expressing retrovirus is generated by transfection of Bosc23 cells with respective retroviral cDNA. (2) Bone marrow cells are isolated from donor mice primed with 5-FU. (3) Cultured bone marrow cells are infected with control or TRAIL-expressing retrovirus. (4) Bone marrow recipient mice are lethally irradiated and (5) are transplanted with the gene-transduced bone marrow cells by trail vein injection. (6) This leads to bone marrow reconstituted mice.

3. BONE MARROW CELL PREPARATION

Before isolating bone marrow cells from donor mice, the mice must be primed with 5-fluorouracil (5-FU) (Sigma-Aldrich, St. Louis, MO). 5-FU treatment serves to kill proliferating cells by interfering with the activities of DNA and RNA, thereby inhibiting bone marrow cell proliferation and thus enriching hematopoietic progenitor cells in the bone marrow. 5-FU is prepared in a sodium chloride solution, sterilized by filtration through a 0.22-μm filter (Millipore, Bedford, MA), and administered in a single dose of 150 mg/kg by tail vein injection into 6- to 8-week old wild-type BALB/c female mice.

Bone marrow isolation is performed on day 4 after 5-FU treatment. First, a mouse is euthanized with CO_2. The fur skin is soaked with 70% ethanol. The mouse is laid dorsal side down on a surgery board. The femurs and tibias are dissected out and placed in a petri dish on ice. An opening is made at both ends of the femurs and tibias with surgery scissors, and the bone marrow is flushed out into a collection tube with a 21–G2 syringe needle attached to a 3-ml syringe containing 2 ml of Eagle's MEM supplemented with 2% FBS (Stem Cell Technology, Vancouver, British Columbia, Canada) by inserting the tip of the needle into one opened end of the bone and then pressing the plunger of the syringe to force the 2% FBS cell medium through the bone.

The ensuing steps are carried out in the tissue culture hood. The bone marrow is resuspended in the 2% FBS cell medium by use of a 5-ml pipette, washed by adding fresh Eagle's MEM for a total volume of 10 ml, pipetting up and down, and then centrifuged in a SORVALL® RT600D centrifuge (DuPONT, Wilmington, DE) at 1500 rpm for 5 min at 4 °C. After centrifugation, the supernatant is removed, and the cell pellet is resuspended in Eagle's MEM containing 15% FBS. A small aliquot of the cell suspension is removed, mixed with a 9-fold volume of the red blood cell lysis solution ACK (0.15 mol/L NH_4CI, 1.0 mmol/L $KHCO_3$, 0.1 mmol/L Na_2EDTA, pH 7.3) and used to estimate the cell count (Kruisbeek, 1993). In our experience, 1 to 3×10^6 bone marrow cells can be obtained from the femurs and tibias of a single mouse prepared as described previously. The cells are plated at 10×10^6 per petri dish (100-mm, nontissue culture treated) in 10 ml of Eagle's MEM containing 15% FBS, 2 mM glutamine, 50 units/ml penicillin, 50 μg/ml streptomycin, 1 mM pyruvate, 6 μg/ml murine IL-3, 10 μg/ml murine IL-6, and 100 μg/ml murine stem cell factor (PeproTech Inc., Rocky, New Jersey). The culture is maintained in a cell incubator at 37 °C under 5% CO_2 for 48 h.

4. RETROVIRAL TRANSDUCTION OF BONE MARROW CELLS

In this step, retroviral transduction is enhanced by use of RetroNectin® (Takara Bio Inc., distributed by Fisher Scientific, Pittsburgh, PA), a recombinant human fibronectin fragment CH–296 that serves to colocalize target cells and virions on the RetroNectin® molecules. The RetroNectin®-enhanced transduction can be performed according to the manufacturer's instructions or with some minor modifications as follows.

4.1. Coating plate with RetroNectin®

A stock solution of RetroNectin® is prepared by dissolving RetroNectin® in sterile ddH_2O to a concentration of 1 mg/ml and filtering the solution through a 0.22-μm MILLEX®–GV filter (Millipore, Bedford, MA). The stock solution can be stored at −20 °C. On the day of transduction, nontissue culture–treated polystyrene 6-well plates (Becton Dickinson Labware, Franklin Lakes, NJ) are coated with RetroNectin®. To prepare the RetroNectin® coated plates, 2 ml of RetroNectin® working solution at 10 μg/ml in PBS is added to each well, and the plate is allowed to stand at room temperature for 2 h. The RetroNectin® solution is then removed, and 2 ml of the blocking solution, 2% bovine serum albumin (BSA) in PBS, is added to each well. Again, the plate is allowed to stand at room temperature for 30 min. The wells should include those for samples as GFP negative control that will not be infected and an aliquot will be used as blank for cell sorting by flow cytometry.

4.2. Cell preparation

The cell supernatant from the bone marrow cell culture in a petri dish is collected into a 15-ml or 50-ml conical tube. Three to 5 ml of Eagle's MEM is added to the harvested petri dish, and the remaining cells are scraped into the medium by use of a Costar® cell lifter (Corning Incorporated, Corning, NY) and collected into the conical tube. Harvested cells are centrifuged in a SORVALL® RT600D centrifuge at 1500 rpm at 4 °C for 5 min. The supernatant is removed, and the cells are resuspended in Eagle's MEM containing 60% FBS. Cell concentration is counted with a small aliquot (10 μl) of the cell suspension mixed with a 9-fold volume (90 μl) of red blood cell lysis solution ACK. The cell concentration is adjusted to a range of 3 to 4 × 10^6 cells/ml. While this step proceeds, stock retroviral infection medium is thawed in a 37 °C waterbath, then kept on ice.

4.3. Adding retroviral medium

The BSA blocking solution is removed from the RetroNectin® coated 6-well plate. The wells are washed twice with 5 ml of Hanks' balanced salt solution (HBSS) containing 25 mM HEPES. Certain wells are reserved for control samples, such as a negative GFP control that will not be infected and a negative TRAIL control (humanized GFP). The well for the negative GFP control sample will contain the same number of cells as well as the same concentrations of FBS and growth factors as described in the following but will lack the retroviral infection medium. The following components are dispensed in each infection well: 0.5 ml cell suspension containing 60% FBS (1 to 1.5 × 10^6 cells) and 1.5 ml of stock retroviral infection medium (pMIG as negative TRAIL control or pMIG-mTRAIL as TRAIL transduction) to make a total volume of 2 ml with final concentrations of 15% for FBS, 6 ng/ml for IL-3, 10 ng/ml for IL-6, and 100 ng/ml for SCF. The cells are incubated at 37 °C under 5% CO_2 for 2 days (including repetition of infection the next day, as described later).

4.4. Repetition of infection

Twenty-four hours after the initial infection, 1 ml of stock retroviral infection medium is added to each infection well, and 1 ml of Eagle's MEM is added to the GFP negative control well. Appropriate amounts of FBS and growth factors are added to each well to maintain the concentrations at the levels described previously. The culture is incubated at 37 °C under 5% CO_2 for another 24 h.

4.5. Washing cells

The cells in each well are resuspended by use of a pipette and transferred to a conical tube. To collect the remaining adherent cells, 1 ml of Eagle's MEM containing 15% FBS is added to cover the bottom of each well, the cells are suspended in solution by use of a pipette, and this fraction of the cells is combined with the previously collected portion of cells from each well. For each tube of cells, corresponding to an individual well, 5 ml of Eagle's MEM is added, and the tubes are centrifuged at 1500 rpm in a SORVALL® RT600D centrifuge for 5 mins at 4 °C. After the supernatant is removed, the cells are resuspended in 2 ml of Eagle's MEM containing 15% FBS and plated back to their original well. FBS and growth factors are added to each well to adjust the concentrations to the same levels as described previously. The cells are incubated at 37 °C under 5% CO_2 for 2 days. We typically achieve an infection efficiency of approximately 30% with primary bone marrow cells (unpublished observations).

5. Bone Marrow Transplantation

In this step, transduced cells are harvested and sorted by fluorescence-activated cell sorting analysis (FACS) (Becton Dickinson Immunocytometry System, FACS Calibur, San Jose, CA), and bone marrow recipient mice are lethally irradiated. Although these two preparations are described separately later, they are carried out almost simultaneously. Sorted GFP$^+$ cells are then injected into the tail vein of the irradiated syngeneic mice.

5.1. Cell preparation

The cultured cells are resuspended in each well with a pipette and transferred to a conical tube. The remaining adherent cells are collected by adding 4 ml of PBS to each well in the PBS and by use of a Costar® cell lifter to suspend the cells in solution. These cells are combined with the cell suspension previously collected from the same well. The cells are spun in a

SORVALL® RT600D centrifuge at 1500 rpm for 5 min at 4 °C, and the supernatant is removed. For lysing red blood cells, the centrifuged cells are resuspended in1 ml of the ACK solution and kept at room temperature for 5 min. The cell suspension is diluted by adding 9 ml of PBS, and the cell number is counted with a small aliquot of the diluted cell suspension. Once again, the cells are centrifuged and the supernatant is removed. The cells are resuspended in PBS at 10 to 20×10^6 cells/ml and placed on ice. Cell sorting is carried out immediately. GFP^+ cells are sorted along with untransduced bone marrow cells (negative GFP control) by FACS and collected in Eagle's MEM (without serum). Sorted cells ($GFP^+ >90\%$) are washed twice in cold PBS by centrifugation and kept on ice. While the cell preparation steps proceed, the irradiation step (following) is performed almost simultaneously.

5.2. Irradiation of bone marrow recipient mice

Mice that are the same strain as the bone marrow donors are used as bone marrow recipients. A gamma irradiator GammaCell® 40 Exactor (MDS Nordion, Ottawa, ON, Canada) is used to generate γ-radiation. The minimum lethal dosage of irradiation is chosen and may vary depending on the specific strain of mice. For example, two doses of 450 rad, with each dose 4 h apart, are administered to BALB/c female 6- to 8-week-old mice, whereas one dose of 950 rad is given to B57BL/6 mice. Determination of the minimum lethal dosage can be obtained experimentally. For each experiment, two to four mice that are lethally irradiated but not transplanted with syngeneic bone marrow cells. These mice are used as an irradiation control mice to evaluate the success of the irradiation.

5.3. Tail vein injection

Transduced cells suspended in PBS are transplanted to bone marrow recipient mice by I.V. injection of 0.5 to 1×10^6 cells in 200 μl into the lateral tail vein. Negative control (negative for GFP or TRAIL transduction) mice are injected with either nontransduced cells (GFP negative control) or blank vector transduced cells (GFP-positive negative TRAIL control). A 1-ml syringe with a 30-G{1/²} needle can be used for tail vein injection.

 # 6. ASSESSMENT OF BONE MARROW TRANSPLANTATION

Successful bone marrow transplantation (BMT) can be evaluated first by observing the death of irradiated mice to which no syngeneic bone marrow cells were transplanted after irradiation and the survival of those irradiated mice that were subjected to bone marrow transplantation. Survival of irradiated mice that did not receive donor bone marrow cells may suggest that the irradiation is not successful and/or the dosage of irradiation below the minimum lethal amount required. The death of irradiated mice that were reconstituted with donor bone marrow cells may be attributed to various factors, including 1) the dosage of irradiation is too high, 2) low viability of the transplanted bone marrow cells, or 3) pathologic conditions of the mice. In our study, irradiated mice without bone marrow transplantation usually die in a week, and we consider the survival of irradiated mice with BMT beyond 3 weeks an indication of successful BMT.

To detect transduced gene expression, visualization of GFP in the white blood cells (WBC) of peripheral blood samples can serve as an initial index for successful BMT and transduced gene expression *in vivo*. This can be achieved without antibody staining by performing FACS analysis on a WBC suspension. The WBC suspension is prepared by withdrawing whole blood samples from the animal and lysing the red blood cells with ACK buffer. Peripheral whole blood samples can be obtained from the orbital sinus or by tail bleeding into an anticoagulant tube. In our experience, the cell sorting step performed during cell preparation for bone marrow transplantation, as described previously, yields a WBC population in which >90% of cells are GFP$^+$. Levels of TRAIL mRNA transcription *in vivo* can be detected by quantitative real-time RT-PCR by use of total WBC RNA from the animals. Primers and probes for the RT-PCR analysis of mouse TRAIL expression have been described previously (Ghaffari *et al.*, 2003). TRAIL protein can be detected by an immunoblotting assay that uses WBC lysates and the rat anti-mouse TRAIL antibody N2B2 (BD Biosciences PharMingen, San Diego, CA). TRAIL expression *in situ* in tissue such as spleen can be detected by immunohistochemistry staining on tissue slides, which have been prepared from 10% formalin-preserved and then paraffin-embedded samples, by use of a TRAIL antibody as the primary antibody (ICL labs Inc., Newberg, Oregon) (Song *et al.*, 2006).

7. Pathologic Evaluation of TRAIL-Transduced Mice

Physical examination is the first step in assessing any abnormalities of the transduced mice. This is carried out 1 and 2 months after bone marrow transplantation, allowing a sufficient amount of time for the reconstituted bone marrow to replace the original bone marrow function. The routine physical examination includes body weight, body temperature, heart rate, respiratory rate, and activity.

In addition, morphologic examination of a bone marrow cytospin preparation is helpful in identifying general features such as the presence of different cell lineages at various stages of maturation. This is commonly done by suspending a small aliquot of the diluted cell suspension (50 μl)m before cell lysis, and staining with May–Grunwald Giemsa, Wright Giemsa, or Diff-Quik (least preferred to assess cytologic details) stains. Identification of myeloid, erythroid, and megakaryocytic series with assessing maturity will give a fair overview of cellularity, maturation, and presence and absence of toxicity.

Because TRAIL transduction is carried out with bone marrow cells, an extensive analysis of hematopoietic cells is one of the major aspects of the evaluation of any potential toxic effect by TRAIL overexpression *in vivo*. A complete blood cell count and a differential leukocyte count of peripheral blood samples provide valuable information regarding the quality and quantity of each cell population of the whole blood. These analyses are performed by use of an automated hematology analyzer.

It has been shown that systemic administration of TNF or the Fas ligand (FasL) can cause lethal toxic effects (Nagata, 1997) and intraperitoneal (i.p.) administration of an anti-Fas antibody into wild-type mice results in severe damage to the liver by apoptosis and rapid death (Ogasawara *et al.*, 1993). To study the potential toxicity of TRAIL overexpression by bone marrow transduction and transplantation, we have administered a Fas agonist antibody (clone Jo2) (BD Biosciences PharMingen, San Diego, CA) at 50 μg per mouse by intravenous (I.V.) injection into wild-type control mice to generate toxicity-positive samples and compared tissue samples from these mice with samples from the TRAIL-transduced mice. Five hours after the injection of the Jo2 anti-Fas antibody, the major organs, including the liver, spleen, kidney, lung, heart, and brain, were collected and immediately preserved in 10% formalin and processed into paraffin-embedded samples. The organs from age- and sex-matched TRAIL-transduced mice were also processed during the same day and in the same fashion. We used hematoxylin and eosin (H&E) staining for histopathologic analysis and immunohistochemical staining for evaluation of apoptosis by the terminal deoxynucleotidyl transferase-mediated dUTP nick end-labeling (TUNEL) assay on 5-μm thick formalin-fixed, paraffin-embedded tissue sections. In our study, we found that TRAIL-transduced mice demonstrated TRAIL overexpression in WBC and spleen samples and had normal complete blood counts and differential counts. The results of H&E and immunohistochemical staining also showed no visible toxic evidence in their organs, as shown in Fig. 19.2. The TRAIL-transduced mice had a normal life span; they lived 1.5 years before being euthanized ($n = 3$).

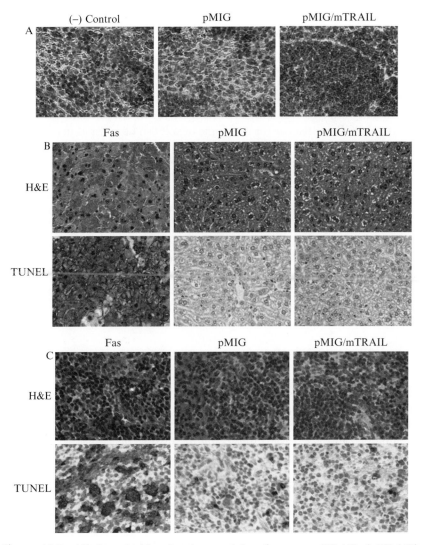

Figure 19.2 (A) Immunohistochemistry staining for mouse TRAIL (mTRAIL) protein in spleen tissue from mice 1 month after BMT. (−) Control mice received bone marrow without retroviral infection. pMIG mice received pMIG retroviral-infected bone marrow. pMIG/mTRAIL mice received bone marrow infected with the pMIG/ mTRAIL retrovirus. Positive immunohistochemistry staining for mTRAIL in spleen sample of mTRAIL-transduced mouse is shown. (B) Histologic and pathologic examination on liver samples from bone marrow recipient mice. Samples from (top labels left to right) Fas: mice that received 50 μg Fas agonist antibody by I.V. injection. The samples were harvested and processed for pathologic examination study 5 h after the injection; other sample labels are indicated in Fig. 19.2A. Top panel, H&E staining. Bottom panel, TUNEL staining. Liver hemorrhage can be seen in H&E staining of the Fas sample. (C) Spleen samples labeled in the same order as in Fig. 19.2B. (See color insert.)

8. METHODS FOR IMMUNOHISTOCHEMICAL STAINING FOR TRAIL EXPRESSION

Immunohistochemistry (IHC) is a commonly used technique in pathology for detection and characterization of antigens of interest. Immunohistochemistry has an added advantage over Western blots, flow cytometry, or PCR in that the morphologic details can be assessed in addition to detecting antigen expression. In the bone marrow transduction and transplantation model described, IHC analysis permits detecting the cell lineage in which the antigen of interest is expressed.

Many commercial kits are available for optimal immunohistochemical staining (such as VectaStain). We used a Dako immunostain kit for detecting TRAIL antigen staining. The following are the general steps in the IHC staining of paraffin-embedded bone marrow tissue sections. Tissue sections are cut in silanized charged slides (SuperFrost, Fisher Scientific or BBC Biochemical, Stanwood, WA) at 4 to 5-μm thick. The slides are baked (45 to 60 min at 56 to 60 °C) in a drying oven and subsequently cooled to room temperature.

8.1. Deparaffinization

The slides are labeled with an appropriate marker (the process removes alcohol-based ink). The slides are immersed in a Coplin jar or glass jar in the following steps: xylene, three changes 5 min each; 100% ethanol, two changes 5 min each; 95% ethanol, repeat two times. After this, the slides are immersed in water for 5 min or till the next step.

8.2. Antigen retrieval

Antigen retrieval is an essential step to expose epitopes of the antigens that have been partially denatured by the formalin treatment and paraffin-embedding process. Antigen retrieval can be done in many ways, including protease digestion to digest the cross bonds created between proteins during formalin processing, mild heat or steam, EDTA (ethylene diamine tetra-acetic acid), Tris, or citrate retrievals. TRAIL immunostains were performed by use of pressure retrieval with citrate buffer. This is commercially available or can be prepared (2.94 g sodium citrate dihydrate in 1000 ml of double distilled water, 500 μl Tween 20. The pH is adjusted to 6.0 with HCL). Pressure retrieval is performed by use of the Pascal pressure cooker (Dako, Carpinteria, CA). The slides are brought to room temperature, washed with TBS-T (Tris-buffered saline with Tween-20: 0.05 M Tris (6.06 g), 0.3 M NaCl (17.53 g), 0.1% Tween 20 (0.1 ml). Add ddH$_2$O to a total volume of 100 ml. Adjust pH to 7.2 to 7.6 with HCl). After this, the slides are wiped with a clean tissue paper, making sure the tissue is not rubbed. The tissue is marked with a circle drawn with a PAP-pen (that creates an hydrophobic barrier for reagents).

8.3. Serum block

Add 3 drops of stock rabbit serum (serum from the animal from which the antibody was made) to 10 ml of TBS buffer in a mixing bottle. Use 3 to 5 drops to cover the entire tissue. Incubate in room temperature for 10 min. Wash with TBST; two times 5 min each.

8.4. Peroxidase block

This step is crucial specifically for marrow or splenic tissue because bone marrow cells are endogenous peroxidase rich. The slides are treated with 100 μl (2 to 5 drops to cover entire tissue) of hydrogen peroxide (30% hydrogen peroxide in 0.9% normal saline—100 μl of hydrogen peroxide in 900 μl of normal saline). If the resulting IHC slide has a lot of background because of endogenous peroxidase staining, as a troubleshooting step, this solution is made fresh. Incubate for 10 to 30 min. Wash with TBST three times 5 min each after this blocking step.

8.5. Primary antibody stains

Rabbit anti–mouse polyclonal TRAIL antibodies (ICL Labs Inc., Newburg, OR) were used in a dilution of 1:50 (diluted in TBS-bovine serum albumin diluent). Approximately 100 μl of antibody solution is used to cover the entire specimen and incubated for 1 h in room temperature or overnight at 4 to 8 °C. Alternately, the entire slide can also be immersed in a slide chamber with 2.5 ml of TBS-BSA with 2 μl of primary antibody solution and placed in a rocker at room temperature for 1 h. An inexpensive slide chamber for antibody incubation purposes is marketed by a vendor (Pro-Histo, Columbia, SC).

8.6. Secondary antibody and detection

Rinse the slides with PBS-T twice, 5 mins each; 100 μl of a horseradish peroxidase–conjugated secondary antibody targeting rabbit (anti-rabbit) IgG is used and incubated for 30 min at room temperature (Dako Envision Plus kit). The Envision plus kit from Dako uses a proprietary dextran-based polymer that helps in reducing background stain and increases primary staining intensity. Similar polymer-based kits are also available from other vendors, including VectaStain. After this, wash twice with TBST and incubate with the DAB-chromogen (this is freshly prepared. In the avidin biotin chromogen method, the biotinylated secondary antibodies interact with avidin biotin–peroxidase complex to develop a brown immunostain. To avoid overstaining, this step is closely monitored, and the slides are dipped in double distilled water as the color starts emerging. This can be done near the microscope to be able to monitor the staining under a low-power objective. Once the slides develop a positive stain, they are washed in water to remove the excess stain and dehydrated by reversing the preceding hydration steps: 95% ethanol—100% ethanol—xylene. The slides can be incubated in xylene overnight to ensure complete dehydration and removal of residual ethanol. After this step, they are coverslipped with toluene-based mounting medium and can be stored indefinitely.

9. Applications that Use the Bone Marrow Transplantation Approach

We used our TRAIL-transduced mouse model to study the potential pathologic effect of TRAIL overexpression on normal tissues and the capability of this model to enhance antitumor surveillance. Our results have shown that TRAIL overexpression in hematopoietic cells can inhibit the growth of syngeneic tumor grafts and that TRAIL overexpression is not toxic to normal tissues (Song *et al.*, 2006).

The bone marrow transplantation approach can be further explored as a means to study the roles of other molecules in the regulation of TRAIL function *in vivo*. Some intracellular proteins are thought to be key mediators of TRAIL-induced apoptosis. For example, overexpression of Bcl-2 or Bcl-XL in several human cancer cell lines blocks TRAIL-induced cell death (Fulda *et al.*, 2002; Guo and Xu, 2001; Lamothe and Aggarwal, 2002; Munshi *et al.*, 2001). TRAIL sensitivity is also associated with the expression of cellular FLICE inhibitory protein (c-FLIP) in several human cell lines (Griffith *et al.*, 1998; Kim *et al.*, 2000; 2002; Zhang *et al.*, 2004). NFκB, Akt, and the FOXO family of Forkhead transcription factors have also been implicated in the regulation of TRAIL sensitivity of some cancer cells (Ghaffari *et al.*, 2003; Modur *et al.*, 2002; Nesterov *et al.*, 2001; Ravi and Bedi, 2002; Thakkar *et al.*, 2001). The role of these proteins in mediating TRAIL-induced apoptosis seems to be cell type–dependent, because studies on some other cell types failed to show a similar effect (Keogh *et al.*, 2000; LeBlanc and Ashkenazi, 2003; Mitsiades *et al.*, 2002; Zhang *et al.*, 2001). Most of the studies of the potential regulators were performed *in vitro*; thus, it will be necessary to determine whether those regulators have the same functions *in vivo*. To test the role of a given candidate molecule in the regulation of TRAIL function *in vivo*, a chosen cancer cell line may be transduced to overexpress the candidate gene, and these cancer cells may then be inoculated in TRAIL-transduced syngeneic mice for assessing the role of the candidate gene in TRAIL-mediated tumor inhibition. Similarly, another gene of interest, other than TRAIL, may be transduced into hematopoietic cells of mice through this BMT approach. The resulting mice may then be inoculated with syngeneic tumor cells for the evaluation of the role of gene of interest in tumor growth.

In summary, a BMT approach can serve as a useful tool for the study of TRAIL-mediated anticancer activity *in vivo* and can also be exploited to study the role of other molecules in the regulation of cancer growth *in vivo*. Considering that the efficiency of retroviral-mediated gene transfer in cultured primary cells is usually approximately 30%, and if a high percentage of the transduced cells are required in transduced mice, cell sorting by FACS is necessary to enrich the population of transduced cells. In our experience, approximately two bone marrow donor mice are required to generate one bone marrow–transduced (recipient) animal. Consequently, if a large quantity of transduced mice were required for an *in vivo* study, twice the number of bone marrow donor mice would be needed to provide sufficient bone marrow cells. In addition to the large number of mice, this procedure is usually a labor-consuming process. This could be a limiting factor for such large-scale studies.

ACKNOWLEDGMENTS

We thank members of the Khosravi-Far laboratory, past and present, for their scientific contributions, and special thanks to Jessica Platti for her comments and suggestions on the manuscript. R. K. F. is supported by NIH grants CA105306 and HL080192. R. K. F. is an American Cancer Society Scholar. Immunohistochemical methodology was supported by the Specialized Histopathology Core Lab of the Dana Farber/Harvard Cancer Center (NIH-P30CA6516).

REFERENCES

Ashkenazi, A., Pai, R. C., Fong, S., *et al.* (1999). Safety and antitumor activity of recombinant soluble Apo2 ligand. *J. Clin. Invest.* **104(2),** 155–162.

Fulda, S., Meyer, E., and Debatin, K. M. (2002). Inhibition of TRAIL-induced apoptosis by Bcl-2 overexpression. *Oncogene* **21(15),** 2283–2294.

Ghaffari, S., Jagani, Z., Kitidis, C., Lodish, H. F., and Khosravi-Far, R. (2003). Cytokine and BCR-ABL mediate suppression of TRAIL-induced apoptosis through inhibition of forkhead FOXO3a transcription factor. *Proc. Natl. Acad. Sci. USA* **100,** 6523–6528.

Griffith, T. S., Chin, W. A., Jackson. G. C., Lynch, D. H., and Kubin, M. Z. (1998). Intracellular regulation of TRAIL-induced apoptosis in human melanoma cells. *J. Immunol.* **161(6),** 2833–2840.

Guo, B. C., and Xu, Y. H. (2001). Bcl-2 over-expression and activation of protein kinase C suppress the trail-induced apoptosis in Jurkat T cells. *Cell Res.* **11(2),** 101–106.

Kayagaki, N., Yamaguchi, N., Nakayama, M., Eto, H., Okumura, K., and Yagita, H. (1999). Type I interferons (IFNs) regulate tumor necrosis factor–related apoptosis-inducing ligand (TRAIL) expression on human T cells: A novel mechanism for the antitumor effects of type I IFNs. *J. Exp. Med.* **189(9),** 1451–1460.

Keogh, S. A., Walczak, H., Bouchier-Hayes, L., and Martin, S. J. (2000). Failure of Bcl-2 to block cytochrome c redistribution during TRAIL-induced apoptosis. *FEBS Lett.* **471(1),** 93–98.

Kim, K., Fisher, M. J., Xu, S. Q., and el-Deiry, W. S. (2000). Molecular determinants of response to TRAIL in killing of normal and cancer cells. *Clin. Cancer Res.* **6(2),** 335–346.

Kim, Y., Suh, N., Sporn, M., and Reed, J. C. (2002). An inducible pathway for degradation of FLIP protein sensitizes tumor cells to TRAIL-induced apoptosis. *J. Biol. Chem.* **277(25),** 22320–22329.

Kruisbeek, A. M. (1993). In vitro assays for mouse lymphocyte function. In "Current Protocols in Immunology" (J.E Coligan, A.M Kruisbeek, D.H Margulies, E.M Shevach, and W. Strober, eds.) pp. 3.1.1–3.1.5, John Wiley & Sons, Inc. New York.

Lamothe, B., and Aggarwal, B. B. (2002). Ectopic expression of Bcl-2 and Bcl-xL inhibits apoptosis induced by TNF-related apoptosis-inducing ligand (TRAIL) through suppression of caspases-8, 7, and 3 and BID cleavage in human acute myelogenous leukemia cell line HL-60. *J. Interferon Cytokine Res.* **22(2),** 269–279.

LeBlanc, H. N., and Ashkenazi, A. (2003). Apo2L/TRAIL and its death and decoy receptors. *Cell Death Differ.* **10(1),** 66–75.

Lee, H. O., Herndon, J. M., Barreiro, R., Griffith, T. S., and Ferguson, T. A. (2002). TRAIL: A mechanism of tumor surveillance in an immune privileged site. *J. Immunol.* **169(9),** 4739–4744.

Mitsiades, N., Mitsiades, C. S., Poulaki, V., Anderson, K. C., and Treon, S. P. (2002). Intracellular regulation of tumor necrosis factor-related apoptosis-inducing ligand-induced apoptosis in human multiple myeloma cells. *Blood* **99(6),** 2162–2171.

Modur, V., Nagarajan, R., Evers, B. M., and Milbrandt, J. (2002). FOXO proteins regulate tumor necrosis factor–related apoptosis inducing ligand expression. Implications for PTEN mutation in prostate cancer. *J. Biol. Chem.* **277(49),** 47928–47937.

Munshi, A., Pappas, G., Honda, T., McDonnell, T. J., Younes, A., Li, Y., and Meyn, R. E. (2001). TRAIL (APO-2L) induces apoptosis in human prostate cancer cells that is inhibitable by Bcl-2. *Oncogene* **20(29),** 3757–3765.

Nagata, S. (1997). Apoptosis by death factor. *Cell* **88,** 355–365.

Nesterov, A., Lu, X., Johnso, M., Miller, G. J., Ivashchenko, Y., and Kraft, A. S. (2001). Elevated AKT activity protects the prostate cancer cell line LNCaP from TRAIL-induced apoptosis. *J. Biol. Chem.* **276(14),** 10767–10774.

Ogasawara, J., Watanabe-Fukunaga, R., Adachi, M., Matsuzawa, A., Kasugai, T., Kitamura, Y., Itoh, N., Suda, T., and Nagata, S. (1993). Lethal effect of the anti-Fas antibody in mice. *Nature* **364(6440),** 806–809.

Ravi, R., and Bedi, A. (2002). Requirement of BAX for TRAIL/Apo2L-induced apoptosis of colorectal cancers: Synergism with sulindac-mediated inhibition of Bcl-x(L). *Cancer Res.* **62(6),** 1583–1587.

Smyth, M. J., Takeda, K., Hayakawa, Y., Peschon, J. J., van den Brink, M. R., and Yagita, H. (2003). Nature's TRAIL—On a path to cancer immunotherapy. *Immunity* **18(1),** 1–6.

Song, K., Benhaga, N., Anderson, R. L., and Khosravi-Far, R. (2006). Transduction of tumor necrosis factor-related apoptosis-inducing ligand into hematopoietic cells leads to inhibition of syngeneic tumor growth *in vivo*. *Cancer Res.* **66,** 6304–6311.

Thakkar, H., Chen, X., Tyan, F, Gim, S., Robinson, H., Lee, C., Pandey, S. K., Nwokorie, C., Onwudiwe, N., and Srivastava, R. K. (2001). Pro-survival function of

Akt/protein kinase B in prostate cancer cells. Relationship with TRAIL resistance. *J. Biol. Chem.* **276(42),** 38361–38369.

Walczak, H., Miller, R. E., Ariail, K., *et al.* (1999). Tumoricidal activity of tumor necrosis factor–related apoptosis-inducing ligand *in vivo. Nat. Med.* **5(2),** 157–163.

Wiley, S. R., Schooley, K., Smolak, P. J., *et al.* (1995). Identification and characterization of a new member of the TNF family that induces apoptosis. *Immunity* **3(6),** 673–682.

Zhang, X., Jin, T. G., Yang, H., DeWolf, W. C., Khosravi-Far, R., and Olumi, A. F. (2004). Persistent c-FLIP(L) expression is necessary and sufficient to maintain resistance to tumor necrosis factor–related apoptosis-inducing ligand-mediated apoptosis in prostate cancer. *Cancer Res.* **64(19),** 7086–7091.

Zhang, X. D., Zhang, X. Y., Gray, C. P., Nguyen, T., and Hersey, P. (2001). Tumor necrosis factor–related apoptosis-inducing ligand-induced apoptosis of human melanoma is regulated by smac/DIABLO release from mitochondria. *Cancer Res.* **61(19),** 7339–7348.

CHAPTER TWENTY

Overcoming Resistance to Trail-Induced Apoptosis in Prostate Cancer by Regulation of c-FLIP

Xiaoping Zhang, Wenhua Li, *and* Aria F. Olumi

Contents

Abstract

Using Tumor necrosis factor Related Apoptosis Inducing Ligand (TRAIL) for cancer therapy is attractive, because TRAIL is effective against cancer cells without inducing significant cytotoxicity, making it an ideal cancer drug. However, some cancer cells evade TRAIL-induced apoptosis and become resistant. We have been investigating the molecular mechanisms that differentiate between TRAIL-resistant and TRAIL-sensitive prostate cancer cells. We have found that transcriptional regulation of the anti-apoptotic molecule, c-FLIP(L), can regulate sensitivity of cancer cells to TRAIL. We have found that c-Fos, represses expression of c-FLIP(L), and promotes TRAIL-induced apoptosis.

Department of Urology, Massachusetts General Hospital, Harvard Medical School, Boston, Massachusetts

Methods in Enzymology, Volume 446
ISSN 0076-6879, DOI: 10.1016/S0076-6879(08)01620-0

Identifying molecular mechanisms that differentiate between sensitive and resistant cancer cells will help improve pro-apoptotic cancer therapies.

1. INTRODUCTION

Apoptosis is a cellular response that regulates important processes such as homeostasis, immunosurveillance, and elimination of unwanted cells. It has become clear that most, but not all, types of apoptosis require activation of a class of cysteine proteases, the caspases. There are two major signaling pathways of apoptosis: the extrinsic pathway and intrinsic pathway. The extrinsic pathway activates executive caspase-8 and caspase-10 through death receptors on the cellular surface, whereas intrinsic pathway activates caspase-9 by releasing cytochrome *c* and activating other mediators from mitochondria.

Caspases are activated in a hierarchical order. Activated caspase-8 triggers the extrinsic apoptotic pathway by directly activating effectors such as caspase-3 and caspase-7. Caspase-8 can also initiate the intrinsic apoptotic pathway through the activation of Bid (Sinicrope and Penington, 2005; Suliman *et al.*, 2001). Both pathways lead to the activation of caspase-3 and eventual apoptotic cell death (Suliman *et al.*, 2001).

Shortly after the identification of caspase-8 and caspase-10, a structurally related protein was cloned independently by nine groups and, therefore, originally had eight different names (Tschopp *et al.*, 1998). The most widely used name is c-FLIP, shortened from cellular FLICE inhibitor protein (FLICE: FADD-like interleukin-1b-converting enzyme).

2. STRUCTURE AND MOLECULAR MECHANISMS OF c-FLIP IN APOPTOSIS

c-FLIP is a human cellular homolog of viral FLICE-inhibitory proteins(v-FLIPs) (Thome *et al.*, 1997) and contains tandem death-effector domains and caspase-like domain similar to pro-caspase-8 and pro-caspase-10 but lacks amino acid residues that are critical for caspase activity, most notably the cysteine in the catalytic center (Chang and Yang, 2000; Irmler *et al.*, 1997). When ligands such as TNFα, FasL, or TRAIL interact with specific death domain receptors, the interaction will induce intracellular cytoplasmic formation of the DISC (death inducing signaling complex) (Bodmer *et al.*, 2000; Chang*et al.*, 2006; Kischkel *et al.*, 2000; Micheau and Tschopp, 2003; Schneider *et al.*, 1997; Sheridan *et al.*, 1997; Sprick *et al.*, 2000). DISC formation involves recruitment of caspase-8/10 through an adaptor protein (FADD, TRADD, and RIP) to the death effector domain (DED) of the activated receptor (Pan *et al.*, 1997; Zhang *et al.*, 2005). c-FLIP protein homologs interrupt apoptotic signaling by competing with caspase-8 for binding to the DED domains of FADD and also regulate apoptosis through their interference with the recruitment of caspase-8 to FADD (Irmler *et al.*, 1997; Medema *et al.*, 1997; Wajant *et al.*, 2000).

c-FLIP mRNA gives rise to multiple protein isoforms. There are three protein isoforms: a long c-FLIP form (c-FLIP[L]) (Irmler *et al.*, 1997), a short c-FLIP form (c-FLIP[s]) (Irmler *et al.*, 1997; Krueger *et al.*, 2001a; 2001b), and a third recently identified form, called FLIP(R) (Golks *et al.*, 2005). All three isoforms contain DED domains and can, therefore, remain bound to FADD and interrupt complete caspase-8/10 processing and activation. However, the exact roles of different c-FLIP isoforms remain controversial (Peter, 2004). As an example, most published reports involving ectopic expression of c-FLIP(L) suggest that c-FLIP(L) or its caspase-cleaved 43-kDa form has an anti-apoptotic role. Moreover, c-FLIP$^{-/-}$ mouse embryonic fibroblasts have been shown to be more sensitive to FasL-induced apoptosis (Yeh *et al.*, 2000b), which strongly suggests that c-FLIP(L) has an anti-apoptotic function. However, two reports have proposed that c-FLIP(L) may have a dual function, a pro-apoptotic function at low physiologic concentrations and an anti-apoptotic function at high cellular concentrations (Chang *et al.*, 2002; Micheau *et al.*, 2002).

3. EXPRESSION OF C-FLIP

c-FLIP is predominantly expressed in the heart, skeletal muscle, and peripheral blood leukocytes (Irmler *et al.*, 1997). Surprisingly, some viruses encode the homolog of c-FLIP, which also controls sensitivity toward death–receptor–mediated apoptosis (Thome *et al.*, 1997). Moreover, c-FLIP–deficient mice do not survive past day 10.5 of embryogenesis and exhibit impaired heart development (Yeh *et al.*, 2000a). Hence, c-FLIP is thought to be involved in the regulation of the immune system and development. The expression of c-FLIP proteins and their role in tumor progression are still under investigation. Two recent reports showed that c-FLIP in Hodgkin's lymphomas could protect lymphoma cells from autonomous FasL-mediated cell death while preserving their ability to evade immunosurveillance (Dutton *et al.*, 2004; Mathas *et al.*, 2004), which indicates that c-FLIP has a crucial role in regulation of cell death, a potential role in malignant transformation, proliferation, and metastasis, and the levels of intracellular c-FLIP, therefore, may determine the sensitivity of cancer cells to apoptotic triggers. We collected microarray data from Oncomine

Table 20.1   c-FLIP expression in normal and cancerous tissues

	Normal vs. Normal	Cancer vs. Normal	Cancer vs. Cancer	Tumor Grade
Appendix	1			
Blood	2			
Head–Neck	1			
Heart	1			
Lung	1	1	1 1	
Lymph Node	1			
Muscle	1			
Others	1			
Ovarian	1 1			
Placenta	1			
Prostate	1			
Spinal Cord	1 1			
Thymus	2			
Tongue	1			
Tonsil	1			
Trachea	1			
Umbilical	1			
Brain	3	1	1 1	1
CellLine			1	
Lymphoma			3 4	
Multi-cancer			1 1	
Renal			1 2	
Breast				
Bladder				
Leukemia		1	1	
Normal				
Myeloma				
Sarcoma			1	

P-Value Threshold: 1E-6; Outlier Rank Threshold: 50; Numbers stands for the number of studies
■ High level; ■ Medium level; □ Low level; ■ High level; ■ Medium level; □ Low level

(www.oncomine.org) and analyzed the expression of c–FLIP in different normal and tumor tissues. We found that c–FLIP expression was relatively higher in blood, head–neck, lung, lymph node, muscle, tonsil, trachea, and umbilical tissues, whereas c–FLIP levels were lower in brain tissue (Table 20.1). Therefore, microarray gene data suggest that expression of c–FLIP may vary in different organs. Moreover, the expression of c–FLIP may be increased in some malignancies compared with normal tissue, whereas its expression may be decreased in other malignancies compared with normal tissues (Fig. 20.1). Immunohistochemical analyses in bladder (Korkolopoulou *et al.*, 2004) and prostate cancer (Dr. A. P. Kumar,

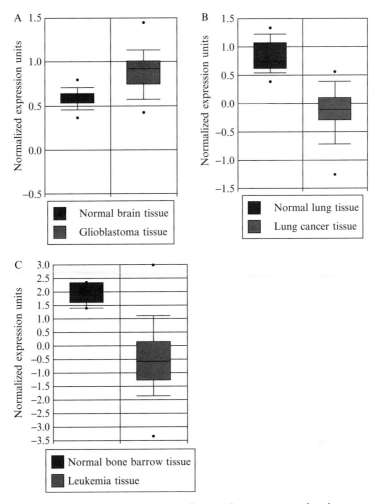

Figure 20.1   The differential expression of c-FLIP between normal and cancerous tissues. (A) c-FLIP level in glioblastoma is higher than that of normal brain tissues. (B) c-FLIP level in lung cancer is lower than that of normal lung tissues. (C) c-FLIP level in leukemia is lower than that of normal bone marrow tissues.

University of Texas, personal communication) suggest that c-FLIP's expression may correlate with more advanced tumors.

4. REGULATION OF C-FLIP EXPRESSION

Intracellular c-FLIP(L) can be regulated at the transcriptional, translational, or posttranslational levels (Kim, 2002; Zhang *et al.*, 2004). We have shown in the past that persistent expression of c-FLIP(L) is necessary and sufficient to maintain resistance to TRAIL-induced apoptosis. Expression of c-FLIP(L) has been shown to be modulated by NF-κB (Benoit et al., 2004; Okamoto et al., 2006), Akt (Namet al., 2003; Skurk et al., 2004), c-Myc (Ricci et al., 2004), p53 (Fukazawa et al., 2001), and E3-ubiquitin ligase (Chang et al., 2006).

To determine whether alterations in transcription can affect TRAIL-induced apoptosis, TRAIL-resistant (PC3-TR and LNCaP) and TRAIL-sensitive (PC3) cells (Zhang et al., 2004; Zhang et al., 2007b) were treated with TRAIL/Apo-2L (100 ng/ml) in the presence or absence of actinomycin D (Fig. 20.2A), a general inhibitor of transcription. We found that actinomycin D had little effect on the cell viability of the TRAIL/Apo-2L-sensitive cells, yet a combination of actinomycin D and TRAIL/Apo-2L enhanced apoptosis in TRAIL-resistant PC3-TR and LNCaP, suggesting that inhibition of cellular transcription can enhance TRAIL-induced apoptosis (Fig. 20.2B). Microarray analysis of differentially expressed genes after TRAIL treatment suggested that expression of c-Fos was significantly upregulated in the TRAIL-sensitive PC3 cells. In contrast, expression of c-Fos was significantly down regulated in the TRAIL-resistant PC3-TR and LNCaP cells (Table 20.2).

Because c-FLIP(L) is partially regulated transcriptionally (Gao et al., 2005; Li et al., 2007; Roue et al., 2007), its putative promoter region contains multiple c-Fos/AP-1 binding sites (Fig. 20.2C), and c-Fos is differentially expressed in TRAIL-sensitive and TRAIL-resistant cells, we wished to examine whether c-Fos regulates expression of c-FLIP(L). We hypothesized that AP-1 family of proteins may be an important regulator of c-FLIP(L) and, as a result, play a key role in mediating a cell's response to TRAIL-induced apoptosis. We examined the potential AP-1 binding sites in the putative c-FLIP(L) regulatory region (17,000 base pairs upstream of c-FLIP(L)'s ATG start codon [Fig. 20.2C]). We identified and examined binding of c-Fos to 14 AP-1 binding sites in the putative c-FLIP(L) regulatory region by means of chromatin immunoprecipitation (ChIP) assays, which included six AP-1 binding sites upstream of exon 1 (designated sites "a" through "f" in Fig. 20.2C) and eight within intron 1-2. We only detected binding of c-Fos protein to c-FLIP(L)'s AP-1(f) site (Fig. 20.2D). ChIP assays demonstrated that binding of c-Fos to the c-FLIP(L) AP-1(f) site increased in the TRAIL-sensitive PC3 cells, whereas c-Fos binding to the c-FLIP(L) AP-1(f) site was reduced in the TRAIL-resistant PC3-TR and LNCaP cells after treatment with TRAIL/Apo-2L. To confirm the importance of c-Fos/AP-1 binding AP-1(f) site on regulating c-FLIP(L) expression, we deleted this AP-1(f) site in our c-FLIP(L) promoter luciferase reporter. We found that deletion of c-FLIP(L)'s AP-1(f) site abolished the ability of c-Fos to suppress c-FLIP(L) expression (Fig. 20.2D). More detailed studies have shown that TRAIL treatment in TRAIL-sensitive cancer cells promote c-Fos to translocate from the cytoplasm to the nucleus and upregulate AP-1 activity. We have found that

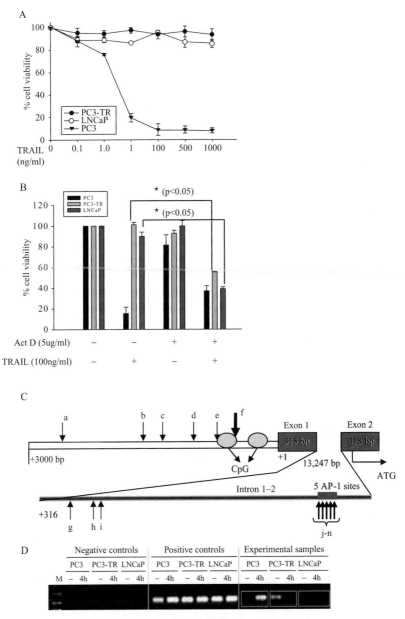

Figure 20.2 c-FLIP in prostate cancer cells can be regulated at the transcriptional level. (A) Cell viability of prostate cancer cells measured by MTT assay. (B) Cell viability was assessed in prostate cancer cells that were pretreated with RNA synthesis inhibitor, Actinomycin D (Act D), for 1 h followed by treatment with TRAIL (100 ng/ml) for another 24 h. (C) Potential AP-1 binding sites in the cFLIP(L) promoter and regulatory region. (D) AP-1 (f) binding to c-FLIP(L) promoter analyzed by CHIP assay. Error bars indicate SD of three replicate experiments. "★" Represents significant differences between controls and experimental samples. (C and D were reproduced with permission from *Cancer Res.* **67**, 9425; 2007.)

Table 20.2 Differentially expressed genes in TRAIL-sensitive and TRAIL-resistant cells

Genes	PC3 U	PC3 T	LCB	PC3-TR U	PC3-TR T	LCB	LNCaP U	LNCaP T	LCB
SOCS box-containing WD protein SWiP-1	91.61	155.32	1.54	210.78	102.14	−1.85	307.6	103.79	−2.67
Notch homolog 3 (Drosophila)	95.16	231.15	2.22	90.97	37.69	−1.63	126.91	49.19	−2.2
v-fos FBJ murine osteosarcoma viral oncogene homolog	−0.07	53.02	**2.57**	93.08	−3.34	**−4.55**	61.81	7.13	**−2.2**
collagen, type VI, alpha 1	718.18	1790.2	2.16	235.85	136.6	−1.55	401.67	240.52	−1.51
myosin VIIB	33.06	92.81	2.53	32.59	17.05	−1.52	52.44	20.92	−1.92
hypothetical protein FLJ14360	14.36	66.69	3.79	27.88	5.79	−2.26	25.29	9.5	−1.53

upregulation and cytoplasmic to nuclear translocation of c–Fos is necessary, but insufficient, for cancer cells to be sensitive to TRAIL. We postulate that one of the mechanisms that c–Fos/AP-1 primes cancer cells to TRAIL is through direct binding of c–Fos protein to the c–FLIP(L) putative promoter region and repressing the expression of c–FLIP(L) (Zhang *et al.*, 2007b).

5. TARGETING c-FLIP TO ENHANCE APOPTOSIS IN CANCER CELLS

Because c–FLIP is an important modulator of apoptosis, and its expression and activity can be regulated at multiple levels, targeting c–FLIP(L) as a cancer therapeutic agent can be attractive. For example, the combination of TRAIL with the chemotherapeutic agent, doxorubicin, has been shown to effectively decrease the expression of c–FLIP(L), thus sensitizing prostate cancer cells to TRAIL-induced apoptosis (Kelly *et al.*, 2002). Others have shown that CDDO (a novel triterpenoid, 2-cyano-3,12-dioxooleana-1,9-dien-28-oic acid) compounds may sensitize cells to pro-apoptotic agents by downregulation of c–FLIP(L) or upregulation of the cell surface TRAIL receptors, TRAIL-R1 and TRAIL-R2 (Hyer *et al.*, 2005). Previously ,we had also found that c–Fos/AP-1 functions as a pro-apoptotic molecule by directly repressing the anti-apoptotic gene, c–FLIP(L). Other groups have shown that expression of c–FLIP(L) could be modulated by NF-κB (Benoit *et al.*, 2004; Okamoto *et al.*, 2006). These findings suggest that strategies to potentiate c–Fos/AP-1 activation and/or inhibit NF-κB may repress the expression of c–FLIP(L) and enhance the efficacy of apoptotic inducers, such as TRAIL, for treatment of various malignancies.

12-O-Tetradecanoylphorbol-13-acetate (TPA) is a strong inducer of c–Fos/AP-1. We have demonstrated that TRAIL or a TRAIL-R2 agonist antibody combined with low-dose TPA upregulates AP-1 proteins and its activity, reduces c–FLIP(L) levels, and potentiates apoptosis in TRAIL-resistant LNCaP cells in *in vitro* and *in vivo* experiments (Zhang *et al.*, 2007a). Therefore, TPA, when combined with the pro-apoptotic agent TRAIL, is effective in changing the phenotype of some TRAIL-resistant prostate cancers to a TRAIL-sensitive phenotype.

We have also used the proteosome inhibitor, MG-132, to inhibit NF-κB. We have found that MG-132 not only inhibited NF-κB activity (Fig. 20.3A), but it also increased AP-1 activity by promoting nuclear translocation of c–Fos and c–Jun and their heterodimerization (Fig. 20.3B) (Li *et al.*, 2007). Treatment of the cells with MG-132 alone did not affect the level of c–FLIP(L) protein levels, however, when combined with TRAIL, c–FLIP(L) level decreased at both the mRNA and protein level (Fig. 20.3C, D). Therefore, MG-132 significantly increases sensitivity of PC3-TR cells by concomitant

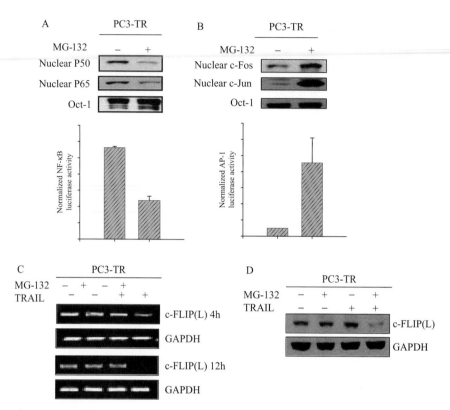

Figure 20.3 MG–132 sensitizes resistant prostate cancer cells to TRAIL. (A) MG–132 decreased NF-κB–related proteins (p50 and p65) (*top panel*) and inhibited NF-κB's activity (*bottom panel*). (B) MG–132 upregulated nuclear protein levels of AP-1 family members, c-Fos and c-Jun, (*top panel*) and activated AP-1 activity (*bottom panel*). TRAIL combined with MG–132 represses c-FLIP(L) as demonstrated in the semiquantitative reverse transcription-PCR (C) and Western blot (D) analyses (C and D were reproduced with permission from *Cancer Res.* **67**, 2247; 2007).

activation of AP-1 and repression of NF-κB to prime cancer cells to undergo TRAIL-induced apoptosis.

6. METHODS AND MATERIALS

6.1. Cell culture and production of PC3-TR

All cell culture materials were obtained from Cellgro (Herndon, VA) and plasticware was from Becton Dickinson Labware (Bedford, MA). PC3, DU145, and LNCaP prostate cancer cell lines were obtained from the American Type Culture Collection (ATCC, Manassas, VA). PC3-TR was a TRAIL–resistant subline established from parental PC3 cells by TRAIL selection. PC3 cells were treated with TRAIL (100 ng/ml). After 24 h, viable cells were rescued by removing TRAIL and replenishing the cells with full medium. When the plates reached 80% confluency, the cells were again treated with TRAIL (100 ng/ml) for 24 h. The cycle was repeated, and PC3-TR cells were generated after 2 months and maintained in medium with TRAIL. PC3-TR cells were released from TRAIL at least one passage before use. All cells were cultured in RPMI-1640 tissue culture medium supplemented with 2 mM L-glutamine, 10% fetal bovine serum, and 1% penicillin-streptomycin (each at 50 μg/ml) at 37 °C with 5% CO_2.

6.2. Cell viability assays

Cell viability was determined by MTT method in accordance with the manufacturer's instructions (Roche Diagnostics, Indianapolis, IN). In brief, 5×10^4 PC3, DU145 cells and 7.5×10^4 LNCaP cells were seeded in 96-well plates and cultured for 24 h before treatment. Cells were then treated with various concentrations of TRAIL for 24 h. MTT was added followed by solubilization buffer 4 h later. Absorbance was measured at 590 nm (630 nm was the reference wavelength) by use of a microtiter plate reader. Viability of untreated cells was set at 100%, and absorbance of wells without cells was set at zero. All results were from at least triplicate experiments.

6.3. cDNA microarray assays

Total RNA was isolated with the RNeasy Mini Kit (Qiagen, Chatsworth, CA). The RNA yield and purity were evaluated by measuring A_{260}/A_{280} and agarose gel electrophoresis. cDNA Microarray was performed by use of the human gene arrays GeneChip (Affymetrix, Santa Clara, CA) and 21,625 genes were analyzed according to the manufacturer's instructions. Arrays were scanned by use of an Affymetrix confocal scanner and analyzed by the Microarray software (Affymetrix). Intensity values were scaled so that the overall fluorescence intensity of each chip of the same type was equivalent. If the 90% lower confidence bound (LCB) of the fold change (FC) between the experiment and the baseline was >1.5, the corresponding genes were considered to be differentially expressed. In our analysis, LCB readings were considered to be more reliable than FC readings for analysis of differential gene expression (Ramalho-Santos et al., 2002).

6.4. Luciferase assays

Cells were seeded into 24-well plates. When the cells were 80% confluent, both AP-1 luciferase reporter (25 ng/well) and Ranilla reporter (5 ng/well) from Stratagene (La Jolla, CA) or NF-κB reporter and Ranilla reporter from

Stratagene (La Jolla, CA) were cotransfected into cells. Here, Ranilla served as an internal control for transfection efficiency. After 24 h of transfection, cells were treated with TRAIL (100 ng/ml) for 4 h, and then both attached and floating cells were collected and centrifuged at 1000 rpm for 5 min at 4 °C. Pellets were rinsed twice with phosphate-buffered saline (PBS), and the cell pellet was prepared in the presence of $1\times$ passive lysis buffer (Dual-Luciferase Assay System Kit, Promega, Madison, WI). Samples were stored at $-20\,^\circ$C until detection. The activities of NF-κB and AP-1 luciferase and Renilla luciferase were determined following the dual-luciferase reporter assay protocol recommended by Promega (Madison, WI, USA). Twenty microliters of cell lysate was transferred into the luminometer tube containing 100 μl luciferase assay reagent (LAR), and firefly luciferase activity (M_1) was first measured then Renilla luciferase activity (M_2) was measured after adding 100 μ; of Stop & Glo Reagent. The results were calculated and expressed as the ratio of M_1/M_2. The experiments were carried out three times with duplicate samples. The data are presented as mean $\pm$SD.

6.5. ChIP assays

ChIP assay was performed by the ChIP Assay Kit (Upstate Cell Signaling Solutions, Lake Placid, NY). Cells were cultured in 10-cm dishes treated with or without TRAIL for 4 h. Fixation of cross-linked DNA and proteins was carried out by adding formaldehyde for final concentration of 1% and incubated for 10 min at 37 °C. Both attached and floating cells were collected, washed, and resuspended in 200 μl of SDS lysis buffer for 10 min and then sonicated for 10 sec 10 times on ice. Selecting the appropriate time and intensity of sonication was very important for successful CHIP results. In pilot experiments, samples were prepared and sonicated for different times and intensity, and then agarose gels were run to evaluate size of the DNA fragment. The DNA gel showed DNA ladder and size of most of the DNA was between 500 bps and 1000 bps. Samples were centrifuged for 10 min at 13,000 rpm at 4 °C, and the supernatant was harvested. The concentration of each sample was quantitated by use of BCA protein assay. Positive controls were 10% of each DNA sample, which did not include the immunoprecipitation step. The remainder of the samples was equally divided into two groups. The experimental group was immunoprecipitated with specific c-Fos (D-1) antibody, whereas the negative control group was immunoprecipitated with general mouse IgG antibody. After immunoprecipitation, protein-DNA crosslinking was reversed. The isolated DNA was first purified, then amplified by PCR, by use of specific primers encompassing the c-FLIP(L) AP-1(f) binding site (GeneBank). The PCR conditions were denaturation at 94 for 50 sec, annealing at 56 °C for 50 sec, and polymerization at 72 °C for 1 min (total number of cycles = 30), final extension at 72 °C for 10 min for 35 cycles. The primers for the experiments in Fig. 20.4A were 5'-CCT GTG ATC CCA GCA CTT TG-3' (forward primer) and 5'- CAC CAT GCC CGA CTA ATT TT-3' (reverse primer).

6.6. Semiquantitative reverse transcription-PCR analysis

Total RNA was isolated with the RNeasy Mini Kit (Qiagen, Valencia, CA) according to the manufacturer's instructions. The RNA yield and purity were evaluated by measuring A_{260}/A_{280} and agarose gel electrophoresis. RT-PCR was performed by use of a Superscript One-Step RT–PCR kit (Invitrogen Life Technologies, Carlsbad, CA); 0.4 μg of the total RNA was used for RT-PCR in 25 μl of total volume. cDNA synthesis was performed at 50 for 30 min with the following cycle temperatures and times: denaturation at 94 °C for 50 sec, annealing at 56 °C for 50 sec, and polymerization at 72 °C for 2 min (total number of cycles = 30), final extension at 72 °C for 10 min. In each reaction, the same amount of GAPDH was used as an internal control. The primers used for PCR were as follows: c-FLIP(L), 5'-GTC TGCTGA AGT CAT CCA TCAG-3' (forward) and 5'-CTT ATG TGT AGG AGA GGA TAA G-3' (reverse); c-Fos, 5'-GAA TAA GAT GGC TGC AGC CAA ATG C-3' (forward) and 5'-AAG GAA GAC GTG TAA GCA GTG CAG C-3' (reverse); GAPDH, 5'-TCC ACC ACC CTG TTG CTG TA-3' (forward); and 5'-ACC ACA GTC CAT GCC ATC AC-3' (reverse). The PCR products were resolved on 1% agarose gels, stained with ethidium bromide, and then photographed.

6.7. Cell extracts and western blot analysis

Cells were harvested for total cell lysates with RIPA buffer (1% NP-40, 50 mM Tris-HCl (pH 8.0), 150 mM NaCl, 0.5% deoxycholate, and 0.1% SDS) containing a mixture of protease inhibitors (cocktail 1×, 1 mM PMSF, 20 mM, 40 mM NaF, and 3 mM Na_3VO_4). After sonication for 15 sec, cell debris was discarded by centrifugation at 12,000g for 10 min at 4 °C, and the protein concentration was determined by BCA protein assay reagent (Pierce, Rockford, IL). The procedure for the nuclear protein extraction was carried out according to the manufacturer's instructions (NE-PER nuclear and cytoplasmic extraction reagents Kit [Pierce Biotechnology, Rockford, IL]). The cell pellet was harvested in Eppendorf tubes, and the supernatant was carefully removed and discarded; 200 μl of ice-cold CER I was added to the cell pellet (per 40 mg). The tube was vigorously vortexed on the highest setting for 15 sec to fully resuspend the cell pellet. The tube was incubated on ice for 10 min, then 11 μl of ice-cold CER II was added to the tube. The mixture was mixed in the tube with vortex for 5 sec on the highest setting and then incubated on ice for 1 min. The reaction was again vortexed for 5 sec on the highest setting, then centrifuged for 5 min at maximum speed in a microcentrifuge ($\sim$16,000g). Immediately, the supernatant (cytoplasmic extract) fraction was transferred to a clean pre-chilled tube. This tube was placed on ice until further use or storage. The insoluble (pellet) fraction was resuspended, which contains nuclei, in 100 μl of ice-cold NER. The nuclear fraction was vortexed on the highest setting for 15 sec and then iced for 10 min. Vortexing and icing of the nuclear fraction was repeated every 10 min five times. The nuclear fraction was centrifuged at maximum speed ($\sim$16,000g) in a microcentrifuge for 10 min. Immediately, the supernatant (nuclear extract) was transferred to a clean prechilled tube and iced. Extracts were placed in $-80\,^{\circ}$C storage until use.

The amount of nuclear protein was quantitated by immunoblot analysis, with anti–Oct-1 or GAPDH as controls. Protein extracts were resolved by 10 to 12% SDS-PAGE and transferred to nitrocellulose membranes by electroblot analysis. Nitrocellulose blots were blocked with 5% (w/v) nonfat dry milk or 3% BSA in Tris-buffered saline/Tween buffer, and incubated with the indicated primary antibody in Tris-buffered saline/Tween containing 2% milk or 1% BSA overnight at 4 $^{\circ}$C. The blots were stained with the appropriate horseradish peroxidase–conjugated secondary antibody. Immunostained proteins were visualized on X-ray film by use of the enhanced chemiluminescence detection system (Amersham Pharmacia Biotech, Piscataway, NJ).

ACKNOWLEDGEMENTS

Grants were received from the Department of Defense (W81XWH-05-1-0080), NIH (DK64062), and Howard Hughes Medical Institute/SPORE grant to the Biomedical Research Support Program at Harvard Medical School (53000234-0006) to A. F. O.

REFERENCES

Benoit, V., Chariot, A., Delacroix, L., Deregowski, V., Jacobs, N., Merville, M. P., and Bours, V. (2004). Caspase-8–dependent HER-2 cleavage in response to tumor necrosis factor alpha stimulation is counteracted by nuclear factor kappaB through c-FLIP-L expression. *Cancer Res.* **64,** 2684–2691.

Bodmer, J. L., Holler, N., Reynard, S., Vinciguerra, P., Schneider, P., Juo, P., Blenis, J., and Tschopp, J. (2000). TRAIL receptor-2 signals apoptosis through FADD and caspase-8. *Nat. Cell Biol.* **2,** 241–243.

Chang, D. W., Xing, Z., Pan, Y., Algeciras-Schimnich, A., Barnhart, B. C., Yaish-Ohad, S., Peter, M. E., and Yang, X. (2002). c-FLIP(L) is a dual function regulator for caspase-8 activation and CD95-mediated apoptosis. *EMBO J.* **21,** 3704–3714.

Chang, H. Y., and Yang, X. (2000). Proteases for cell suicide: Functions and regulation of caspases. *Microbiol. Mol. Biol. Rev.* **64,** 821–846.

Chang, L., Kamata, H., Solinas, G., Luo, J. L., Maeda, S., Venuprasad, K., Liu, Y. C., and Karin, M. (2006). The E3 ubiquitin ligase itch couples JNK activation to TNF alpha–induced cell death by inducing c-FLIP(L) turnover. *Cell* **124,** 601–613.

Dutton, A., O'Neil, J. D., Milner, A. E., Reynolds, G. M., Starczynski, J., Crocker, J., Young, L. S., and Murray, P. G. (2004). Expression of the cellular FLICE-inhibitory protein (c-FLIP) protects Hodgkin's lymphoma cells from autonomous Fas-mediated death. *Proc. Natl. Acad. Sci. USA* **101**, 6611–6616.

Fukazawa, T., Fujiwara,, T., Uno,, F., Teraishi, F., Kadowaki, Y., Itoshima, T., Takata, Y., Kagawa, S.,, Roth, J. A., Tschopp, J., et al. (2001). Accelerated degradation of cellular FLIP protein through the ubiquitin-proteasome pathway in p53-mediated apoptosis of human cancer cells. *Oncogene.* **20**, 5225–5231.

Gao, S., Lee, P., Wang, H., Gerald, W., Adler, M., Zhang, L., Wang, Y. F., and Wang, Z. (2005). The androgen receptor directly targets the cellular Fas/FasL-associated death domain protein-like inhibitory protein gene to promote the androgen-independent growth of prostate cancer cells. *Mol. Endocrinol.* **19**, 1792–1802.

Golks, A., Brenner, D., Fritsch, C., Krammer, P. H., and Lavrik, I. N. (2005). c-FLIPR, a new regulator of death receptor-induced apoptosis. *J. Biol. Chem.* **280**, 14507–14513.

Hyer, M. L., Croxton, R., Krajewska, M., Krajewski, S., Kress, C. L., Lu, M., Suh, N., Sporn, M. B., Cryns, V. L., Zapata, J. M., et al. (2005). Synthetic triterpenoids cooperate with tumor necrosis factor–related apoptosis-inducing ligand to induce apoptosis of breast cancer cells. *Cancer Res.* **65**, 4799–4808.

Irmler, M., Thome, M., Hahne, M., Schneider, P., Hofmann, K., Steiner, V., Bodmer, J. L., Schroter, M., Burns, K., Mattmann, C., et al. (1997). Inhibition of death receptor signals by cellular FLIP. *Nature* **388**, 190–195.

Kelly, M. M., Hoel, B. D., and Voelkel-Johnson, C. (2002). Doxorubicin pretreatment sensitizes prostate cancer cell lines to TRAIL induced apoptosis which correlates with the loss of c-FLIP expression. *Cancer Biol. Ther.* **1**, 520–527.

Kim, Y., Suh, N., Sporn, M., and Reed, J. C. (2002). An inducible pathway for degradation of FLIP protein sensitizes tumor cells to TRAIL-induced apoptosis. *J. Biol. Chem.* **277**, 22320–22329.

Kischkel, F. C., Lawrence, D. A., Chuntharapai, A., Schow, P., Kim, K. J., and Ashkenazi, A. (2000). Apo2L/TRAIL-dependent recruitment of endogenous FADD and caspase-8 to death receptors 4 and 5. *Immunity* **12**, 611–620.

Korkolopoulou, P., Goudopoulou, A., Voutsinas, G., Thomas-Tsagli, E., Kapralos, P., Patsouris, E., and Saetta, A. A. (2004). c-FLIP expression in bladder urothelial carcinomas: Its role in resistance to Fas-mediated apoptosis and clinicopathologic correlations. *Urology* **63**, 1198–1204.

Krueger, A., Baumann, S., Krammer, P. H., and Kirchhoff, S. (2001a). FLICE-inhibitory proteins: Regulators of death receptor-mediated apoptosis. *Mol. Cell. Biol.* **21**, 8247–8254.

Krueger, A., Schmitz, I., Baumann, S., Krammer, P. H., and Kirchhoff, S. (2001b). Cellular FLICE-inhibitory protein splice variants inhibit different steps of caspase-8 activation at the CD95 death-inducing signaling complex. *J. Biol. Chem.* **276**, 20633–20640.

Li, W., Zhang, X., and Olumi, A. F. (2007). MG-132 sensitizes TRAIL-resistant prostate cancer cells by activating c-Fos/c-Jun heterodimers and repressing c-FLIP(L). *Cancer Res.* **67**, 2247–2255.

Mathas, S., Lietz, A., Anagnostopoulos, I., Hummel, F., Wiesner, B., Janz, M., Jundt, F., Hirsch, B., Johrens-Leder, K., Vornlocher, H. P., et al. (2004). c-FLIP mediates resistance of Hodgkin/Reed–Sternberg cells to death receptor–induced apoptosis. *J. Exp. Med.* **199**, 1041–1052.

Medema, J. P., Scaffidi, C., Kischkel, F. C., Shevchenko, A., Mann, M., Krammer, P. H., and Peter, M. E. (1997). FLICE is activated by association with the CD95 death-inducing signaling complex (DISC). *EMBO J.* **16**, 2794–2804.

Micheau, O., Thome, M., Schneider, P., Holler, N., Tschopp, J., Nicholson, D. W., Briand, C., and Grutter, M. G. (2002). The long form of FLIP is an activator of caspase-8 at the Fas death-inducing signaling complex. *J. Biol. Chem.* **277**, 45162–45171.

Micheau, O., and Tschopp, J. (2003). Induction of TNF receptor I–mediated apoptosis via two sequential signaling complexes. *Cell* **114,** 181–190.

Nam, S. Y., Jung, G. A., Hur, G. C., Chung, H. Y., Kim, W. H., Seol, D. W., and Lee, B. L. (2003). Upregulation of FLIP(S) by Akt, a possible inhibition mechanism of TRAIL-induced apoptosis in human gastric cancers. *Cancer Sci.* **94,** 1066–1073.

Okamoto, K., Fujisawa, J., Reth, M., and Yonehara, S. (2006). Human T-cell leukemia virus type-I oncoprotein tax inhibits Fas-mediated apoptosis by inducing cellular FLIP through activation of NF-kappaB. *Genes Cells* **11,** 177–191.

Pan, G., Ni, J., Wei, Y. F., Yu, G., Gentz, R., and Dixit, V. M. (1997). An antagonist decoy receptor and a death domain–containing receptor for TRAIL. *Science* **277,** 815–818.

Peter, M. E. (2004). The flip side of FLIP. *Biochem. J.* **382,** e1–e3.

Ramalho-Santos, M., Yoon, S., Matsuzaki, Y., Mulligan, R. C., and Melton, D. A. (2002). "Stemness": Transcriptional profiling of embryonic and adult stem cells. *Science* **298,** 597–600.

Ricci, M. S., Jin, Z., Dews, M., Yu, D., Thomas-Tikhonenko, A., Dicker, D. T., and El-Deiry, W. S. (2004). Direct repression of FLIP expression by c-myc is a major determinant of TRAIL sensitivity. *Mol. Cell. Biol.* **24,** 8541–8555.

Roue, G., Perez-Galan, P., Lopez-Guerra, M., Villamor, N., Campo, E., and Colomer, D. (2007). Selective inhibition of IkappaB kinase sensitizes mantle cell lymphoma B cells to TRAIL by decreasing cellular FLIP level. *J. Immunol.* **178,** 1923–1930.

Schneider, P., Thome, M., Burns, K., Bodmer, J. L., Hofmann, K., Kataoka, T., Holler, N., and Tschopp, J. (1997). TRAIL receptors 1 (DR4) and 2 (DR5) signal FADD-dependent apoptosis and activate NF-kappaB. *Immunity* **7,** 831–836.

Sheridan, J. P., Marsters, S. A., Pitti, R. M., Gurney, A., Skubatch, M., Baldwin, D., Ramakrishnan, L., Gray, C. L., Baker, K., Wood, W. I., *et al.* (1997). Control of TRAIL-induced apoptosis by a family of signaling and decoy receptors. *Science* **277,** 818–821.

Sinicrope, F. A., and Penington, R. C. (2005). Sulindac sulfide-induced apoptosis is enhanced by a small-molecule Bcl-2 inhibitor and by TRAIL in human colon cancer cells overexpressing Bcl-2. *Mol. Cancer Ther.* **4,** 1475–1483.

Skurk, C., Maatz, H., Kim, H. S., Yang, J., Abid, M. R., Aird, W. C., and Walsh, K. (2004). The Akt-regulated forkhead transcription factor FOXO3a controls endothelial cell viability through modulation of the caspase-8 inhibitor FLIP. *J. Biol. Chem.* **279,** 1513–1525.

Sprick, M. R., Weigand, M. A., Rieser, E., Rauch, C. T., Juo, P., Blenis, J., Krammer, P. H., and Walczak, H. (2000). FADD/MORT1 and caspase-8 are recruited to TRAIL receptors 1 and 2 and are essential for apoptosis mediated by TRAIL receptor 2. *Immunity* **12,** 599–609.

Suliman, A., Lam, A., Datta, R., and Srivastava, R. K. (2001). Intracellular mechanisms of TRAIL: Apoptosis through mitochondrial-dependent and -independent pathways. *Oncogene* **20,** 2122–2133.

Thome, M., Schneider, P., Hofmann, K., Fickenscher, H., Meinl, E., Neipel, F., Mattmann, C., Burns, K., Bodmer, J. L., Schroter, M., *et al.* (1997). Viral FLICE-inhibitory proteins (FLIPs) prevent apoptosis induced by death receptors. *Nature* **386,** 517–521.

Tschopp, J., Irmler, M., and Thome, M. (1998). Inhibition of fas death signals by FLIPs. *Curr. Opin. Immunol.* **10,** 552–558.

Wajant, H., Haas, E., Schwenzer, R., Muhlenbeck, F., Kreuz, S., Schubert, G., Grell, M., Smith, C., and Scheurich, P. (2000). Inhibition of death receptor–mediated gene induction by a cycloheximide-sensitive factor occurs at the level of or upstream of Fas-associated death domain protein (FADD). *J. Biol. Chem.* **275,** 24357–24366.

Yeh, W. C., Itie, A., Elia, A. J., Ng, M., Shu, H. B., Wakeham, A., Mirtsos, C., Suzuki, N., Bonnard, M., Goeddel, D. V., *et al.* (2000a). Requirement for Casper (c-FLIP) in regulation of death receptor–induced apoptosis and embryonic development. *Immunity* **12,** 633–642.

Yeh, W. C., Itie, A., Elia, A. J., Ng, M., Shu, H. B., Wakeham, A., Mirtsos, C., Suzuki, N., Bonnard, M., Goeddel, D. V., *et al.* (2000b). Requirement for Casper (c-FLIP) in regulation of death receptor–induced apoptosis and embryonic development. *Immunity* **12,** 633–642.

Zhang, X., Li, W., and Olumi, A. F. (2007a). Low-dose TPA enhances TRAIL-induced apoptosis in prostate cancer cells. *Clin. Cancer Res.* In press.

Zhang, X., Cheung, R. M., Komaki, R., Fang, B., and Chang, J. Y. (2005). Radiotherapy sensitization by tumor-specific TRAIL gene targeting improves survival of mice bearing human non-small cell lung cancer. *Clin. Cancer Res.* **11,** 6657–6668.

Zhang, X., Jin, T. G., Yang, H., DeWolf, W. C., Khosravi-Far, R., and Olumi, A. F. (2004). Persistent c-FLIP(L) expression is necessary and sufficient to maintain resistance to tumor necrosis factor–related apoptosis-inducing ligand-mediated apoptosis in prostate cancer. *Cancer Res.* **64,** 7086–7091.

Zhang, X., Zhang, L., Yang, H., Huang, X., Otu, H., Libermann, T., DeWolf, W. C., Khosravi-Far, R., and Olumi, A. F. (2007b). c-Fos as a proapoptotic agent in tumor necrosis factor–related apoptosis-inducing ligand-induced apoptosis in prostate cancer cells. *Cancer Res.* **67,** 9425–9437.

CHAPTER TWENTY-ONE

Caspase Assays: Identifying Caspase Activity and Substrates *In Vitro* and *In Vivo*

Cristina Pop, Guy S. Salvesen, *and* Fiona L. Scott

Contents

Abstract

The measurement of general caspase activity and the quantification of purified recombinant caspases *in vitro* can be accomplished with relative ease. But the determination of which caspases are active in a cellular context is much more challenging. This is because commercially available small molecule substrates and inhibitors do not display sufficient specificity to dissect the complex interplay of caspase pathways. Here we describe procedures that can be used to validate which caspases are active in cell culture models and determine which caspases are responsible for specific cleavage events. We also recommend methods for working with recombinant initiator caspases *in vitro* and suggest ways to accurately assess the cleavage efficiency of natural caspase substrates.

Program in Apoptosis and Cell Death Research, The Burnham Institute for Medical Research, La Jolla, California

Methods in Enzymology, Volume 446

ISSN 0076-6879, DOI: 10.1016/S0076-6879(08)01621-2

1. INTRODUCTION

Caspases are a family of cysteine proteases (proteases that use a cysteine side-chain as the catalytic nucleophile) that are irreversibly activated in the cell during apoptosis, inflammation, and keratinocyte differentiation (reviewed in Fuentes-Prior and Salvesen [2004]). Once activated, caspases cleave substrates after an aspartic acid residue. For the apoptotic caspases-3, -6, -7, -8, -9, and -10 (and possibly caspase-2), a proteolytic cascade amplifies the cell death signal, and substrate cleavage results in dismantling of the cell. Activation of the inflammatory caspases-1, -4, and -5 leads to proteolytic maturation of proinflammatory cytokines (Martinon and Tschopp, 2004). Finally, the barrier function of the epidermis requires profilaggrin processing in a caspase-14–dependent manner in differentiating keratinocytes (Denecker *et al.*, 2007).

The mechanism of caspase activation has been extensively studied and is understood at both the biochemical and atomic level (reviewed in Boatright and Salvesen [2003]). Apical or initiator apoptotic caspases (caspases-8, -9, and -10) reside in the cytoplasm as inactive monomers. On a cell death signal, they are recruited to higher order protein platforms, where they are activated by dimerization (Boatright and Salvesen, 2003). Cleavage of the single chain caspase occurs as a consequence of activation and probably stabilizes the dimeric conformation, maintaining a productive active site and the capacity for substrate catalysis (Pop *et al.*, 2007). In contrast, the effector apoptotic caspases-3, -6, and -7 exist as inactive dimers in resting cells. To mature a catalytic active site, these enzymes are cleaved in the inter-subunit linker by active apical caspases or other immune cell derived proteases.

After a cell death trigger, caspase activity can be readily detected by use of small peptidic-based substrates (fluorogenic or colorimetric) and active site directed inhibitors. Many companies sell these reagents under the premise that they are specific for one caspase over the rest. Examples include DEVD for caspase-3 and -7, LEHD for caspase-9, IETD for caspase-8, VEID for caspase-6, and VDVAD for caspase-2. These are the optimal sequences for the respective caspases, but they are certainly not specific. Over the past 10 years there have been hundreds of publications in which investigators have used one or more of these substrates to try to define which caspase(s) is/are active in a complex mixture such as a cell lysate. A very important point ignored at an investigator's peril is that caspases have overlapping substrate specificities (Stennicke *et al.*, 2000; Talanian *et al.*, 1997; Thornberry *et al.*, 1997). Although experts in the field have been aware of this since the initial characterization of caspase specificity, a recent study has highlighted the pitfalls of the use of small-peptidic substrates and inhibitors in complex samples (McStay *et al.*, 2007). It is now clear that the commonly used small peptidic substrates and inhibitors are totally unreliable and usually target caspase-3 in whole cells and cell lysates simply because it is the most active and abundant caspase (McStay *et al.*, 2007; Stennicke *et al.*, 2000).

On the basis of the activation mechanism of the initiator caspases and the lack of specificity of small peptide substrates, demonstrating activation and, consequently, activity of initiator caspases in complex samples is not trivial. In this chapter we describe in detail some recently developed methods to: (1) assist researchers when working with recombinant caspases *in vitro*, (2) delineate which caspase is active in cells, and (3) demonstrate which caspase is responsible for a cleavage event in whole cells.

 ## 2. Caspase Methods

2.1. Activation of recombinant initiator caspases with kosmotropes

Because of their activation mechanism, recombinant initiator caspases-8 and -9 exist as a mixture between inactive monomers and active dimers in solution (Boatright *et al.*, 2003). This is explained by the fact that the K_d for dissociation is approximately 50 μM for cleaved caspase-8 and 100 μM for caspase-9, whereas the protein concentration of an average preparation of recombinant material purified from *E. coli* expression is approximately 10 μM (Blanchard *et al.*, 1999; Donepudi *et al.*, 2003; Renatus *et al.*, 2001). This provides a problem when working with recombinant caspase-8 and -9 in standard caspase assay buffer, because it is hard to know the exact concentration of the active form of the enzyme. In addition, when performing an active-site titration of initiator caspases by use of an irreversible inhibitor such as zVAD-fmk, the very nature of the irreversible reaction drives the equilibrium toward dimerization, overestimating the actual "active" caspase concentration (Renatus *et al.*, 2001). To overcome these issues, we have developed a buffer system to achieve maximal activity from recombinant caspase-8, -9, and -10 with kosmotropic salts (Boatright *et al.*, 2003; Pop *et al.*, 2007).

Kosmotropes stabilize proteins in an aqueous environment. They act either by ordering the structure of water or by compensating for the loss of hydrogen bonding in proteins at high salt concentration; either way, the overall effect is "order" or less entropy. Examples of ionic kosmotropic salts are citrate and related organic acid salts (malic, malonic, succinic), phosphates, ammonium, and sulfates. Examples of nonionic kosmotropes are polyethylene glycol, proline, glycine, betaine, and trehalose.

At high concentrations (0.5 to 1.5 M) kosmotropic salts can have two effects: (1) they induce dimerization of monomeric caspases, in effect increasing the number of active sites in solution (i.e., increase in k_{cat}); and

(2) order the loops in the active site of the dimeric caspase, promoting substrate binding (i.e., decrease in K_M; Gouvea *et al.*, 2006; Schmidt and Darke, 1997). In reality, at least for the initiator caspases, both effects occur simultaneously.

For initiator caspases the most effective kosmotrope is Na citrate. To emphasize the efficiency of the kosmotropes, we purified caspase-8, -9, and -10 monomers by gel filtration. We then added 1 M Na citrate and measured activity. In this case the activation is more than 100-fold (Fig. 21.1A). If we start with a mix of monomer/dimer, as is found during normal purification of initiators from *E. coli*, the activation is between 10- and 100-fold (Boatright *et al.*, 2003). By contrast, for recombinant effector caspases-3 and -7, which are 100% dimeric in solution, Na citrate increases the activity by only 2- to 3-fold (Fig. 21.1A) because of the stabilization of the catalytic loops (Pop *et al.*, 2007). Other efficient kosmotropes for apical caspases are the salts of malonic, malic, and succinic acid (Fig. 21.1B), whereas aspartic and maleic acid salts are less efficient activators.

Interestingly, it has been shown that 100 mM Na malonate can inhibit caspase-1 by binding to the active site pocket (Romanowski *et al.*, 2004). However, caspase-1 activity is maintained during overnight incubation in 1 M malonate (Scheer *et al.*, 2005). In our hands, 1 M malonate is a weak activator of caspase-1. It is, therefore, possible that some kosmotropes may have multiple effects on enzyme activity, depending on the enzyme type, salt concentration, protein solubility in kosmotrope, or kosmotrope interaction with the catalytic loops.

Active-site titration emphasizes the mechanism of initiator caspase activation by kosmotrope is dimerization. When caspase-8 is titrated with zVAD-fmk, in caspase assay buffer versus 1 M Na citrate (Stennicke and

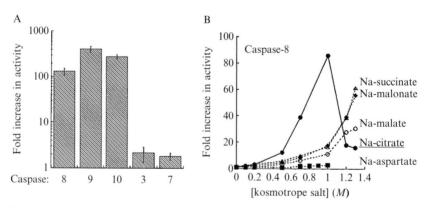

Figure 21.1 Activation of caspases by Na citrate. (A) Increase in the caspase enzymatic activity by 1 M Na citrate after incubation at 37° C for 15 min (caspase-3 and -7) or 1 h (caspase-8, -9, and -10). (B) Activation of caspase-8 by kosmotropes related to Na citrate at 37° C. Activation was in low salt buffer containing 10 mM DTT (*no CHAPS*) *and the indicated kosmotrope concentration.*

Salvesen, 2000), the concentration is closest to the predicted, on the basis of the A_{280} estimation in the presence of kosmotrope (Fig. 21.2). This is also true for caspase-9 (data not shown).

2.1.1. Reagents

Recombinant caspases-3, -7, -8, -9, and -10 were expressed in *E. coli* and purified as previously described (Stennicke and Salvesen, 1999).

Na citrate stock: 1.4 M Na citrate, 50 mM Tris or 50 mM NaH$_2$PO$_4$/Na$_2$HPO$_4$, pH 7.4. The stock is made by dissolving the buffer crystals in distilled water, followed by slow addition of solid Na citrate or citric acid and then adjustment of the pH with concentrated HCl or NaOH, respectively. *Note*: the solubility limit of Na citrate in water is 1.7 M; therefore, correct adjustment of the solution volume is important to avoid citrate crystallization. The solution is sterile filtered to prevent bacterial growth and is stable for several months at room temperature.

DTT: 1 M in water, stored in aliquots at −20 °C.

CHAPS stock: 10% (w/v) in water, stable at room temperature for several months.

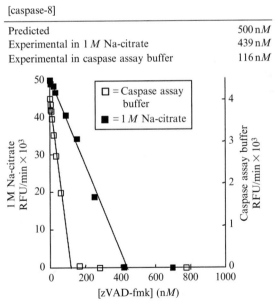

[caspase-8]	
Predicted	500 nM
Experimental in 1 M Na-citrate	439 nM
Experimental in caspase assay buffer	116 nM

Figure 21.2 Active-site titration of 500 nM caspase-8 with zVAD-fmk in caspase assay buffer or buffer containing 1 M citrate. Kosmotrope increased the number of active sites by dimerization, generating a caspase-8 solution that approaches 100% active. In caspase assay buffer recombinant caspase-8 is only 23% active.

Low salt buffer: 50 mM Tris, pH 7.4, or 50 mM NaH$_2$PO$_4$/ Na$_2$HPO$_4$, pH 7.4.

Caspase assay buffer: 20 mM PIPES, 0.1 M NaCl, 5% (w/v) sucrose, 0.1% (w/v) CHAPS, 10 mM DTT (freshly added), pH 7.4. This buffer is the general optimal assay buffer for executioner caspases (Stennicke and Salvesen, 1997).

Caspase substrates: 10 mM stocks in DMSO. Ac-IETD-AFC for caspases-8 and -10, Ac-LEHD-AFC for caspase-9, and Ac-DEVD-AFC for caspase-3, -6, and -7. Although we use fluorogenic substrates, colorimetric substrates are also compatible with this procedure.

Other kosmotrope stocks: 1.4 M Na malonate, 1.4 M Na malate, 1.4 M Na succinate, 1.1 M Na aspartate in 25 mM NaH$_2$PO$_4$/ Na$_2$HPO$_4$, pH 7.4. (These have not been tested for long-term stability).

2.1.2. Procedure

Use 96-well plates for the assays. From the stock solutions, prepare 1 M Na citrate, 50 mM Tris, pH 7.4, containing 10 mM DTT and 0.05 % CHAPS, by diluting in low salt buffer. *Note*: CHAPS precipitates in 1 M Na citrate; therefore, use concentrations less than 0.1%. CHAPS can be omitted if the experiments last less than 2 h.

Add the caspase of interest at final concentrations of 10 to 20 nM for caspase-8, -9, -10 or 0.5 to 1 nM for caspase-3 and -7 and mix by pipetting.

Incubate at 37 °C. For the effector caspases-3 and -7, the activation process is rapid, so the incubation time should not exceed 15 to 20 min. For the initiator caspases-8 and -9, the kinetics of activation is slower and depends on both the enzyme concentration and temperature. As an example, caspase-8 at high concentration (112 nM) is maximally activated after only 1 h incubation in Na citrate, whereas it takes 7 h for caspase-8 at low concentration (7.5 nM) to be come fully activated (Fig. 21.3).

For determination of the catalytic activity, add the appropriate fluorogenic substrate to a final concentration of 100 μM. The initial velocity is monitored by use of a plate reader equipped with filters that excite at 405 nm and detect emission at 510 nm.

As a control, the experiment is repeated in caspase assay buffer.

Notes: Cleavage of natural substrates, either purified or in complex mixtures, can also be monitored in the presence of kosmotropes, as long as they do not precipitate under the assay conditions (Pop *et al.*, 2006). If cleavage is monitored by SDS-PAGE, then the proteins need to be separated from kosmotrope by use of TCA precipitation. High concentrations of salts are not compatible with electrophoresis.

In general, Na citrate increases the long-term stability of initiator caspases compared with caspase assay buffer. Nanomolar concentrations of caspases-8 and -10 maintain their activity overnight at ambient temperature,

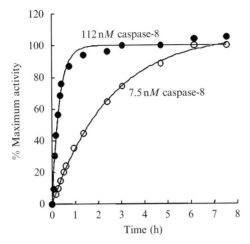

Figure 21.3 Kinetics of caspase-8 activation in 1 *M* citrate at 25 °C. The kinetics of kosmotrope-induced activation of caspase-8 is directly related to the concentration of the protease.

in the presence of 0.05 % CHAPS and kosmotrope. We do not observe precipitation of either caspase-8 or -10 in 1 *M* Na citrate, even at micromolar concentrations. Caspase-9 at submicromolar concentration loses approximately 20% activity after overnight incubation. At concentrations above 1 μM it precipitates in 1 *M* Na citrate. In Na citrate, caspase-3 and -7 are only stable for few hours, whereas caspase-6 loses approximately 50% activity because of precipitation. Caspase-7 at concentrations in the micromolar range precipitates in sodium citrate. Although we do not know why caspase-3 is unstable in kosmotropes, we suspect microprecipitation may be a factor.

Caspase-2 is 100% dimeric in solution and is activated approximately 5- to 10-fold in Na citrate. Caspases 1, -4, and -5 show 2- to 4-fold increases in the activity in 1 *M* Na citrate (C. Pop and F. L. Scott, unpublished results).

When preparing the working solution containing kosmotropes, it is important to adjust the pH to 7.4 with compatible buffers that are stable in the neutral range.

The concentration of "active" initiator caspase by active-site titration in kosmotrope is only true for the conditions containing the same amount of kosmotrope.

Activation by kosmotropes is reversible. Removal of kosmotrope by dilution or dialysis is sufficient to inactivate apical caspases. Also, addition of chaotropic salts, like $NaClO_4$, $NaNO_3$, or $MgCl_2$, to a mixture of monomer-dimer will shift the equilibrium to the inactive species, because of dimer dissociation (Pop *et al.*, 2007).

2.2. Determining the catalytic efficiency of caspases for natural substrates

An estimation of the k_{cat}/K_M parameter for cleavage of natural protein substrates has been described (Stennicke and Salvesen, 2000). The assay is based on the concentration of protease required to cleave half of the substrate ($[E]_{1/2}$) for a given time (t). We routinely quantify $[E]_{1/2}$ for a given substrate by in-gel densitometry, quantitative immunoblotting, or radiography. This parameter is a measure of how good a protease is for that substrate; therefore, the $[E]_{1/2}$ values of a protease vary widely for different substrates. For this method to return an accurate measure of k_{cat}/K_M, the substrate concentration must be below the K_M for that enzyme/substrate pair. The caveat is that more often than not, the K_M for a protease/substrate pair is unknown. Taking this into consideration, we want to emphasize that the best strategy is to set up the assay with the lowest substrate concentration that can be reliably quantified with the most sensitize detection system available. We highlight this by analyzing cleavage of a p35 variant containing an inactivating mutation that converts this potent caspase inhibitor into substrate (Riedl *et al.*, 2001). The $[E]_{1/2}$ for cleavage of p35 C2A by caspase-3 varies when the substrate concentration is 5, 1 and 0.2 μM (Fig. 21.4). Substrate concentrations above the K_M underestimate the efficiency of the cleavage.

$$k_{cat}/K_{M(app)} = \ln2/[E]_{1/2} \cdot t \qquad (21.1)$$

where the $k_{cat}/K_{M\,(app)}$ is the apparent catalytic efficiency, $[E]_{1/2}$ is the concentration of protease for which half the substrate is consumed, and t is the incubation time.

2.3. Determining which caspase cleaves a substrate in whole cells

So your favorite protein is cleaved during apoptosis or another interesting biologic situation. You are pretty sure a caspase is responsible, but you want to know which one. Given the extremely close specificity shared by members of the apoptotic caspase family, this becomes a challenging question. Contrary to popular belief, small cell-permeable peptidic-based inhibitors (e.g., DEVD, LEHD, IETD, VEID, and VDVAD-fmks) sold by many companies are not specific for any given caspase, making them ineffective in complex samples such as whole cells or cell lysates (see preceding).

We have developed a technique that takes advantage of natural caspase inhibitors, and other apoptosis regulators, that have evolved exquisite potency and specificity (Table 21.1). These natural proteins have overcome

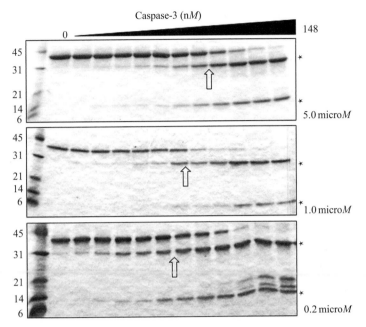

Figure 21.4 Cleavage of p35 C2A by caspase-3. The p35 C2A substrate (in 50 mM Tris, 0.1 M NaCl, and 250 mM imidazole, pH 8) was mixed 1:1 with serial dilutions of recombinant caspase-3 (in 50 mM HEPES, pH 7.4, 0.1 M NaCl, 0.04% CHAPS, 2 mM DTT, 2% sucrose) to give the indicated final concentrations. Reactions were incubated for 1 h at 37 °C. Reactions containing 5 and 1 μM substrate were stopped by addition of SDS-loading buffer, whereas reactions containing 0.2 μM substrates were precipitated with 10% TCA and the pellets dissolved in SDS loading buffer. All samples were separated on 8 to 18% SDS-PAGE and proteins detected with Blue staining. *Cleaved p35 C2A; *open arrows,* 50% cleavage of p35 C2A.

the problem of nonspecificity by evolving unusual inhibitory mechanisms (Stennicke *et al.*, 2002). The basic concept is that the inhibitors are exogenously overexpressed in the cell system you are interested in, apoptosis induced with your agent of interest, and the substrate of interest is monitored by immunoblot.

To use this procedure, it is important that the cell system you are looking at can be transfected with exogenous plasmids with reasonable efficiency or transduced with retrovirus, adenovirus, or lentivirus. We routinely use transfection with commercially available cationic lipid-based transfection reagents and HEK293 cells. However, the basic procedure can be adapted to most cell lines. This method can also be used to study any cellular insult that generates a caspase-dependent response. It is useful for analyzing cleavage of endogenous substrates, if good antibodies are available, or exogenously overexpressed substrates, with or without epitope tags. Care

Table 21.1 Natural caspase and apoptosis inhibitors

Protein	Caspase or pathway inhibited
Caspase-8 dominant negative catalytic mutant (C285A according to the caspase-1 numbering system)	Caspase-8
Caspase-9 dominant negative catalytic mutant (C285A according to the caspase-1 numbering system)	Caspase-9
XIAP	Caspase-3, -7, and -9
CrmA	Caspase-1 and -8
Bcl-xL or Bcl2	Inhibits mitochondria apoptosis pathway upstream of caspase-9 activation
p35	All caspases

should be taken when deciding whether to use an N- or C-terminal tag, because some cleavage products may be unstable, particularly those containing PEST sequences (Rogers *et al.*, 1986).

As an example, we have dissected the caspase-8 initiated extrinsic apoptosis pathway. We used tumor necrosis factor–related apoptosis-inducing ligand (TRAIL), although any extrinsic apoptotic stimuli can be used. The example substrate is directly cleaved by caspase-8 in cells undergoing TRAIL-induced apoptosis (Fig. 21.5). Proteins that directly inhibit caspase-8 activation or activity (caspase-8 dominant negative, CrmA and p35) blocked substrate cleavage, whereas XIAP, which inhibits caspase-3, -7, and -9 did not. Bcl-xL is a negative control, because it regulates caspase-9–dependent apoptosis at the mitochondria.

2.3.1. Reagents
The reagents described are available from many sources.

Plasmids are available from Addgene. All cDNAs were cloned into pcDNA3 for mammalian expression systems: p35-FLAG (Ryan *et al.*, 2002), Myc-XIAP (Takahashi *et al.*, 1998), Bcl-xL-HA (Chang *et al.*, 1997), caspase-8 dominant negative catalytic mutant C285A-HA (Boatright *et al.*, 2003), CrmA (Ryan *et al.*, 2002).

TRAIL (Alexis Biochemicals)

Antibodies: mouse anti-XIAP and anti-HSP90 (BD Transduction Laboratories), mouse anti-FLAG (M2; Sigma), rabbit anti-HA (Santa Cruz) and CrmA (kind gift from Dr. David Pickup, Duke University, Durham, NC). Secondary antibodies were HRP conjugated donkey anti-rabbit IgG and donkey anti-mouse IgG (Amersham Biosciences)

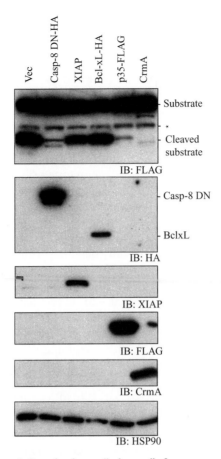

Figure 21.5 Caspase-8 directly cleaves "substrate" after engagement of the extrinsic apoptosis pathway. HEK293 cells cotransfected with 0.2 μg FLAG-tagged substrate and 0.8 μg of the indicated cDNA were treated with 100 ng/ml TRAIL for 2 h. Cell lysates were normalized for protein concentration, resolved by SDS-PAGE, and immunoblotted with the indicated antibodies. *Nonspecific protein detected by anti-FLAG antibody.

PBS: 136 mM NaCl, 2.6 mM KCl, 10 mM Na_2HPO_4, 1.76 mM KH_2PO_4, pH 7.4

Modified radioimmunoprecipitation buffer (mRIPA): 10 mM Tris, 150 mM NaCl, 1% (v/v) NP-40, 0.5% (v/v) deoxycholate, 0.1% (w/v) SDS, 5 mM EDTA, pH 7.4 containing protease inhibitors (10 μM zVAD-fmk, 10 μM E-64, 10 μM leupeptin, 150 μg/ml PMSF, and 1 μM pepstatin).

2.3.2. Procedure

Plate HEK293 cells into 6-well trays at 1 to 3×10^5 cells/well and let attach for 18 to 24 h in a humidified tissue culture incubator.

Transfect cells with 0.2 μg of FLAG-tagged substrate and 0.8 μg of the indicated plasmid, according to the manufacturers instructions.

After 24 h, add fresh media containing 100 ng/ml TRAIL and incubate the cells for 2 h.

Harvest cells by collecting media containing apoptotic cells into a 15-ml tube. Wash remaining cells with 1 ml of PBS, detach with a cell scraper, and add to the media containing the apoptotic cells. Wash the wells once more with 0.5 ml of PBS to remove any remaining cells and add to the tube containing the rest of the cell sample.

Centrifuge cells at 1000 rpm (180g) in a benchtop centrifuge for 5 min and remove the supernatant by aspiration. Wash cells twice with 1 ml of PBS, spinning at 1500g in an Eppendorf benchtop centrifuge between each wash. The final wash is aspirated, being sure to remove all of the remaining wash fluid.

The cell pellet can either be stored at $-20\,°C$ (we have stored pellets for as long as 1 week but would suggest storing at $-70\,°C$ for longer periods of time). Alternately, the cell pellet can be directly lysed on ice with 50 to 100 μl ice-cold mRIPA lysis buffer containing protease inhibitors.

Clarify lysates by centrifugation at 16,000g at 4 $°C$ for 10 min.

Determine the protein concentration by use of a detergent compatible protein assay (D_C Protein Assay kit, Biorad), separate equal amounts of protein on an 8 to 18% SDS-PAGE gel, transfer to PVDF membrane, and immunoblot as previously described (Denault and Salvesen, 2003).

Notes: We routinely save 1/5th of the cell sample before lysis, for apoptosis analysis either by Annexin-V staining (Bossy-Wetzel and Green, 2000) or DNA cleavage (Nicoletti *et al.*, 1991).

2.4. Determining which caspases are active in whole cells

As we have emphasized earlier in this chapter, initiator caspases-8 and -9 are activated by dimerization and are subsequently cleaved in an autocatalytic manner. However, in many instances, initiator caspases are cleaved in cells as a bystander event during apoptosis. For example, caspase-8 is cleaved in cells undergoing granzyme B–induced apoptosis, yet no ligand-stimulated death-inducing signaling complex (DISC) was formed, and caspase-8 was never activated (Boatright *et al.*, 2003; Duan *et al.*, 1996). In contrast, for executioner caspases-3, -6, and -7, cleavage is the activating event. Therefore, demonstrating "active" caspase-3, -6, or -7 is relatively straightforward, with detection of the cleaved caspase by simple immunoblot being the technique of choice. Unfortunately, it is not so easy to demonstrate "active" caspase-8 and -9 in whole cells, and immunoblotting alone is never conclusive. To make matters worse, as described earlier, small peptidic-based substrates are not specific enough to be used in complex samples such as cell lysates (McStay *et al.*, 2007).

To address this technical problem, a number of groups have developed *in vivo* labeling techniques designed to trap the active caspase. The advantage of this strategy is that it harnesses the catalytic machinery of the active protease, so only the active enzyme is detected. The inhibitor usually consists of a "handle" (e.g., biotin), the recognition sequence, and a "warhead" (e.g., a fluoromethyl ketone, chloromethyl ketone, or acyloxymethyl ketone; reviewed in Berger *et al.* [2004]). As a consequence of nucleophilic attack, the catalytic cysteine residue of the caspase becomes covalently bound to the inhibitor and can be purified out of a complex sample by way of the biotin group. For a broad range of caspases, the most commonly used is biotinylated zVAD-fmk (bVAD-fmk). To detect active caspases in whole cells, it is important to use *O*-methylated bVAD-fmk (bVAD(Ome)-fmk), because this version has increased cell permeability. For labeling active caspases in cell lysates, nonmethylated bVAD-fmk is adequate.

A recent publication described a technique specifically designed to label and trap initiator caspases-2, -8, and -9 in whole cells during various apoptotic stimuli (Tu *et al.*, 2006). Unfortunately, we and other groups have not been successful with the technique as described, despite the use of a variety of cell types and apoptotic stimuli. This is probably because a number of technical issues including, but not limited to (1) the abundance of the initiator caspases in cells, (2) the percent of initiator caspase activated during apoptosis, (3) the stability of the caspase after activation, (4) caspase-3, which is activated downstream, is the most catalytically active of all caspases and also the most abundant, effectively competing for the available bVAD-fmk, and (5) most cells contain high amounts of endogenously biotinylated proteins (e.g., carboxylases), which compete for streptavidin beads used to isolate the labeled caspases. We, and others, have only had success labeling executioner caspases-3, -6, and -7 in whole cells (Berger *et al.*, 2006; Denault *et al.*, 2006; Denault and Salvesen, 2003).

Here we describe an optimized procedure for *in vitro* trapping of executioner caspases. We have included a preincubation step with a proteasome inhibitor (MG132) to stabilize caspases at the protein level during apoptosis. Active caspase-3 is readily precipitated from Jurkat cells undergoing caspase-8–dependent apoptosis, yet no active caspase-8 can be detected (Fig. 21.6). When bVAD(Ome)-fmk is preincubated with cells before induction of apoptosis, autocatalytic removal of caspase-3 N-terminal peptide is prevented (p20 large subunit). If bVAD(Ome)-fmk is added after apoptosis is induced, both the p20 and p17 (N-peptide removed) forms of caspase-3 are detected.

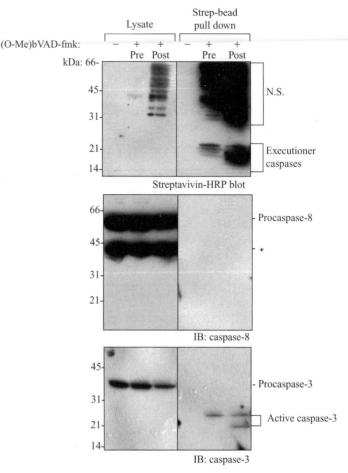

Figure 21.6 *In vivo* labeling of caspase-3 during extrinsic apoptosis. Jurkat cells were treated with 100 ng/ml anti-Fas antibody (CH11) for 4 h, lysed, and precipitated with streptavidin agarose beads. Precipitates were separated on SDS-PAGE and probed with streptavidin-HRP or immunoblotted with the indicated antibody. *Pre,* Cells were pretreated with 50 μM bVAD(Ome)-fmk for 1 h before Fas antibody; *post,* 50 μM bVAD (Ome)-fmk was added to CHAPS buffer before lysis; *N.S.*, proteins that react nonspecifically with bVAD-fmk; *nonspecific protein detected by anti–caspase-8 antibody. All samples were pretreated with 10 μM MG132 to stabilize labeled caspases.

2.4.1. Reagents

Biotinyl-Val-Ala-(O-methyl)Asp-fluoromethyl ketone ((bVAD(Ome)-fmk), MP Bioscience)
MG132 (Calbiochem)
Agonistic anti-Fas antibody, clone CH11 (Millipore)

PBS: 136 mM NaCl, 2.6 mM KCl, 10 mM Na$_2$HPO$_4$, 1.76 mM KH$_2$PO$_4$, pH 7.4

CHAPS buffer: 50 mM HEPES, 150 mM KCl, 0.1% (w/v) CHAPS, pH 7.4

CHAPS buffer containing protease inhibitors (CHAPS buffer with 10 μM zVAD-fmk, 10 μM E-64, 10 μM leupeptin, 10 μM MG132, 150 μg/ml PMSF, and 1 μM pepstatin)

Streptavidin agarose beads (Pierce)

Antibodies: monoclonal caspase-8 antibody (clone C15, generous gift from Dr. Markus Peter, University of Chicago, Chicago, IL) and monoclonal caspase-3 antibody (BD Pharmingen)

Horseradish peroxidase–conjugated streptavidin (Sigma-Aldrich)

2.4.2. Procedure

Wash 5 × 10^7 Jurkat cells in fresh RPMI media, resuspend the cells in 1 ml of RPMI containing 10 μM MG132 and 50 μM bVAD(Ome)-fmk and incubate at 37° C for 1 h.

Add agonistic anti-Fas antibody (CH11) to 100 ng/ml and incubate cells for a further 4 h.

Wash cells twice with PBS, resuspend in 500 μl of CHAPS buffer containing protease inhibitors, and freeze-thaw once to lyse cells. (Lysates can be stored at −20 °C for as long as 1 week, and at −70 °C for longer periods of time).

Add 30 μl of streptavidin agarose beads (washed once in 1 ml of CHAPS buffer) to the lysate and incubate at 4° C for 4 h on a rocking platform. (Samples can be incubated overnight).

Wash beads three times with 1 ml CHAPS buffer, once with 1 ml CHAPS buffer containing 0.5 M NaCl, and once more in 1 ml CHAPS buffer.

Resuspend the beads in 50 μl SDS loading buffer, electrophorese on an 8 to 18% SDS-PAGE gel, transfer to PVDF membrane, and immunoblot as previously described (Denault and Salvesen, 2003).

Notes: This procedure is most useful for suspension cells, because they can be treated in a small volume at high density, cutting down on the amount of apoptotic agent required.

In the protocol published by Tu and colleagues (2006), there was a boiling step after the cells are resuspended in CHAPS buffer. We omitted this step because in our hands more than 95% of the protein precipitated.

3. Conclusion

In conclusion, we hope these protocols and tips help researchers more accurately determine which caspases are active in their biology of interest and which caspases are responsible for specific cleavage events. In addition, we hope an increased awareness about the limited usefulness of small peptide substrates and inhibitors will make researchers more cautious in both the design of their own experiments and the interpretation of others' data.

REFERENCES

Berger, A. B., Vitorino, P. M., and Bogyo, M. (2004). Activity-based protein profiling: Applications to biomarker discovery, *in vivo* imaging and drug discovery. *Am. J. Pharmacogenomics* **4**, 371–381.

Berger, A. B., Witte, M. D., Denault, J. B., Sadaghiani, A. M., Sexton, K. M., Salvesen, G. S., and Bogyo, M. (2006). Identification of early intermediates of caspase activation using selective inhibitors and activity-based probes. *Mol. Cell* **23**, 509–521.

Blanchard, H., Kodandapani, L., Mittl, P. R. E., Di Marco, S., Krebs, J. F., Wu, J. C., Tomaselli, K. J., and Grütter, M. G. (1999). The three-dimensional structure of caspase-8: An initiator enzyme in apoptosis. *Structure* **27**, 1125–1133.

Boatright, K. M., Renatus, M., Scott, F. L., Sperandio, S., Shin, H., Pedersen, I., Ricci, J. E., Edris, W. A., Sutherlin, D. P., Green, D. R., and Salvesen, G. S. (2003). A unified model for apical caspase activation. *Mol. Cell* **11**, 529–541.

Boatright, K. M., and Salvesen, G. S. (2003). Mechanisms of caspase activation. *Curr. Opin. Cell. Biol.* **15**, 725–731.

Bossy-Wetzel, E., and Green, D. R. (2000). Detection of apoptosis by annexin V labeling. *Methods Enzymol.* **322**, 15–18.

Chang, B. S., Minn, A. J., Muchmore, S. W., Fesik, S. W., and Thompson, C. B. (1997). Identification of a novel regulatory domain in Bcl-X(L) and Bcl-2. *EMBO J.* **16**, 968–977.

Denault, J. B., Bekes, M., Scott, F. L., Sexton, K. M., Bogyo, M., and Salvesen, G. S. (2006). Engineered hybrid dimers: Tracking the activation pathway of caspase-7. *Mol. Cell* **23**, 523–533.

Denault, J. B., and Salvesen, G. S. (2003). Human caspase-7 activity and regulation by its N-terminal peptide. *J. Biol. Chem.* **278**, 34042–34050.

Denecker, G., Hoste, E., Gilbert, B., Hochepied, T., Ovaere, P., Lippens, S., Van den Broecke, C., Van Damme, P., D'Herde, K., Hachem, J. P., Borgonie, G., Presland, R. B., *et al.* (2007). Caspase-14 protects against epidermal UVB photodamage and water loss. *Nat. Cell. Biol.* **9**, 666–674.

Donepudi, M., Mac Sweeney, A., Briand, C., and Gruetter, M. G. (2003). Insights into the regulatory mechanism for caspase-8 activation. *Mol. Cell* **11**, 543–549.

Duan, H., Orth, K., Chinnaiyan, A. M., Poirier, G. G., Froelich, C. J., He, W. W., and Dixit, V. M. (1996). ICE-LAP6, a novel member of the ICE/Ced-3 gene family, is activated by the cytotoxic T cell protease granzyme B. *J. Biol. Chem.* **271**, 16720–16724.

Fuentes-Prior, P., and Salvesen, G. S. (2004). The protein structures that shape caspase activity, specificity, activation and inhibition. *Biochem. J.* **384**, 201–232.

Gouvea, I. E., Judice, W. A., Cezari, M. H., Juliano, M. A., Juhasz, T., Szeltner, Z., Polgar, L., and Juliano, L. (2006). Kosmotropic salt activation and substrate specificity of poliovirus protease 3C. *Biochemistry* **45**, 12083–12089.

Martinon, F., and Tschopp, J. (2004). Inflammatory caspases: Linking an intracellular innate immune system to autoinflammatory diseases. *Cell* **117**, 561–574.

McStay, G. P., Salvesen, G. S., and Green, D. R. (2007). Overlapping cleavage motif selectivity of caspases: Implications for analysis of apoptotic pathways. *Cell Death Differ.* **15**(2), 322–331.

Nicoletti, I., Migliorati, G., Pagliacci, M. C., Grignani, F., and Riccardi, C. (1991). A rapid and simple method for measuring thymocyte apoptosis by propidium iodide staining and flow cytometry. *J. Immunol. Methods* **139,** 271–279.

Pop, C., Fitzgerald, P., Green, D. R., and Salvesen, G. S. (2007). Role of proteolysis in caspase-8 activation and stabilization. *Biochemistry* **46,** 4398–4407.

Pop, C., Timmer, J., Sperandio, S., and Salvesen, G. S. (2006). The apoptosome activates caspase-9 by dimerization. *Mol. Cell* **22,** 269–275.

Renatus, M., Stennicke, H. R., Scott, F. L., Liddington, R. C., and Salvesen, G. S. (2001). Dimer formation drives the activation of the cell death protease caspase 9. *Proc. Natl. Acad. Sci. USA* **98,** 14250–14255.

Riedl, S. J., Renatus, M., Snipas, S. J., and Salvesen, G. S. (2001). Mechanism based inactivation of caspases by the apoptotic suppressor p35. *Biochemistry* **40,** 13274–13280.

Rogers, S., Wells, R., and Rechsteiner, M. (1986). Amino acid sequences common to rapidly degraded proteins: The PEST hypothesis. *Science* **234,** 364–368.

Romanowski, M. J., Scheer, J. M., O'Brien, T., and McDowell, R. S. (2004). Crystal structures of a ligand-free and malonate-bound human caspase-1: Implications for the mechanism of substrate binding. *Structure (Camb).* **12,** 1361–1371.

Ryan, C. A., Stennicke, H. R., Nava, V. E., Lewis, J., Hardwick, J. M., and Salvesen, G. S. (2002). Inhibitor specificity of recombinant and endogenous caspase 9. *Biochem. J.* **366,** 595–601.

Scheer, J. M., Wells, J. A., and Romanowski, M. J. (2005). Malonate-assisted purification of human caspases. *Protein Expr. Purif.* **41,** 148–153.

Schmidt, U., and Darke, P. L. (1997). Dimerization and activation of the herpes simplex virus type 1 protease. *J. Biol. Chem.* **272,** 7732–7735.

Stennicke, H. R., Renatus, M., Meldal, M., and Salvesen, G. S. (2000). Internally quenched fluorescent peptide substrates disclose the subsite preferences of human caspases 1, 3, 6, 7 and 8. *Biochem. J.* **350,** 563–568.

Stennicke, H. R., Ryan, C. A., and Salvesen, G. S. (2002). Reprieval from execution: The molecular basis of caspase inhibition. *Trends Biochem. Sci.* **27,** 94–101.

Stennicke, H. R., and Salvesen, G. S. (1997). Biochemical characteristics of caspases-3, -6, -7, and -8. *J. Biol. Chem.* **272,** 25719–25723.

Stennicke, H. R., and Salvesen, G. S. (1999). Caspases: Preparation and characterization. *Meth. Enzymol.* **17,** 313–319.

Stennicke, H. R., and Salvesen, G. S. (2000). Caspase assays. *Methods Enzymol.* **322,** 91–100.

Takahashi, R., Deveraux, Q., Tamm, I., Welsh, K., Assa-Munt, N., Salvesen, G. S., and Reed, J. C. (1998). A single BIR domain of XIAP sufficient for inhibiting caspases. *J. Biol. Chem.* **273,** 7787–7790.

Talanian, R. V., Quinlan, C., Trautz, S., Hackett, M. C., Mankovich, J. A., Banach, D., Ghayur, T., Brady, K. D., and Wong, W. W. (1997). Substrate specificities of caspase family proteases. *J. Biol. Chem.* **272,** 9677–9682.

Thornberry, N. A., Rano, T. A., Peterson, E. P., Rasper, D. M., Timkey, T., Garcia-Calvo, M., Houtzager, V. M., Nordstrom, P. A., Roy, S., Vaillancourt, J. P., Chapman, K. T., and Nicholson, D. W. (1997). A combinatorial approach defines specificities of members of the caspase family and granzyme B. Functional relationships established for key mediators of apoptosis. *J. Biol. Chem.* **272,** 17907–17911.

Tu, S., McStay, G. P., Boucher, L. M., Mak, T., Beere, H. M., and Green, D. R. (2006). *In situ* trapping of activated initiator caspases reveals a role for caspase-2 in heat shock-induced apoptosis. *Nat. Cell. Biol.* **8,** 72–77.

CHAPTER TWENTY-TWO

SYNTHESIS AND BIOPHYSICAL CHARACTERIZATION OF STABILIZED α-HELICES OF BCL-2 DOMAINS

Gregory H. Bird,* Federico Bernal,* Kenneth Pitter, *and* Loren D. Walensky

Contents

Abstract

Rational design of compounds to mimic the functional domains of BCL-2 family proteins requires chemical reproduction of the biologic complexity afforded by the relatively large and folded surfaces of BCL-2 homology (BH) domain peptide α-helices. Because the intermolecular handshakes of BCL-2 proteins are so critical to controlling cellular fate, we undertook the development of a toolbox of peptidic ligands that harness the natural potency and specificity of BH α-helices to interrogate and potentially medicate the deregulated apoptotic pathways of human disease. To overcome the classic deficiencies of peptide

Department of Pediatric Oncology and the Program in Cancer Chemical Biology, Dana-Farber Cancer Institute, and the Division of Hematology/Oncology, Children's Hospital Boston, Harvard Medical School, Boston, Massachusetts
* These authors contributed equally to the work.

Methods in Enzymology, Volume 446
ISSN 0076-6879, DOI: 10.1016/S0076-6879(08)01622-4

reagents, including loss of bioactive structure in solution, rapid proteolytic degradation *in vivo*, and cellular impermeability, we developed a new class of compounds based on hydrocarbon stapling of BH3 death domain peptides. Here we describe the chemical synthesis of Stabilized Alpha-Helices of BCL-2 domains or SAHBs, and the analytical methods used to characterize their secondary structure, proteolytic stability, and cellular penetrance.

1. INTRODUCTION

The mission of chemical genetics is to identify compounds that directly and specifically alter protein function, so that physiologic activities can be investigated, and ultimately manipulated, on a conditional basis in real time. Identifying or generating such reagents to probe the broad range of apoptotic protein functions *in vitro* and specifically manipulate apoptotic pathways *in vivo* has been a significant and worthy challenge. Despite the obvious benefit of adopting "Nature's solution" to target protein–protein interactions, peptides present significant biophysical and pharmacologic drawbacks, which include structural unfolding when taken out of context from the full-length protein and resultant loss of biologic activity, rapid proteolysis *in vivo*, and, except for a minority of peptides, the inability to penetrate intact cells. Because the peptide α-helix participates in such a wide variety of intermolecular biologic recognition events, it is not surprising that a major focus of modern organic chemistry has been to devise synthetic strategies to recreate the architecture and stability of biologically active structures for both basic research and medicinal purposes.

An important strategic breakthrough in stabilizing α-helices derived from installing a covalent bond between amino acids in an attempt to "lock" the peptide structure in place (Fig. 22.1) (Bracken *et al.*, 1994; Jackson *et al.*, 1991; Phelan *et al.*, 1997). Initial successes with covalent helix stabilization, however, involved both polar crosslinks, which impede cell permeability, and labile crosslinks, which are readily hydrolyzed by proteolytic enzymes (amides) or reduced (disulfides) (Fig. 22.1A,B). Grubbs and coworkers circumvented these drawbacks by forming a crosslink between *O*-allyl serine residues on adjacent turns of the α-helix (Fig. 22.1C); the ring closing metathesis (RCM) chemistry uses a ruthenium catalyst to form a covalent bond between non-natural amino acid residues containing terminal double bonds or olefins (Blackwell and Grubbs, 1994). Ironically, although this chemical approach was successful in generating the covalent hydrocarbon crosslink, there was little to no enhancement of peptide α-helicity. Subsequently, Verdine and colleagues developed an alternate approach that used α,α-disubstituted non-natural

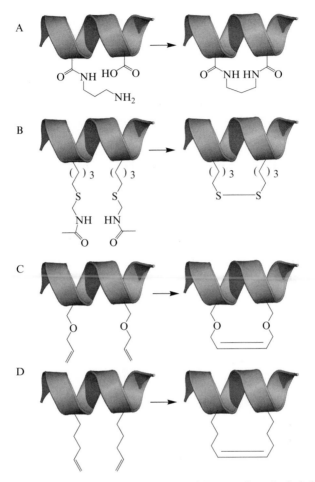

Figure 22.1 Approaches to covalent α-helical stabilization have included the use of (A) lactam bridges, (B) disulfide bridges, (C) ruthenium-catalyzed ring closing metathesis (RCM) of O-allyl serine residues, and (D) RCM of α,α-disubstituted non-natural amino acids bearing alkyl tethers.

amino acids containing alkyl tethers (Fig. 22.1D) (Schafmeister *et al.*, 2000). By experimenting with alternative placement of these non–natural amino acids along the peptide scaffold, in addition to varying stereochemistry and alkyl tether length, the chemical features required to dramatically stabilize a model helical peptide by use of an all-hydrocarbon chain crosslink were revealed. Here we describe the synthesis and biophysical characterization of stabilized α-helices of BCL-2 domains (SAHBs), our first biologic application of the "peptide stapling" technology (Walensky *et al.*, 2004).

2. SYNTHESIS OF NON-NATURAL AMINO ACIDS FOR PEPTIDE STAPLING

The asymmetric synthesis of the α,α-disubstituted amino acids required for peptide stapling is adapted from the method of Williams and colleagues (Fig. 22.2) (Williams and Im, 1991; Williams *et al.*, 1988). The production of the Fmoc-protected crosslinking amino acid "S5" commences with the use of the Williams morpholinone (5*S*,6*R*)-4-*tert*-butoxycarbonyl-5,6-diphenyl-morpholin-2-one (**4**), which is commercially available (e.g., Sigma Aldrich) or can be generated from (1*R*,2*S*)-2-amino-1,2-diphenylethanol (**1**) as illustrated in Fig. 22.2 (**2, 3**) and previously described in detail (Williams *et al.*, 2003). The stereochemistry of the bulky phenyl groups dictates the chirality of the produced α,α-disubstituted amino acids. Installation of the helix-inducing α-methyl group (Banerjee *et al.*, 2002) is performed by deprotonating the α-carbon of the morpholinone with sodium bis(trimethylsilyl)amide and

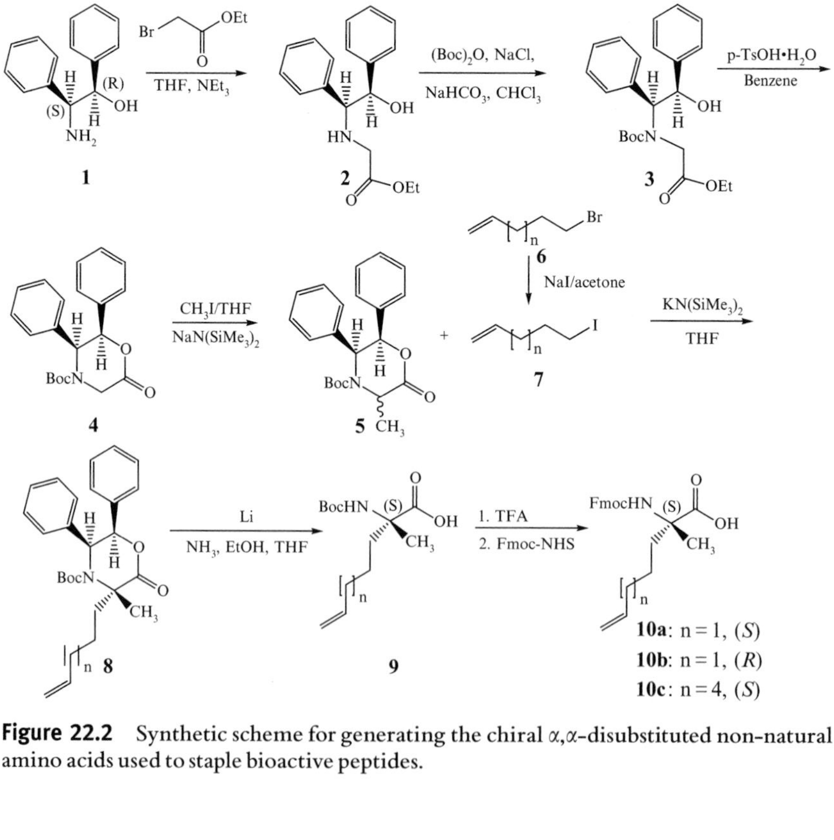

Figure 22.2 Synthetic scheme for generating the chiral α,α-disubstituted non-natural amino acids used to staple bioactive peptides.

alkylating with iodomethane at $-78\,^\circ$C. The α-methylated morpholinone, (5S,6R)-4-*tert*-butoxycarbonyl-5,6-diphenyl-3-methyl-morpholin-2-one (**5**), is obtained as a crystalline solid at a 3:2 ratio of diastereomers. The diastereomeric mixture of morpholinones is then treated with potassium bis(trimethylsilyl) amide in the presence of an alkenyl iodide (**7**) generated from a Finkelstein reaction of the bromide precursor (**6**). The second alkylation is performed at $-42\,^\circ$C and cleanly produces the α, α-bisalkylated morpholinone, 3S,5S,6R)-4-*tert*-butoxycarbonyl-5,6-diphenyl-3-methyl-3-(pent-4-enyl)-morpholin-2-one (**8**; n=1) in nearly quantitative yield with excellent diastereoselectivity (>98% *d.e.*) as determined by chiral HPLC.

After the synthesis of the bis-alkylated morpholinone, the chiral directing group is cleaved by reduction with lithium in liquid ammonia to yield the Boc-protected amino acid (*S*)-2-(*tert*-butoxycarbonylamino)-2-methyl-hept-6-enoic acid (**9**). Several methods exist for the cleavage of the C–N and C–O bonds in the morpholinone (Williams *et al.*, 2003); however, dissolving metal reduction is the only method that is compatible with the compounds synthesized (Greenfield *et al.*, 1954), as the alternative methods requiring reduction with H_2 and palladium on carbon or oxidation with sodium periodate or lead (IV) tetraacetate would destroy the terminal olefin needed for the crosslinking reaction. To complete the synthesis, the Boc-protecting group on the α-amine is converted to Fmoc to ensure compatibility with Fmoc solid-phase peptide synthesis. The Boc amino acid is dissolved in dichloromethane and exposed to excess trifluoroacetic acid to yield the free amine. Upon removal of the volatile components, the residue is dissolved in a 1:1 mixture of water and acetone and treated with sodium carbonate and *N*-(9-fluorenylmethoxycarbonyloxy)-succinimide (Fmoc-OSu) which yields, after column chromatography on silica gel, the desired Fmoc-protected amino acid (*S*)-2-(((9H-fluoren-9-yl)methoxy) carbonylamino)-2-methyl-hept-6-enoic acid (**10a**) in 75% yield. Whereas two S5 amino acids (**10a**) are used to generate (*i, i* + *4*) crosslinks, the R5 (**10b**) and (**10c**) (or S5 and R8) derivatives are used for (*i, i* + *7*) stapling. Once purified, the Fmoc-protected α,α-disubstituted amino acids are ready for use in solid-phase peptide synthesis.

3. DESIGN AND SYNTHESIS OF STABILIZED α-HELICES OF BCL-2 DOMAINS (SAHBs)

The BH3 domains of many BCL-2 family members have been structurally defined as amphipathic α-helices (Chou *et al.*, 1999; Day *et al.*, 2005; Denisov *et al.*, 2003; Hinds *et al.*, 2003; McDonnell *et al.*, 1999; Muchmore *et al.*, 1996; Petros *et al.*, 2000; 2001; Sattler *et al.*, 1997; Suzuki *et al.*, 2000).

Whereas the hydrophobic face of the α-helix contacts a hydrophobic pocket of its interacting partner, charged residues on the hydrophilic face participate in electrostatic pairings with the target protein in addition to being exposed to the aqueous environment (Sattler *et al.*, 1997). When designing the insertion points for hydrocarbon staples, assessment of available structural data is ideal so that native amino acids essential to protein interaction can be preserved. For example, the alanine scan performed by Sattler *et al.* for the interaction between BAK BH3 and BCL-X_L $\triangle$C determined the important hydrophobic and charged amino acids for BH3 engagement (Sattler *et al.*, 1997). Thus, in this example, the S5 amino acid is specifically implanted in (*i, i + 4*) pairings that do not disrupt key interacting residues of the hydrophobic contact surface (Fig. 22.3).

The construction of a stapled BH3 peptide can be performed either manually or by use of an automated peptide synthesizer (Fig. 22.4). In either case, the solid-phase synthesis resin of choice (e.g., Rink amide MBHA resin with loading levels of 0.4 to 0.6 mmol/g resin) is swollen with 1-methyl-2-pyrrolidinone (NMP) for 15 min. After draining, the Fmoc group is deprotected by exposure to a 20% (v/v) solution of piperidine in NMP for 30 min. After extensive washing with NMP, the resin beads are exposed first to a 0.5 *M* solution of amino acid in NMP, followed by a 0.5 *M* solution of 2-(6-chloro-1*H*-benzotriazole-1-yl)-1,1,3,3-tetramethylaminium hexa-fluorophosphate (HCTU) coupling agent (Albericio, 2004). The resin is then treated with *N,N*-diisopropyl ethylamine (DIEA), and the reaction is

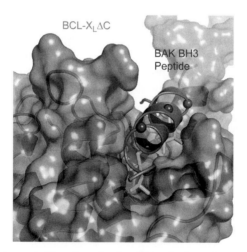

Figure 22.3 Design considerations for installing hydrocarbon staples derive from structural data (e.g., BAK BH3/BCL-X_L $\triangle$C; PDB 1BXL [Sattler *et al.*, 1997]) that highlight key interacting surfaces to be avoided. Several (*i, i + 4*) residues indicated as red balls on the noninteracting surface of the BAK BH3 helix are amenable to replacement with the S5 non-natural amino acid.

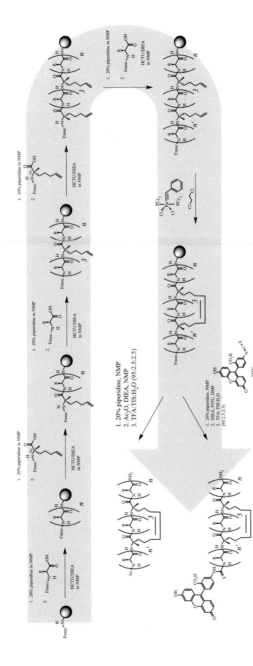

Figure 22.4 Synthetic scheme for the generation of SAHBs by Fmoc–based solid-phase peptide synthesis and ruthenium–catalyzed olefin metathesis.

carried out for 30 to 60 min with either shaking or bubbling under nitrogen. Typically, 4 equivalents of Fmoc–amino acid, 3.9 equivalents of coupling agent, and 8 equivalents of DIEA are used, and double couplings are performed. After completion of the peptide synthesis, the amino terminus can either be acetylated by reaction with acetic anhydride and DIEA or subjected to alternative derivatization (e.g., FITC-β-Ala, biotin-β-Ala). For amino terminal modifications that contain such sulfur atoms, the β-Ala can be installed at this stage but FITC or biotin capping is deferred until after metathesis to avoid sulfur-based poisoning of the ruthenium catalyst.

The olefin metathesis step is carried out by first swelling the resin with 1,2-dichloroethane followed by exposure to a 10 mM solution of bis(tricyclohexylphosphine)-benzylidene ruthenium (IV) dichloride (Grubbs' first generation catalyst) in 1,2-dichloroethane (0.20 mol% on the basis of resin substitution) for 2 h. The stapling reaction is carried out twice with constant bubbling under nitrogen. The resin-bound peptide is then washed with 1,2-dichloroethane three times and dried under a stream of nitrogen. Sulfur-containing moieties can be N-terminally appended at this stage after Fmoc deprotection in piperidine. The completed peptide is cleaved from the resin and deprotected by exposure to trifluoroacetic acid (TFA)–based cleavage cocktails such as reagent K (82.5% TFA, 5% thioanisole, 5% phenol, 5% water, 2.5% 1, 2-ethanedithiol) or TFA/triisopropyl silane (TIS)/water (95%, 2.5%, 2.5%), and precipitated with methyl-*tert*-butyl ether at 4 °C followed by lyophilization.

Lyophilized SAHB peptides are purified by reverse-phase HPLC by use of a C$_{18}$ column. The compounds are characterized by LC/MS, with mass spectra obtained either by electrospray in positive ion mode or by MALDI-TOF. Quantitation is achieved by amino acid analysis on a Beckman 6300 high-performance amino acid analyzer. Working stock solutions are generated by dissolving the lyophilized material in DMSO at 1 to 10 mM. Lyophilized powder and DMSO stock solutions are stored at −20 °C.

4. STRUCTURAL ASSESSMENT OF SAHBs BY CIRCULAR DICHROISM (CD)

To evaluate secondary structure improvements of hydrocarbon-stapled BH3 peptides, CD spectra are recorded and analyzed. Generally, short peptides do not exhibit significant α-helical structure in solution because the entropic cost of maintaining a conformationally restricted structure is not overcome by the enthalpic gain from hydrogen bonding of the peptide backbone. Indeed, we find that unmodified BH3 peptides, ranging in length from 16 to 25 amino acids, display α-helical propensities of less than 25% (Letai *et al.*, 2002), whereas installation of a chemical staple typically enhances α-helicity by 3- to 5-fold (Fig. 22.5) (Walensky *et al.*, 2004; 2006).

CD spectra are recorded on an Aviv Biomedical spectrometer (Model 410) equipped with a Peltier temperature controller and a thermoelectric sample changer with 5-position rotor. A total of five scans from 190 to 260 nm in 0.5-nm increments with 0.5 sec averaging time are collectively averaged to obtain each spectrum using a 1-mm path length cell. For temperature scans, three scans are averaged at temperature increments of 5 °C from 1 °C to 91 °C. The target peptide concentration is 25 to 50 μM, and exact concentrations are confirmed by quantitative amino acid analysis of two CD sample dilutions. It is important to carefully assess compound solubility by first dissolving the peptide in buffer followed by high-speed tabletop centrifugation to ensure that there is no pellet. SAHBs are generally reconstituted in 5 mM potassium phosphate (pH 7.5) or Milli-Q deionized water, but alternate aqueous buffers spanning a wider pH range or organic co-solvents may be required to optimize solubility.

The CD spectra are initially plotted as wavelength vs millidegree, the default output of the instrument. Once the precise peptide concentration is confirmed, the mean residue ellipticity [θ], in units of degree·cm^2·dmol^{-1}·residue^{-1}, is derived from Eq. (22.1):

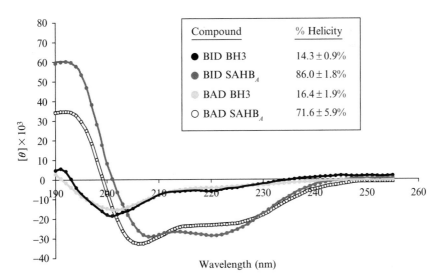

Figure 22.5 Circular dichroism spectra demonstrate the increased α-helicity of SAHBs compared to the corresponding unmodified BH3 peptides (Walensky *et al.*, 2006).

$$[\theta] = \text{millidegree/molar concentration/number} \atop \text{of amino acid residues} \tag{22.1}$$

Once converted to mean residue ellipticity, percent α-helicity can be calculated with Eqs. (22.2) and (22.3)(Forood *et al.*, 1993):

$$\%\text{Helicity} = 100 \times [\theta]_{222}/[\theta]_{222}^{\text{max}} \text{where} \tag{22.2}$$

$${}^{\text{max}}[\theta]_{222} = -40,000 \times [1 - (2.5/\text{number of amino acid residues}) \tag{22.3}$$

Alternately, most CD instruments are bundled with curve-fitting software programs, such as CDDN (Bohm *et al.*, 1992), that deduce the relative fractions of secondary structure including α-helix, parallel and antiparallel β-sheet, β-turn, and random coil.

5. PROTEASE RESISTANCE TESTING OF SAHBs

Protein degradation *in vivo* is a natural metabolic process that activates and deactivates biologically active peptides in a regulated, homeostatic fashion. Peptide proteolysis, however, remains a major hurdle for the conversion of synthetic peptides to pharmaceuticals. One of the benefits of enforcing peptide α-helicity is the shielding of the vulnerable amide bond from *in vivo* proteolysis. Because proteases require that peptides adopt an extended conformation to hydrolyze amide bonds, the structural constraint afforded by the hydrocarbon staple renders the crosslinked peptides protease-resistant. To measure and optimize SAHB stability, we use several proteolytic stability assays.

5.1. *In vitro* trypsin/chymotrypsin degradation assay

To assess protease resistance *in vitro*, proteolytic enzymes are selected based on the sequence composition of the peptide substrate, such that effective fragmentation can be achieved. Trypsin, which recognizes Arg and Lys, and chymotrypsin, which predominantly cleaves after Phe, Tyr, Trp, Leu, Met, are commonly used. Several experimental and analytical approaches can be undertaken. For example, relative stability of a FITC-labeled unmodified and stapled peptide can be compared by exposing the compounds (e.g., 5 μg) to trypsin agarose (Pierce) (substrate/enzyme $\approx$125) over time (e.g., 0, 10, 20, 90, and 180 min) (Fig. 22.6A). Reactions are quenched by tabletop

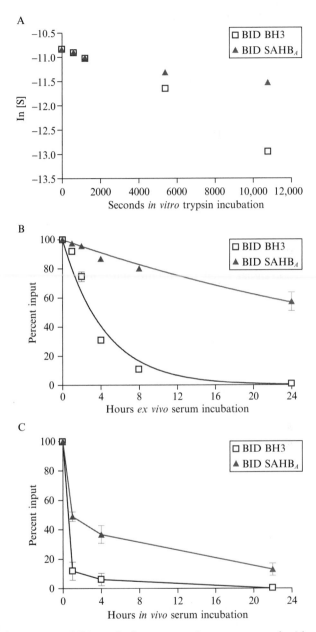

Figure 22.6 SAHBs exhibit marked protease resistance compared with unmodified BH3 peptides as assessed by (A) *in vitro* trypsin degradation assay, (B) *ex vivo*, and (C) *in vivo* serum stability assays (Walensky *et al.*, 2004).

centrifugation at high speed, and remaining full-length substrate in the isolated supernatant is analyzed by HPLC-based quantitation with fluorescence detection at excitation/emission settings of 495/530 nm.

Alternately, mass spectrometry-based quantitation can be used to increase sensitivity and avoid the use of N-terminally derivatized FITC SAHBs. MS/MS-based detection and quantitation is typically used in pharmacokinetic studies of stability and metabolism to eliminate signal suppression from complex mixtures; however, a standard single-quadripole mass spectrometer detector found in most academic institutions is also capable of quantifying intact peptide in an *in vitro* assay consisting of enzyme and peptide dissolved in a phosphate-buffered calcium chloride solution. For example, we use an Agilent model 1200 LC/MS with the following settings: 20 μl injection, 0.6 ml flow rate, a gradient of water (0.1% formic acid) to 20 to 80% acetonitrile (0.075% formic acid) over 10 min, with 4 min at completion to revert to starting gradient conditions and 0.5 min post-time, allowing for an overall 15-min run time. The diode-array detector (DAD) signal is set to 280 nm with an 8 nm bandwidth. Mass spectrometry detector (MSD) settings are in scan mode with one channel gated on $(M + 2H)/2$, $\pm$ 1 mass unit and the other on $(M + 3H)/3$, $\pm$ 1 mass unit. Integration of each MSD signal typically yields counts for area under the curve of $\sim 10^8$. Because each run takes 15 min, four duplicate samples can be staggered such that four data points are acquired per hour, with the proteolytic enzyme promptly added to the sample once the zero-hour time point is autoinjected. Plotting of MSD area vs time generates an exponential decay curve with error bars typically 5% or less. An internal control of acetylated tryptophan carboxamide at a concentration of 100 μM can be used with absorbance at 278 nm to normalize each MSD data point.

5.2. Serum stability assays

To extend the proteolytic analysis to *in vitro* and *in vivo* serum stability, peptides (e.g., 2.5 to 5 μg) are incubated with fresh mouse serum (e.g., 25 μl) at 37 °C or injected (e.g., 10 mg/kg) by tail vein into mice, respectively. *Ex vivo* serum samples or serum isolated from serial tail bleeds (e.g., 25 μL) are collected at various time intervals (e.g., 0, 1, 2, 4, 8, and 24 h), flash frozen, lyophilized, and then extracted with acidified organic solvent solution (e.g., 50:50 acetonitrile/water containing 0.1% trifluoroacetic acid). Levels of intact peptide are detected and quantified as described in section 5.1 (Fig. 22.6B,C).

6. Cell Permeability Screening of SAHBs

Despite the exquisite biologic specificity of native peptides for their protein targets, a major limitation of peptides as therapeutics is their general inability to cross lipid membranes to access the intracellular environment. The charge of amino acid side chains and the polarity of the peptide backbone account for the impenetrability. However, a discrete subclass of cationic peptides (e.g., HIV-TAT, penetratin sequence from the *Antennapedia* homeodomain, poly-Arg), termed cell penetrating peptides or CPPs, have been shown to enter cells (Derossi *et al.*, 1994; Fawell *et al.*, 1994; Schwarze *et al.*, 1999) by an energy-dependent fluid-phase macropinocyto-mechanism (Wadia *et al.*, 2004). Hydrocarbon-stapled BH3 peptides were likewise found to be cell permeable through a pinocytotic pathway, although in contrast to cationic CPPs, initial contact with the plasma membrane does not seem to be mediated by interactions with glycosaminoglycans (Console *et al.*, 2003; Walensky *et al.*, 2004). The cell penetrability of SAHBs is believed to derive from the reinforced α-helical structure of the peptides, as peptide helices are common membrane-interacting protein motifs; the hydrophobicity of the hydrocarbon staple itself may additionally contribute to membrane tropism. To evaluate the cellular uptake of SAHB compounds, we use two assays in combination: FACS analysis and confocal microscopy of FITC-SAHB treated cells. In general, neutral to cationic SAHBs consistently display efficient cellular uptake (Console *et al.*, 2003; Walensky *et al.*, 2004; 2006), whereas anionic species may require sequence modification (e.g., point mutagenesis, sequence shift) to dispense with negative charge (Bernal *et al.*, 2007).

6.1. FACS analysis of FITC-SAHB–treated cells

SAHBs are diluted in stepwise fashion from 100% DMSO stock solutions (1 to 10 mM) into water, with the experimental concentration (e.g., 1 to 10 μM) achieved by final dilution into serum-free tissue culture media. It is important to monitor the treatment solutions for any sign of compound precipitation, which can be avoided by modifying the media solution (e.g., increasing final DMSO concentration to 1 or 2%). Nonadherent cells are washed in serum-free media and diluted to a concentration of 1×10^6 cells/ml, and typically 50 μl (50,000 cells) are combined with 50 μl of FITC-SAHB solution (e.g., 5, 10, and/or 20 μM) concentration and incubated at 37 °C. The corresponding unmodified FITC-peptide is used as a negative control. After 4 h or serial time points, the cells are pelleted at 1400 rpm on a tabletop swinging bucket centrifuge for 4 min, and the supernatant is gently aspirated. The cells are incubated in 50 μl of tissue-culture grade trypsin solution (e.g., 0.25%, Gibco) to cleave surface proteins and thereby help eliminate any nonspecifically bound peptide from the cell surface. After 5 min, the trypsin is quenched with 300 μl of media containing 10% FBS. The cells are then pelleted again, washed in PBS, pelleted, and resuspended in 200 μl of FACS buffer. The procedure is identical for adherent cells, except that 50,000 cells are preplated in serum-containing media and then, at the outset of the experiment, washed twice in serum-free media; the cells are detached from the culture plates during the trypsinization step described previously. Cellular fluorescence is analyzed by use of a FACSCalibur flow cytometer (Becton Dickinson) and FlowJo software (Tree Star) (Fig. 22.7A). FITC intensity is detected on a log scale, and 10,000 cells are counted per treatment replicate. To exclude membrane disruption as the mechanism for cellular entry and fluorescence, cell permeabilization can be evaluated by adding 0.5 $\mu g/ml$ propidium iodide (BD Biosciences) to the analyzed samples. To evaluate an endocytic mechanism of import, the

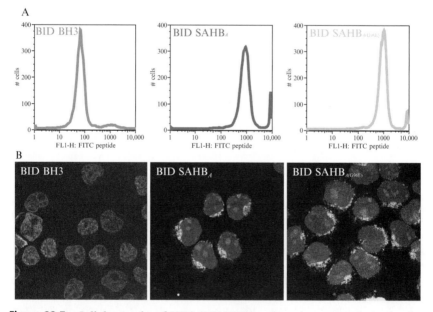

Figure 22.7 Cellular uptake of FITC–BID SAHB and a point mutant derivative, but not the unmodified FITC–BID BH3 peptide, is readily demonstrated by (A) FACS analysis and (B) confocal microscopy of FITC-peptide treated cells in culture. The cellular fluorescence of FITC-SAHB treated cells, as reflected by a shift of the FACS profile to the right compared to FITC–BID BH3–treated cells, corresponds to the intracellular cytosolic localization of FITC–BID SAHB observed by confocal microscopy. (Walensky *et al.*, 2004.)

identical experiment can be performed with: (a) 30 min preincubation of cells at 4 °C followed by 4 h incubation with FITC-compounds at 4 °C to assess temperature-dependence of fluorescent labeling; (b) 60 min pretreatment of cells at 37 °C with 10 mM sodium azide and 50 mM 2-deoxy-D-glucose to deplete cellular ATP and thereby assess energy-dependence of fluorescent labeling (Potocky et al., 2003; Richard et al., 2003).

6.2. Confocal microscopy of FITC-SAHB–treated cells

Whereas the FACS-based cell permeability assay is amenable to rapid screening of compounds, it is not meant to stand alone as proof positive of cellular uptake. Confirmatory testing with confocal microscopy of FITC-SAHB–treated cells is used to demonstrate intracellular fluorescence and subcellular localization. Cells are incubated with FITC-labeled compounds in serum-free media for up to 4 h as described previously or with serum replacement (as 2× serum-containing media) followed by additional incubation intervals (e.g., 20 h) at 37 °C. In preparation for fixation, nonadherent cells are washed twice with PBS and then cytospun at 600 rpm for 5 min onto Superfrost plus glass slides (Fisher); media is aspirated from the wells of adherent cells and then washed twice with PBS in situ. Slides are exposed to freshly prepared 4% paraformaldehyde/PBS, washed with PBS, incubated with a nuclear counterstain (e.g., 100 nM TO-PRO-3 iodide [Molecular Probes]), treated with Vectashield mounting medium (Vector), and then imaged by confocal microscopy (BioRad 1024) (Fig. 22.7B). For double-labeling experiments to identify subcellular compartments, fixed cells can be incubated with a primary antibody (e.g., Tom20 for outer mitochondrial membrane labeling [Schleiff et al., 1997]) and rhodamine-conjugated secondary antibody before nuclear counterstaining. Because cationic CPPs have been shown to relocalize during cellular fixation (Drin et al., 2003; Richard et al., 2003), live confocal microscopy can be used to further confirm the intracellular localization of FITC-SAHB compounds. Cells doubly labeled with FITC-SAHB and live cell organellar markers (e.g., 100 nM MitoTracker [Molecular Probes]) are washed and resuspended in PBS, followed by wet mount preparation and prompt confocal microscopy analysis.

7. SUMMARY

Novel chemical approaches that optimize the pharmacology and bioactivity of peptides are rejuvenating the field of peptide therapeutics. Moreover, chemical reconstitution of native peptide structures has yielded new biologic tools to interrogate signal transduction pathways in vitro and in vivo. In this case, the application of hydrocarbon stapling to BH domain peptides has enabled the synthesis of compounds with structural fidelity, proteolytic stability, and cell permeability, thereby facilitating the biochemical analysis and pharmacologic manipulation of BCL-2 family protein interactions. The application of SAHB compounds to apoptosis research is the subject of a separate chapter.

ACKNOWLEDGMENTS

We thank E. Smith for figure design and editorial assistance, Gregory L. Verdine and the late Stanley J. Korsmeyer for their invaluable mentorship, members of the Walensky laboratory, past and present, for their scientific contributions, and C. E. Schafmeister, J. Po, and I. Escher, former members of the Verdine laboratory, for their seminal work on the all-hydrocarbon crosslinking system. L. D. W. is supported by NIH grants K08HL074049, 5R01CA50239, and 5P01CA92625, a Burroughs Wellcome Fund Career Award in the Biomedical Sciences, a Partnership for Cures Charles E. Culpeper Scholarship in Medical Science, a grant from the William Lawrence Children's Foundation, and the Dana-Farber Cancer Institute High-Tech fund. G.H.B. is the recipient of a Harvard University Center for AIDS Research Scholar Award and F.B. is funded by a grant from the Dana. Farber Cancer Institute Pediatric Low-Grade Astrocytoma Program.

REFERENCES

Albericio, F. (2004). Developments in peptide and amide synthesis. *Curr. Opin. Chem. Bio.* **8,** 211–221.

Banerjee, R., Basu, G., Chene, P., and Roy, S. (2002). Aib-based peptide backbone as scaffolds for helical peptide mimics. *J. Pept. Res.* **60,** 88–94.

Bernal, F., Tyler, A. F., Korsmeyer, S. J., Walensky, L. D., and Verdine, G. L. (2007). Reactivation of the p53 tumor suppressor pathway by a stapled p53 peptide. *J. Am. Chem. Soc.* **129,** 2456–2457.

Blackwell, H. E., and Grubbs, R. H. (1994). Highly Efficient Synthesis of Covalently Cross-Linked Peptide Helices by Ring-Closing Metathesis. *Angew Chem. Int. Ed.* **37,** 3281–3284.

Bohm, G., Muhr, R., and Jaenicke, R. (1992). Quantitative-Analysis of Protein Far Uv Circular-Dichroism Spectra by Neural Networks. *Protein Eng.* **5,** 191–195.

Bracken, C., Gulyas, J., Taylor, J. W., and Baum, J. (1994). Synthesis and nuclear magnetic resonance structure determination of an alpha-helical, bicyclic, lactam-bridged hexapeptide. *J. Am. Chem. Soc.* **116,** 6431–6432.

Chou, J. J., Li, H., Salvesen, G. S., Yuan, J., and Wagner, G. (1999). Solution structure of BID, an intracellular amplifier of apoptotic signaling. *Cell* **96,** 615–624.

Console, S., Marty, C., Garcia-Echeverria, C., Schwendener, R., and Ballmer-Hofer, K. (2003). Antennapedia and HIV transactivator of transcription (TAT) "protein transduction domains" promote endocytosis of high molecular weight cargo upon binding to cell surface glycosaminoglycans. *J. Biol. Chem.* **278,** 35109–35114.

Day, C. L., Chen, L., Richardson, S. J., Harrison, P. J., Huang, D. C., and Hinds, M. G. (2005). Solution structure of prosurvival Mcl-1 and characterization of its binding by proapoptotic BH3-only ligands. *J. Biol. Chem.* **280,** 4738–4744.

Denisov, A. Y., Madiraju, M. S., Chen, G., Khadir, A., Beauparlant, P., Attardo, G., Shore, G. C., and Gehring, K. (2003). Solution Structure of Human BCL-w: Modulation of ligand binding by the C-terminal helix. *J. Biol. Chem.* **278,** 21124–21128.

Derossi, D., Joliot, A. H., Chassaing, G., and Prochiantz, A. (1994). The third helix of the Antennapedia homeodomain translocates through biological membranes. *J. Biol. Chem.* **269,** 10444–10450.

Drin, G., Cottin, S., Blanc, E., Rees, A. R., and Temsamani, J. (2003). Studies on the internalization mechanism of cationic cell-penetrating peptides. *J. Biol. Chem.* **278,** 31192–31201.

Fawell, S., Seery, J., Daikh, Y., Moore, C., Chen, L. L., Pepinsky, B., and Barsoum, J. (1994). Tat-mediated delivery of heterologous proteins into cells. *Proc. Natl. Acad. Sci. USA* **91,** 664–668.

Forood, B., Feliciano, E. J., and Nambiar, K. P. (1993). Stabilization of Alpha-Helical Structures in Short Peptides Via End Capping. *Proc. Nat Acad. Sci. USA* **90,** 838–842.

Greenfield, H., Friedel, R. A., and Orchin, M. (1954). The reduction of simple olefins with sodium and methanol in liquid ammonia. **76,** 1258–1259.

Hinds, M. G., Lackmann, M., Skea, G. L., Harrison, P. J., Huang, D. C., and Day, C. L. (2003). The structure of Bcl-w reveals a role for the C-terminal residues in modulating biological activity. *EMBO J.* **22,** 1497–1507.

Jackson, D. Y., King, D. S., Chmielewski, J., Singh, S., and Schultz, P. G. (1991). General approach to the synthesis of short α-helical peptides. *J. Am. Chem. Soc.* **113,** 9391–9392.

Letai, A., Bassik, M. C., Walensky, L. D., Sorcinelli, M. D., Weiler, S., and Korsmeyer, S. J. (2002). Distinct BH3 domains either sensitize or activate mitochondrial apoptosis, serving as prototype cancer therapeutics. *Cancer Cell* **2,** 183–192.

McDonnell, J. M., Fushman, D., Milliman, C. L., Korsmeyer, S. J., and Cowburn, D. (1999). Solution structure of the proapoptotic molecule BID: A structural basis for apoptotic agonists and antagonists. *Cell* **96,** 625–634.

Muchmore, S. W., *et al.* (1996). X-ray and NMR structure of human Bcl-xL, an inhibitor of programmed cell death. *Nature* **381,** 335–341.

Petros, A. M., Medek, A., Nettesheim, D. G., Kim, D. H., Yoon, H. S., Swift, K., Matayoshi, E. D., Oltersdorf, T., and Fesik, S. W. (2001). Solution structure of the antiapoptotic protein bcl-2. *Proc. Natl. Acad. Sci. USA* **98,** 3012–3017.

Petros, A. M., Nettesheim, D. G., Wang, Y., Olejniczak, E. T., Meadows, R. P., Mack, J., Swift, K., Matayoshi, E. D., Zhang, H., Thompson, C. B., and Fesik, S. W. (2000). Rationale for Bcl-xL/Bad peptide complex formation from structure, mutagenesis, and biophysical studies. *Protein Sci.* **9,** 2528–2534.

Phelan, J. C., Skelton, N. J., Braisted, A. C., and McDowell, R. S. (1997). A general method for constraining short peptides to an α-helical conformation. *J. Am. Chem. Soc.* **119,** 455–460.

Potocky, T. B., Menon, A. K., and Gellman, S. H. (2003). Cytoplasmic and nuclear delivery of a TAT-derived peptide and a beta-peptide after endocytic uptake into HeLa cells. *J. Biol. Chem.* **278,** 50188–50194.

Richard, J. P., Melikov, K., Vives, E., Ramos, C., Verbeure, B., Gait, M. J., Chernomordik, L. V., and Lebleu, B. (2003). Cell-penetrating peptides. A reevaluation of the mechanism of cellular uptake. *J. Biol. Chem.* **278,** 585–590.

Sattler, M., *et al.* (1997). Structure of Bcl-xL-Bak peptide complex: Recognition between regulators of apoptosis. *Science* **275,** 983–986.

Schafmeister, C., Po, J., and Verdine, G. (2000). An all-hydrocarbon cross-linking system for enhancing the helicity and metabolic stability of peptides. *J. Am. Chem. Soc.* **122,** 5891–5892.

Schleiff, E., Shore, G. C., and Goping, I. S. (1997). Interactions of the human mitochondrial protein import receptor, hTom20, with precursor proteins *in vitro* reveal pleiotropic specificities and different receptor domain requirements. *J. Biol. Chem.* **272,** 17784–17789.

Schwarze, S. R., Ho, A., Vocero-Akbani, A., and Dowdy, S. F. (1999). *In vivo* protein transduction: Delivery of a biologically active protein into the mouse. *Science* **285,** 1569–1572.

Suzuki, M., Youle, R. J., and Tjandra, N. (2000). Structure of Bax: Coregulation of dimer formation and intracellular localization. *Cell* **103,** 645–654.

Wadia, J. S., Stan, R. V., and Dowdy, S. F. (2004). Transducible TAT-HA fusogenic peptide enhances escape of TAT-fusion proteins after lipid raft macropinocytosis. *Nat. Med.* **10,** 310–315.

Walensky, L. D., Kung, A. L., Escher, I., Malia, T. J., Barbuto, S., Wright, R. D., Wagner, G., Verdine, G. L., and Korsmeyer, S. J. (2004). Activation of apoptosis *in vivo* by a hydrocarbon-stapled BH3 helix. *Science* **305,** 1466–1470.

Walensky, L. D., Pitter, K., Morash, J., Oh, K. J., Barbuto, S., Fisher, J., Smith, E., Verdine, G. L., and Korsmeyer, S. J. (2006). A stapled BID BH3 helix directly binds and activates BAX. *Mol. Cell* **24,** 199–210.

Williams, R. M., and Im, M. N. (1991). Asymmetric synthesis of monosubstituted and alpha, alpha-disubstituted amino acids via diastereoselective glycine enolate alkylations. *J. Am. Chem. Soc.* **113,** 9276–9286.

Williams, R. M., Sinclair, P. J., Zhai, D., and Chen, D. (1988). Practical asymmetric syntheses of .alpha.-amino acids through carbon-carbon bond constructions on electrophilic glycine templates. *J. Am. Chem. Soc.* **110,** 1547–1557.

Williams, R. M., Sinclair, P. J., DeMong, D. E., Chen, D., and Zhai, D. (2003). Asymmetric Synthesis of N-tert-butoxycarbonyl Alpha-Amino Acids: Synthesis of (5S, 6R)-4-tert-butoxycarbonyl-5,6-diphenylmorpholin-2-one. *Organic Syntheses* **80,** 18–30.

CHAPTER TWENTY-THREE

Dissection of the BCL-2 Family Signaling Network with Stabilized α-Helices of BCL-2 Domains

Kenneth Pitter, Federico Bernal, James LaBelle,
and Loren D. Walensky

Contents

Abstract

The BCL-2 family of apoptotic proteins regulates the critical balance between cellular life and death and, thus, has become the focus of intensive basic science inquiry and a fundamental target for therapeutic development in oncology and other diseases. Classified based on the presence of conserved α-helical motifs and pro- and anti-apoptotic functionalities, BCL-2 proteins participate in a complex interaction network that determines cellular fate. The identification of BCL-2 homology domain 3 (BH3) as a critical death helix that engages and regulates BCL-2 family proteins has inspired the development of molecular tools to decode and drug the interaction network. Stabilized Alpha-Helices of BCL-2 domains (SAHBs) are structurally reinforced, protease-resistant, and cell-permeable compounds that retain the specificity of native BH3 death ligands and, therefore, serve as ideal reagents to dissect BCL-2 family interactions

Department of Pediatric Oncology and the Program in Cancer Chemical Biology, Dana-Farber Cancer Institute, and the Division of Hematology/Oncology, Children's Hospital Boston, Harvard Medical School, Boston, Massachusetts

Methods in Enzymology, Volume 446
ISSN 0076-6879, DOI: 10.1016/S0076-6879(08)01623-6

in vitro and *in vivo*. Here, we describe the *in vitro* and cell-based methods that exploit SAHB compounds to determine the functional consequences of BH3 interactions in regulating apoptosis.

1. INTRODUCTION

The discovery of BCL-2 at the chromosomal breakpoint of t(14;18) (q32;q21) lymphomas (Bakhshi *et al.*, 1985; Cleary and Sklar, 1985; Tsujimoto *et al.*, 1985) led to a paradigm shift in our understanding of the origins of cancer. BCL-2 was initially defined as a survival protein capable of prolonging cellular life by evading programmed cell death or apoptosis (Hockenbery *et al.*, 1990; Nunez *et al.*, 1990; Vaux *et al.*, 1988). In follicular lymphoma, the aberrant subjugation of BCL-2 to the transcriptional control of the immunoglobulin heavy chain locus leads to BCL-2 overexpression, a primary oncogenic event responsible for pathologic B-cell survival (McDonnell *et al.*, 1989; Seto *et al.*, 1988). Since this seminal discovery, a growing family of BCL-2–like proteins has been identified and subclassified based on the presence of conserved helical motifs and pro- and anti-apoptotic functionalities (Danial and Korsmeyer, 2004; Youle and Strasser, 2008) (Fig. 23.1). What has emerged is a complex protein-interaction network of guardians and executioners that determine cellular fate. Structural and biochemical studies identified the BCL-2 homology domain 3 (BH3) as a critical α-helical motif that engages BCL-2 family targets to regulate their activities (Cheng *et al.*, 2001; Sattler *et al.*, 1997; Zha *et al.*, 1996). Thus, understanding the selectivities and functional activities of discrete BH3 domains remains essential to decoding the BCL-2 family network. To that end, we have developed a chemical toolbox of Stabilized Alpha Helices of BCL-2 domains (SAHBs) that preserve the primary and secondary structural fidelity of native BH3 domains for biologic study *in vitro* and *in vivo* (Walensky *et al.*, 2004; 2006). Here we describe the application of SAHB compounds to cell death studies focused on discerning the functional consequences of selective BH3 domain interactions.

2. BCL-2 FAMILY BINDING MEASUREMENTS BY FLUORESCENCE POLARIZATION ASSAY

Despite the major advances stemming from apoptosis studies, many critical facets of BCL-2 family death signaling—including the regulation of essential cell death executioners such as BAX—remain mechanistic mysteries. *In vitro* binding assays serve as a starting point for defining BH3 domain interactions. The resultant quantitative affinity data provide a framework for hierarchical ranking of interaction pairs, which in turn forms the basis for hypotheses regarding potential functions of BH3 interactions *in situ*.

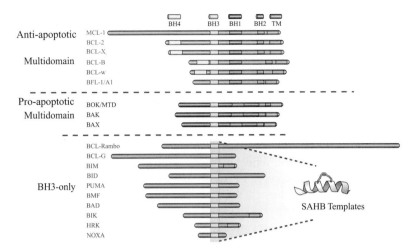

Figure 23.1 The BCL-2 family of proteins is composed of anti-apoptotic and pro-apoptotic members that share conserved BCL-2 homology (BH) domains. The pro-apoptotic BH3 domains serve as sequence templates for the generation of SAHB compounds.

The quality and reproducibility of *in vitro* binding assays explicitly depend on the design and purity of expressed proteins. For example, the potential impact of fusion proteins and tags should be considered and evaluated independently in negative control binding studies; ideally, tagless proteins should be used, although this may not be feasible in all circumstances because of protein solubility, stability, or recovery challenges. In the case of BCL-2 family proteins, the carboxy termini of proteins are typically removed (ΔC) to facilitate solubilization and purification; however, results with such constructs must be interpreted with care, because these sequences could participate in the regulation of native protein interactions. Whenever comparing binding affinities of a particular BH3 domain across distinct BCL-2 family proteins, ideally, truncated constructs should be compared with truncated constructs, and full-length proteins compared with full-length proteins. It is also worth noting that different types of binding assays can produce quantitatively distinct data sets. For example, surface plasmon resonance (SPR), which involves immobilization of either ligand or target, typically yields lower K_d values than fluorescence polarization assays (FPA), which are performed in solution. By orienting immobilized ligand or protein target on a surface, binding interactions may be facilitated compared with FPA, in which there is a greater entropic cost to aligning ligand and target in solution.

For ΔC anti-apoptotic proteins, we use pGEX vectors to express GST-BCL-2, BCL-X_L, BCL-w, MCL-1, and BFl1/A1, followed by thrombin cleavage and FPLC-based gel filtration chromatography. For full-length anti-apoptotic proteins, pET22 vectors are constructed and expressed protein purified by Ni^{2+} column chromatography and gel filtration.

For pro-apoptotic BAX, we use the pTYB1 vector to generate ΔC and full-length BAX chitin-binding fusion proteins, followed by chitin affinity chromatography, DTT-based chitin cleavage, and FPLC purification (Suzuki *et al.*, 2000).

The fluorescence polarization binding assay uses N-terminal fluoresceinated SAHBs to determine the binding affinities of SAHBs for multi-BH domain BCL-2 family member proteins (Fig. 23.2). The technique measures the change in polarization of light that results from a freely mobile and tumbling fluoresceinated molecule engaging a larger protein, which then slows and orients the tumbling of the fluoresceinated moiety in solution. When comparing the affinity of a ligand to a panel of proteins, it is important to note that larger proteins will yield greater absolute changes in polarization than smaller proteins, but this does not affect affinity calculations, unless the target protein is too small to generate a change in polarization when bound.

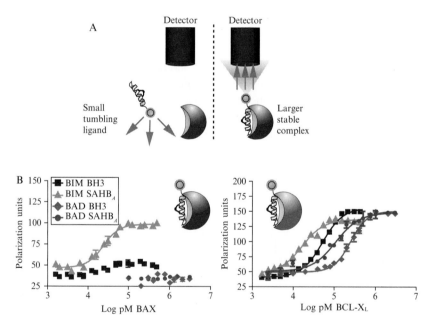

Figure 23.2 Fluorescence polarization binding assay. (A) The interaction between a FITC-ligand at fixed concentration with increasing doses of target protein results in dose-dependent polarization of light, which is measured by the detector. (B) FPA studies demonstrate the high affinity of BIM and BAD BH3 peptides for BCL-X_L, with SAHB derivatives exhibiting enhanced binding activity compared to the corresponding unmodified peptides (Walensky *et al.*, 2006). The selectivity of BH3 interactions is highlighted by BAX binding studies. Whereas BAD BH3 and BAD SAHB peptides display no interactions with BAX even at μM doses of protein, the α-helical BIM SAHB, but not the unmodified (and structurally unfolded) BIM BH3 peptide, binds BAX with nanomolar affinity.

Serial dilutions of protein in 50 mM Tris pH 8, 100 nM NaCl are instilled in 96-well black Costar plates (Costar #3915) to assay ligand binding over a broad range of protein concentrations. FITC-SAHB stocks are stored as a lyophilized powder at $-20\,^\circ$C in tinted containers, and aliquots reconstituted in 100% DMSO to yield a 1 mM stock. For a final peptide concentration of 25 nM, stock FITC-SAHB compound is diluted stepwise into water to yield a 0.5 μM solution (0.1% DMSO) and then 10 μl is repeat-pipetted into each well for a final assay volume of 200 μl. Depending on the brightness of the FITC-ligand, lower or higher concentrations may be used, but the typical range is 10 to 50 nM. Each condition is run in at least triplicate, including replicates that contain no protein so that the fluorimeter can be calibrated against the FITC-ligand alone. Once the FITC-SAHB is added, the plates are incubated in the dark at room temperature until equilibrium is reached. The time to equilibrium is initially determined by monitoring the binding isotherms over time to assess stabilization of binding activity. Fluorescence polarization (mP units) is measured on a Perkin-Elmer LS50B luminescence spectrophotometer equipped with cuvette containing a stir bar, a Spectramax M5 Microplate Reader (Molecular Devices), a BMG POLARstar Optima, or similar device. EC_{50} and K_d values are calculated by nonlinear regression analysis of dose-response curves with Prism software 4.0 (Graphpad). Care must be taken to provide sufficient data points to clearly define both the baseline and maximal mP values, so that the curve fit is statistically meaningful. When the total concentration of fluorescent ligand, L_T, is less than K_d and the assumption $L_T \approx L_{free}$, applied, binding isotherms are fitted to Eq. (23.1):

$$P = P_f + \left[(P_b - P_f) \times \frac{R_T}{K_D + R_T} \right] \qquad (23.1)$$

where P is the measured polarization value, P_f is the polarization of free fluorescent ligand, P_b is the polarization of bound ligand, and R_T is the receptor/protein concentration. However, when $L_T > K_d$, the assumption that $L_T \approx L_{free}$ does not hold because of ligand depletion. As such, binding isotherms are fitted to the more explicit Eq. (23.2):

$$P = P_f + (P_b - P_f) \left[\frac{(L_T + K_D + R_T) - \sqrt{(L_T + K_D + R_T)^2 - 4L_T R_T}}{2L_T} \right]$$

$$(23.2)$$

where P is the measured polarization value, P_f is the polarization of free fluorescent ligand, P_b is the polarization of bound ligand, L_T is the total concentration of fluorescent ligand, and R_T is the receptor/protein concentration (Copeland, 2000). Each data point represents the average of an experimental condition performed in at least triplicate.

3. *In Vitro* Release Assays as a Measure of Pro-Apoptotic Activity

Mitochondrial apoptosis is induced by the oligomerization of BAX and BAK, which are believed to form a yet uncharacterized pore that enables release of apoptogenic mitochondrial contents (Annis *et al.*, 2005; Goping *et al.*, 1998; Gross *et al.*, 1998). How BAX and BAK are activated, either directly by selective BH3 engagement ("direct activation") or indirectly by BH3-mediated inhibition of anti-apoptotics ("derepression"), or both, remains actively debated in the apoptosis field (Kim *et al.*, 2006; Kuwana *et al.*, 2005; Letai *et al.*, 2002; Walensky *et al.*, 2006; Willis *et al.*, 2007). Thus, release assays that probe the biophysical and biochemical characteristics of BAX/BAK are important tools for apoptosis researchers. Like any *in vitro* assay, the release assays presented here have benefits and limitations, yet provide a practical means for probing the capacity of select BCL-2 family proteins to activate or inhibit "release" by directly or indirectly regulating BAX/BAK.

3.1. Liposomal release assay

The purpose of the liposomal release assay is to simulate mitochondrial release but only with the minimal essential components for pro-apoptotic activation: lipid, release agent, and BCL-2 family protein/ligand of interest (Fig. 23.3A). The clear benefit of this reductionist assay is the ability to monitor membrane pore formation by a singular protein in response to a stimulus and in the absence of potentially confounding cellular factors, known or unknown. Despite the clarity of interpretation that this assay provides, it is certainly a giant step away from an isolated mitochondria, which in turn is many more giant steps away from an intracellular mito-chondria. Nevertheless, the *in vitro* liposomal release assay has provided important insights into BAX/BAK physiology (Kuwana *et al.*, 2002; 2005; Oh *et al.*, 2006; Terrones *et al.*, 2004; 2008; Walensky *et al.*, 2006; Yethon *et al.*, 2003), just as *in vitro* structural biology studies have shed enormous light on how BH3 helices engage multi-BH anti-apoptotic grooves (Sattler *et al.*, 1997).

Liposomes are prepared from a mixture of lipids that approximates the lipid content of the outer mitochondrial membrane (Ardail *et al.*, 1990; Kuwana *et al.*, 2002; Lutter *et al.*, 2000; Oh *et al.*, 2006) as indicated in

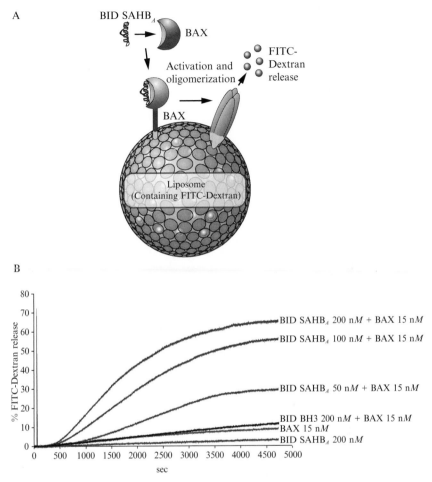

Figure 23.3 Liposomal release assay. (A) Recombinant BAX and SAHB ligand are added to liposomes containing entrapped FITC-dextran. Direct activation and oligomerization of BAX leads to release of liposomal FITC-dextran, which is detected by a fluorimeter. (B) BID SAHB$_A$ induces dose-responsive BAX-mediated liposomal release of FITC-dextran (Walensky *et al.*, 2006). BID SAHB$_A$ or BAX alone have no effect; the combination of unmodified BID BH3 peptide and BAX likewise has no effect in this dose range.

Table 23.1. Solubilized lipids (e.g., chloroform) are combined in glass test tubes as 50-mg aliquots and then vortexed thoroughly (10 to 15 times) to ensure complete mixing. Subsequently, the organic solvent is evaporated by a stream of nitrogen, and the tubes are placed in a vacuum desiccator overnight, resulting in a visible lipidic film on the bottom of the glass tube by morning. Tubes containing the dried lipid mixtures can be placed in Kapak pouches, flushed with nitrogen, sealed, and then stored at −80 °C.

Table 23.1 Composition of large unilamellar vesicles that simulate the lipid content of mitochondrial outer membrane contact sites

Lipid	Stock (mg/ml)	Mol wt	Wt%	Wt lipid (mg)	Vol lipid (ml)	Moles Lipid	Mole %	P atom	Moles P
POPC	10	760.1	41	20.5	2.05	.0270	42.7	1	.027
POPE	10	770.99	22	11	1.1	.0143	22.6	1	.0143
PI	25	909.12	9	4.5	0.18	.0049	7.83	1	.0049
CHOL	10	386.66	8	4	0.4	.0103	16.4	0	0
CL	10	1493.91	20	10	1	.0067	10.6	2	.0134
TOTAL			**100%**	**50 mg**	**4.73 mL**	**.0632**			**0.0596**

The indicated values represent the amounts of lipid combined to yield a 50 mg mixture of POPC (1-palmitoyl-2-oleoyl-*sn*-glycero-3-phosphocholine), POPE (1-palmitoyl-2-oleoyl-*sn*-glycero-3-phosphoethanolamine), PI (beef liver phosphatidylinositol), CHOL (cholesterol), and CL (beef heart cardiolipin).

To produce FITC-dextran encapsulated liposomes, an aliquot of mixed lipids is thawed and resuspended in 1 ml of HK buffer (20 mM HEPES, 150 mM KCl, pH 7) with 50 to 100 mg of FITC-dextran (Molecular Probes #D-1821). The resulting slurry is vortexed until the lipid film has completely dissolved and then freeze-thawed 15 to 20 times between liquid nitrogen and a 37 °C waterbath. A portion of the slurry can be preserved for later use (up to a month) by storing at −80 °C in a pouch flushed with nitrogen. To generate large unilamellar vesicles (LUVs), the slurry is passed through an Avanti Mini-Extruder Set (#610000) equipped with a 100-nm filter. LUVs are purified away from residual unencapsulated FITC-dextran by gel filtration with a Sephacryl S-300 HR column (GE Healthcare) at a flow rate of 1 ml/min, with liposomes typically emerging at the ~12 ml fraction. Once purified, the lipid concentration of the solution is assessed by a colorimetric phosphate assay described in detail elsewhere (Böttcher *et al.*, 1961) and performed in duplicate (see Table 23.1 for phosphate content of the lipid mixture). A liposomal sample of 10 μl typically yields a value within the standard curve, which ranges from 0 to 100 nmol/μL. FITC-dextran encapsulated LUVs are stored in foil-wrapped tubes at 4 °C and can be reliably used for 10 to 12 days.

In an experiment that monitors BAX-induced FITC-dextran release, for example, 2 ml of HK buffer containing 10 μg/ml lipid is added to a quartz cuvette under constant stirring at 37 °C and fluorescence monitored over time in a fluorimeter (excitation 488 nm, emission 525 nm). Once the baseline fluorescence measurement stabilizes (typically 15 min), recombinant monomeric BAX (freshly gel filtration purified, final concentration 15 to 50 nM) is added and fluorescence monitored for several minutes to ensure a stable baseline. Subsequently, a dose of SAHB compound is added (typical test range, 0 to 250 nM) and the fluorescence measurements recorded every 2 sec until a stable plateau is reached, at which time the liposomes are quenched with 1% Triton X-100. The data are presented as percent FITC-dextran release, with the average of three replicate Triton values set to 100%, and the average starting baseline value set to 0% (Fig. 23.3B). Of note, a baseline measurement of liposome mixture with SAHB alone is recorded for each dosing level to ensure that the compound itself does not disrupt the liposomes. In general, neutral to negatively charged SAHBs are well tolerated by the liposomes. The liposomal assay is reliable and versatile and can be modified to test inhibition of BH3-induced BAX activation by adding recombinant anti-apoptotic proteins or by preincubating BAX with test ligands or proteins before exposure to liposomes. We have also adapted the assay to generate liposomes containing Ni^{2+}-NTA-lipids (5% DOGS-NTA, Avanti #790404C), which can be used to localize histidine-tagged BH3 peptides or BCL-2 family proteins to the membrane surface to assess the impact of membrane-targeted regulators on BAX release activity.

3.2. Mitochondrial cytochrome *c* release assay

The purpose of this assay is to monitor the effect of ligand- or recombinant protein–induced activation or inhibition of BAX/BAK-mediated cytochrome *c* release in the context of isolated mitochondria (Ellerby *et al.*, 1997; Luo *et al.*, 1998; Scorrano *et al.*, 2002) (Fig. 23.4A). The clear benefit of this approach derives from analysis of the intact organelle, which contains the natural lipid membrane composition and topography, including the cohort of proteins endogenously embedded in the various mitochondrial compartments. Of course, the isolated mitochondria, out of context from the intact cell, are lacking the multitude of factors from other cellular compartments (e.g., endoplasmic reticulum, cytosol) that certainly impact constitutive mitochondrial physiology and signal transduction. Thus, conclusions and generalizations based on such studies must be made with care. Furthermore, meticulous and timely handling of the mitochondria throughout the isolation and experimental procedure is essential to the preparation of suitable mitochondria for meaningful and reproducible assays.

An important feature of this assay is the ability to study and compare the responses of genetically distinct mitochondria that lack one or more of the regulatory BCL-2 family proteins of interest. Because the overall character of genetically modified mitochondria may be quite different from wild-type mitochondria, as recently demonstrated for $Bax^{-/-}Bak^{-/-}$ mitochondria that exhibit altered architecture (Karbowski *et al.*, 2006), data must be interpreted with this caveat in mind. To study endogenous BAK activation, we use wild-type mouse liver mitochondria that naturally lack BAX, which is constitutively cytosolic (Letai *et al.*, 2002). To study recombinant BAX activation, we use either $Bak^{-/-}$ or Alb-cre$^{pos}Bax^{flox/-}Bak^{-/-}$ mouse liver mitochondria, which have yielded essentially identical results (Walensky *et al.*, 2006). Theoretically, if BAX is activated during mitochondrial preparation, it is plausible that endogenous BAX could be present and, therefore, contribute to the measured cytochrome *c* release; however, we do not detect BAX by Western analysis of isolated wild-type or $Bak^{-/-}$ mitochondria (Letai *et al.*, 2002; Walensky *et al.*, 2006). Finally, it is worth mentioning that conflicting results have been reported with regard to the activity of select BH3-only proteins and peptides on mitochondrial activation based on this assay. The finding that select BH3-containing proteins and peptides, such as BID and BIM, activate BAX/BAK-mediated cytochrome *c* release, but others, such as BAD and NOXA do not, led to a subclassification of BH3-only proteins as "activators" (i.e., direct activation of BAX/BAK by BH3) or "sensitizers" (i.e., indirect activation of BAX/BAK by inhibition of anti-apoptotic proteins) of mitochondrial apoptosis (Cheng *et al.*, 2001; Kuwana *et al.*, 2002; Letai *et al.*, 2002). Whereas this model has been supported by subsequent work (Certo *et al.*, 2006; Kim *et al.*, 2006; Kuwana *et al.*, 2005; Walensky *et al.*, 2006), other mitochondrial studies demonstrate activation in response

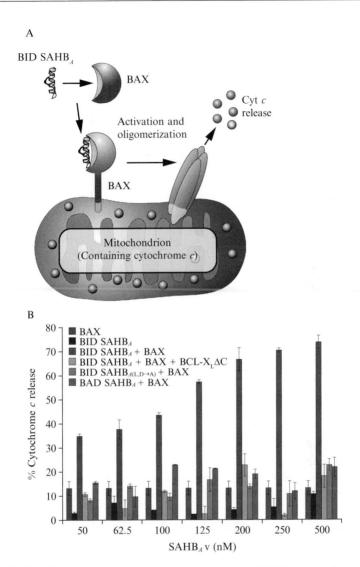

Figure 23.4 Mitochondrial cytochrome c release assay. (A) The molecular interactions that regulate mitochondrial apoptosis are evaluated *in vitro* by treating isolated mitochondria with BH3 ligands and/or recombinant BCL-2 family proteins followed by ELISA-based quantitation of released cytochrome c. (B) BID SAHB$_A$ induced dose-responsive BAX-mediated cytochrome c release from Bak$^{-/-}$ mitochondria (Walensky *et al.*, 2006). Mitochondrial treatment with BAX or BID SAHB$_A$ alone had no such effect. The specificity of BID SAHB$_A$ activity was confirmed by the inability of mutant BID SAHB$_{A(L,D\rightarrow A)}$ or BAD SAHB$_A$ to activate BAX-induced cytochrome c release, and by abrogation of BID SAHB$_A$ activity with BCL-X$_L$ ΔC co-treatment.

to "sensitizers" such as BAD, especially when used in combination with other BH3 peptides that expand anti-apoptotic blockade (e.g., NOXA) (Uren *et al.*, 2007). Such studies, support a model in which BAX and BAK are activated only once BH3-only proteins or peptides target the broad spectrum of anti-apoptotic proteins, thereby eliminating the anti-apoptotic blockade of BAX/BAK (Willis *et al.*, 2007). Indeed, results may vary on the basis of the experimental conditions and the cellular origin of the mitochondria tested (Uren *et al.*, 2007). Ultimately, both mechanisms may contribute, singly or in combination, to the regulation of BAX/BAK activation *in vivo*, depending on the cellular context and apoptotic stimulus.

3.2.1. Buffer preparation

Mitochondrial isolation buffer is composed of 250 mM sucrose (85.6 g/L), 10 mM Tris-HCl (1.57 g/L), and 100 μM EGTA (0.046 g/L), titrated to pH 7.4, sterile filtered, and stored at 4 °C.

The mitochondrial experimental buffer (pH 7.4) contains 125 mM KCl, 10 mM Tris-MOPS, 5 mM glutamic acid, 2.5 mM malic acid, 1 mM K-phos, and 10 μM EGTA-Tris. This buffer is prepared fresh for each experiment from the following stock solutions: 1 M KCl (37.3 g/500 ml); 1 M Tris-MOPS (for 100 ml add 12.1 g of Tris-base, titrate to pH 11, and slowly add 20.9 g of MOPS); 1 M glutamic acid (20.9 g/100 ml of the potassium salt); 1 M malic acid (6.7 g/50 ml); K-phos solution, made by combining 2 ml of 1 M monobasic potassium phosphate with 8 ml of 1 M dibasic potassium phosphate; and 1 M EGTA-Tris (for 5 ml add 0.6 g of Tris-base, titrate to pH 11, and add 2.34 g of tetrasodium-EGTA). Stocks are stored at 4 °C for no longer than 1 month. A 100-ml working buffer is made by combining 12.5 ml of 1 M KCl, 1 ml of 1 M Tris-MOPS, 500 μl of 1 M glutamic acid, 250 μl of 1 M malic acid, 100 μl of 1 M K-phos, 100 μl of 1 M EGTA-Tris, diluted to 100 ml in deionized water, titrated to pH 7.4, and sterile filtered. The buffer should be brought to room temperature before use.

3.2.2. Mitochondrial isolation

A mouse of the desired phenotype is sacrificed and the liver promptly removed and immersed in chilled isolation buffer. The liver is scissor-minced in several changes of isolation buffer, with this repeat rinsing performed to eliminate trace blood. The liver fragments are then dounce homogenized in ~25 ml of chilled isolation buffer until a homogeneous suspension is achieved (typically 3 to 4 dounce cycles at high speed). The liver suspension is maintained on ice and quickly transferred for centrifugation at 800g for 10 min at 4 °C. The supernatant is collected (leaving gross cellular debris behind) and centrifuged at 7000g for 10 min at 4 °C. The supernatant is decanted, and the pellet containing the isolated mitochondria is gently resuspended by pipette trituration in 1 ml of chilled isolation buffer. Enlarging the orifice of the pipette tip by cutting with a sterile razor helps avoid shear forces that could disrupt the mitochondria and cause background cytochrome c release. The resuspended pellet is diluted in 30 to 40 ml of chilled isolation buffer and centrifuged again at $7000g$ for 10 min at 4 °C. After resuspending the pellet in 1 ml isolation buffer, the protein concentration of the mitochondrial suspension is promptly determined (Bio-Rad Protein Assay #500-0006). The liver from a single mouse typically yields ~5 to 10 mg of mitochondrial pellet. Once the protein concentration is determined, the suspension is diluted to a final concentration of 0.5 mg/ml in experimental buffer and promptly added to an already prepared 96-well plate (or Eppendorf tubes) containing the treatment protein(s) and ligand(s) at the desired concentrations (see 3.2.3). Of note, once the mitochondria are immersed in experimental buffer, prolonged storage, even at 4 °C, can lead to background cytochrome c release, and therefore, prompt use of the mitochondria is mandatory.

3.2.3. Plate design and assay

We have adapted the mitochondrial assay to a 96-well format for screening purposes. In this setup, control wells containing vehicle (e.g., 1% DMSO) or 1% Triton X-100 alone represent baseline and maximal cytochrome c release, respectively. The amount of cytochrome c quantitated in the supernatant by ELISA is inserted into Eq. (23.3) to determine percent cytochrome c release:

$$\%\text{Release} = (Abs_x - Abs_{veh}) \times 100/(Abs_{Tr} - Abs_{veh}) \quad (23.3)$$

where Abs_x, Abs_{veh}, and Abs_{Tr}, are the absorbances of the sample, vehicle control, and Triton X-100 control. If fewer samples are run, an Eppendorf tube–based setup can also be used. In this case, the supernatant is separated from the mitochondrial pellet, the pellet is resuspended in 1% Triton X-100, the supernatant of the extracted pellet isolated, and percent cytochrome c release determined by Eq. (23.4):

$$\%\text{Release} = Abs_s \times 100/(Abs_s + Abs_p) \quad (23.4)$$

where Abs_s and Abs_p represent the absorbances of the original supernatant and the supernatant from the detergent-extracted pellet, respectively.

To prepare a plate (clear U-bottom) for recombinant BAX activation analysis, for example, a serial dilution of test SAHB is performed at 4× the desired final concentrations in 25 μl experimental buffer. Recombinant BAX at 4× the desired final concentration is then added by repeat pipette, delivering an additional 25 μl to each well. It is critically important that the recombinant BAX is freshly gel filtration purified to ensure a monomeric solution, thereby avoiding nonspecific BAX-induced release resulting from oligomer-triggered BAX activation and cytochrome c release. The test components for the assay can be modified to accommodate other analyses (e.g., combination SAHB treatments, investigation of endogenous BAK activation). SAHB dilutions are typically prepared before or simultaneous with mitochondrial isolation, and then BAX and mitochondria are added sequentially by repeat pipette once the protein concentration of the mitochondrial preparation is determined. It is critically important that the mitochondrial preparation be homogenously resuspended immediately before repeat pipetting to avoid variability of the suspension that arises from the settling of mitochondria in the stock tube. Experimental conditions are typically run in quadruplicate and important controls include vehicle + mitochondria, BAX + mitochondria, and SAHB + mitochondria to monitor for background and/or nonspecific release. Additional controls include blockade of cytochrome *c* release by added anti-apoptotic proteins and the use of point mutant SAHBs that exhibit impaired BCL-2 family protein interactions (Walensky *et al.*, 2006).

Once the mitochondria have been added, the plate is incubated at room temperature for 40 min. A time course is also beneficial to monitor for changes in release kinetics (Walensky *et al.*, 2006). After the desired incubation time, the plate is centrifuged at 3000 rpm for 10 min at 4 °C and then ~30 μl of each supernatant is carefully removed to another 96-well plate, taking care not to disrupt the pellet, which can be seen at the very bottom of the U-shaped well. For the Eppendorf setup, tubes are spun at 14,000 rpm in a refrigerated table top centrifuge for 10 min, followed by transfer of the total supernatant (a gel loading pipette tip is helpful to avoid disrupting and removing any pellet) to another set of Eppendorf tubes. The pellets are then resuspended (by flicking and vortexing) in 100 μl of 1% Triton X-100, spun at 14,000 rpm in a refrigerated table top centrifuge, and the supernatants transferred to another set of Eppendorf tubes. Samples of supernatant (typically 4 μl) are then added to the wells of an ELISA plate for cytochrome *c* detection according to the manufacturer's protocol (R&D Systems #MCTC0). Of note, residual supernatants can be stored at −20 °C for repeat or deferred analysis. The absorbance data are processed as described at the beginning of this section to determine the percent cytochrome *c* released by the experimental condition (Fig. 23.4B).

4. MEASUREMENT OF CELLULAR APOPTOSIS INDUCTION

As described in our preceding article, a key benefit of SAHBs as tool compounds for dissecting apoptotic pathways is their cell penetrability. Thus, in addition to serving as potential therapeutics for reactivating apoptosis in cancer, the wide variety of SAHBs generated based on the many discrete BH3 sequences can be used to test cancer cell susceptibilities to individual pro-apoptotic death domains. This cell-based survey, combined with identification of the corresponding intracellular SAHB targets, provides a protein interaction–based mechanism for apoptosis induction in a particular cell or tissue.

Because individual cell types (e.g., adherent vs nonadherent, transformed vs nontransformed) may exhibit differential rates and capacities for uptake of cell-penetrating peptides such as SAHBs, an assessment of cellular permeability is first performed for the cell under investigation, as described (Bird et al., 2008). Culture conditions will vary according to the cell type used. For the Jurkat studies described here, we use 1X RPMI-1600, 200 units/ml penicillin/streptomycin, 2 mM L-glutamine, 50 mM HEPES, 50 μM 2-mercaptoethanol, and 10% heat-inactivated fetal bovine serum (FBS). Serum-free media can be used to wash cells before SAHB treatment and during initial exposure to SAHB compounds (typically 2 to 4 h) to maximize cell surface contact and avoid any decrement in activity from serum binding.

A variety of kit-based assays are available to monitor cell viability (e.g., MTT assay [Roche], Cell Titer-Glo^{TM} [Promega]), and apoptosis induction (e.g., caspase-3 activation [Oncogene], annexin-V binding [BD Biosciences]) in reasonably high throughput. Whereas viability assays are useful for screening purposes, active compounds are then assessed in more specific assays for apoptosis induction (Walensky et al., 2004). For example, the annexin-V–binding assay detects the externalization of phosphatidylserine (PS, the binding substrate for annexin V), which is a characteristic event of apoptosis induction (Fig. 23.5). SAHB stocks (1 mM, 100% DMSO) are diluted stepwise into water, and then serially diluted into serum-free media (final volume 50 μl) to achieve the desired dose range (typically 0.15 to 10 μM). Jurkats cells are resuspended and centrifuged twice in serum-free media, diluted to a concentration of 1×10^6 cells/ml in serum-free media, and then 50,000 cells (50 μl) are added to the SAHB solution. After an initial incubation period (typically 2 to 4 h), an additional 100 μl of Jurkat medium containing 20% FBS is added to restore 10% serum conditions and the cells monitored for apoptosis induction over time (e.g., 6, 12, 24, and 48 h). For annexin-V–binding analysis, the cells are pelleted, washed in PBS, re-pelleted, and then resuspended in 200 μl of 1 × annexin-V binding buffer (10 mM HEPES/NaOH, pH7.4, 140 mM NaCl, 2.5 mM CaCl$_2$) containing a 1:500 dilution of fluoresceinated annexin-V according to the manufacturer's protocol (e.g., BD Biosciences). Cellular fluorescence is then detected and quantitated by FACS analysis. Vital dyes such as propidium iodide (PI) or 7-amino-actinomycin (7-AAD) can also be added (1 μg/ml final concentration) to monitor the progression from early apoptosis, at which point cells display externalized PS but retain membrane integrity (annexin-V positive, PI/7-AAD negative), to end-stage apoptosis by which point the dead cells are dye-permeable (annexin-V positive, PI/7-AAD positive) (Fig. 23.5). Each experimental condition is performed in at

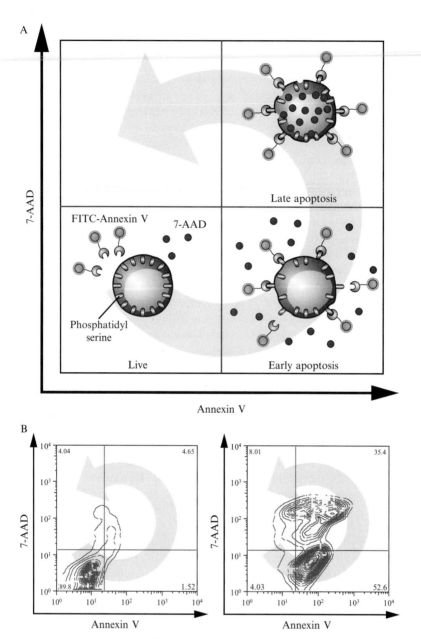

Figure 23.5 Annexin-V apoptosis induction assay. (A) Cells exposed to a pro–apoptotic stimulus and then evaluated by FITC–annexin-V and 7-AAD staining exhibit a characteristic apoptotic progression as detected by FACS analysis. Whereas healthy cells are annexin-V and 7-AAD negative, cells undergoing apoptosis externalize phosphatidylserine and become annexin-V positive. During the early stage of apoptosis induction, membrane integrity is preserved and cells are annexin-V positive but 7-AAD negative. By late-stage

least triplicate, and controls include vehicle (e.g., corresponding serial dilution of DMSO/water), a standard pro-apoptotic stimuli (e.g., staurosporine, 1 μM), and negative control point mutant SAHBs.

5. IDENTIFICATION OF *IN SITU* MECHANISTIC TARGETS OF SAHBS

A major advantage of peptidic compounds is that they are readily derivatized for a host of research applications. For example, the capacity to incorporate a fluorescent label allows for intracellular tracking of SAHBs, which facilitated their visualization in pinocytotic vesicles during uptake and their ultimate localization to the outer mitochondrial membrane in live cells (Walensky *et al.*, 2004). Whereas such FITC-labeled SAHB compounds contributed to our understanding of the import mechanism and organeller targeting, anti-FITC immunoprecipitation from extracts of FITC-SAHB–treated cells enables the identification of intracellular high-affinity SAHB targets and thus provides insight into the mechanism of SAHB-induced activation of apoptosis (Fig. 23.6).

For anti-FITC co-immunoprecipitation experiments, cells (1×10^7) are treated with SAHB peptides (10 to 20 μM) in serum-free media for up to 4 h, followed by serum replacement and timed incubation. Cells are collected by centrifugation and subjected to lysis in 50 mM Tris pH 7.6, 150 mM NaCl, 1% CHAPS, and protease inhibitor cocktail (Sigma). A variety of detergents may be used, but when assaying for BAX/BAK interactions, it is important to use detergent conditions that do not induce BAX/BAK oligomerization (Antonsson *et al.*, 2000). Cellular fractionation protocols that examine specific cellular compartments may also be used. Cellular lysis and all subsequent steps are conducted at 4 °C. Extracts are subjected to centrifugation and supernatants isolated and then incubated with protein A/G-sepharose (Santa Cruz) (50 μl 50% bead slurry/0.5 ml lysate). The precleared supernatants (0.5 ml) are collected after centrifugation, incubated with 10 μl goat-anti-FITC antibody (Abcam) for 1.5 h, and then protein A/G-sepharose (50 μl 50% bead slurry/0.5 ml lysate) added for an additional 1.5-h incubation. After 2-min tabletop centrifugation, the pellets are isolated and then washed three times in lysis buffer containing increasing salt concentrations (150, 300, 500 mM). To liberate precipitated

apoptosis, membrane integrity is lost, and cells become positive for 7-AAD as well. (B) Cultured leukemia cells exposed to serial dilutions of pro-apoptotic SAHBs can be evaluated by FACS analysis over time to identify and quantify the stages of apoptosis induction. Whereas the cells are initially annexin-V and 7-AAD negative (left panel, 90% live), apoptosis induction (right panel) is reflected by a predominant population that is annexin-V positive and 7-AAD negative (bottom right, 53% early apoptosis), and a second population that is both annexin-V and 7-AAD positive (top right, 35% late apoptosis).

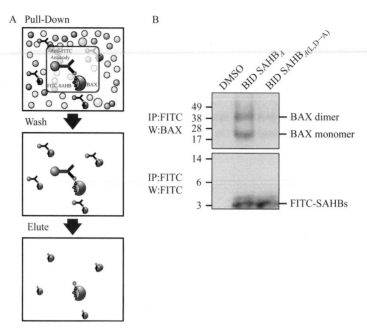

Figure 23.6 *In situ* target identification by SAHB retrieval. (A) Cells treated with FITC-SAHB compounds are lysed and extracts subjected to anti-FITC co-immuno-precipitation with FITC antibody and protein A/G-sepharose pull down. After washing, SAHB-bound protein targets are eluted and identified by BCL-2 family Western analy-sis. (B) Jurkat T cells were incubated with FITC-BID SAHB$_A$ and a mutant derivative for 18 h, followed by cellular lysis in a 1% CHAPS-containing buffer. Anti-FITC pull down co-immunoprecipitated native BAX with FITC-BID SAHB$_A$ but not with the mutant BID SAHB$_{A(L,D \rightarrow A)}$ (Walensky *et al.*, 2006).

protein, the beads are suspended in SDS-containing sample buffer, boiled, and supernatants collected after centrifugation. Samples are electrophoresed (e.g., 4 to 12% gradient Bis–Tris gels [Invitrogen]), and proteins transferred to Immobilon-P membranes (Millipore). After blocking, blots are incu-bated with the specified BCL-2 family protein antibody (e.g., BAX-N20 antibody [Santa Cruz]) or rabbit anti-FITC antibody (Santa Cruz) at 1:500 dilutions in blocking buffer, followed by treatment with secondary antibody for visualization with a chemiluminescent reagent (e.g., Western Lightning^{TM} [Perkin Elmer]). Alternate retrieval strategies, such as streptavidin-based capture of biotinylated SAHBs, are also feasible.

It is important to emphasize that *in situ* target identification studies based on noncovalent interactions between ligand and protein are limited by affinity, which dictates the capacity of the association to withstand cellular lysis and retrieval processing. Whereas high-affinity 1:1 interactions are best suited for this type of analysis, transient interactions, such as enzymatic processes or catalytic "hit and run" phenomena as has been hypothesized for BID–BAX interactions (Eskes *et al.*, 2000; Wei *et al.*, 2000), may require alternate strategies, such as derivitization of SAHBs with covalent trapping moieties (Danial *et al.*, 2008).

6. SUMMARY

SAHBs are cell-permeable compounds that retain the structural and biochemical functionalities of native BH3 domains and, therefore, serve as useful tools to chemically dissect the BCL-2 family interaction network *in vitro* and *in vivo*. Through substituting only two non-natural residues at positions on the noninteracting surface of the BH3 motif, SAHBs preserve the differential binding specificities of discrete BH3 motifs and facilitate the identification of binding preferences among BCL-2 family proteins by use of *in vitro* binding assays and *in situ* co-immunoprecipitation strategies. Mitochondrial and liposomal release assays serve as robust *in vitro* studies to probe the functional impact of BH3-based interactions on activation or inhibition of mitochondrial apoptosis. Functional interaction studies can then be extended to a cellular context owing to the cell permeability of SAHB compounds. Thus, SAHBs are new chemical tools that directly and specifically alter BCL-2 family protein function, enabling the apoptotic network to be investigated, and ultimately manipulated, on a conditional basis in real time.

ACKNOWLEDGMENTS

We thank E. Smith for figure design and editorial assistance, the late Stanley J. Korsmeyer for his indelible mentorship, and former members of the Korsmeyer laboratory, including Kyoung Joon Oh and Joel Morash, and members of the Walensky laboratory for their scientific contributions. L.D.W. is supported by NIH grants K08HL074049, 5R01CA50239, and 5P01CA92625, a Burroughs Wellcome Fund Career Award in the Biomedical Sciences, a Partnership for Cures Charles E. Culpeper Scholarship in Medical Science, a grant from the William Lawrence Children's Foundation, and the Dana-Farber Cancer Institute High-Tech fund. G.H.B. is the recipient of a Harvard University Center for AIDS Research Scholar Award and F.B. is funded by a grant from the Dana-Farber Cancer Institute Pediatric Low-Grade Glioma Program.

REFERENCES

Annis, M. G., Soucie, E. L., Dlugosz, P. J., Cruz-Aguado, J. A., Penn, L. Z., Leber, B., and Andrews, D. W. (2005). Bax forms multispanning monomers that oligomerize to permeabilize membranes during apoptosis. *EMBO J.* **24,** 2096–2103.
Antonsson, B., Montessuit, S., Lauper, S., Eskes, R., and Martinou, J. C. (2000). Bax oligomerization is required for channel-forming activity in liposomes and to trigger cytochrome c release from mitochondria. *Biochem. J.* **345**(Pt 2), 271–278.

Ardail, D., Privat, J. P., Egret-Charlier, M., Levrat, C., Lerme, F., and Louisot, P. (1990). Mitochondrial contact sites. Lipid composition and dynamics. *J. Biol. Chem.* **265,** 18797–18802.

Bakhshi, A., Jensen, J. P., Goldman, P., Wright, J. J., McBride, O. W., Epstein, A. L., and Korsmeyer, S. J. (1985). Cloning the chromosomal breakpoint of t(14;18) human lymphomas: Clustering around JH on chromosome 14 and near a transcriptional unit on 18. *Cell* **41,** 899–906.

Bird, G. H., Bernal, F., Pitter, K., and Walensky, L. D. (2008). Synthesis and Biophysical Characterization of Stabilized Alpha-Helices of BCL-2 Domains. *Methods Enzymol.* **446,** 387–408.

Böttcher, C. J. F., van Gent, C. M., and Pries, C. (1961). A rapid and sensitive sub-micro phosphorus determination. *Anal. Chim. Acta* **24,** 203–204.

Certo, M., Del Gaizo Moore, V., Nishino, M., Wei, G., Korsmeyer, S., Armstrong, S. A., and Letai, A. (2006). Mitochondria primed by death signals determine cellular addiction to antiapoptotic BCL-2 family members. *Cancer Cell* **9,** 351–365.

Cheng, E. H., Wei, M. C., Weiler, S., Flavell, R. A., Mak, T. W., Lindsten, T., and Korsmeyer, S. J. (2001). BCL-2, BCL-X(L) sequester BH3 domain-only molecules preventing BAX- and BAK-mediated mitochondrial apoptosis. *Mol. Cell* **8,** 705–711.

Cleary, M. L., and Sklar, J. (1985). Nucleotide sequence of a t(14;18) chromosomal breakpoint in follicular lymphoma and demonstration of a breakpoint-cluster region near a transcriptionally active locus on chromosome 18. *Proc. Natl. Acad. Sci. USA* **82,** 7439–7443.

Copeland, R. A. (2000). Enzymes: A practical introduction to structure, mechanism, and data analysis **2nd ed**, Wiley-VCH, New York.

Danial, N. N., and Korsmeyer, S. J. (2004). Cell death: Critical control points. *Cell* **116,** 205–219.

Danial, N. N., *et al.* (2008). Dual role of proapoptotic BAD in insulin secretion and beta cell survival. *Nat. Med.* **14,** 144–153.

Ellerby, H. M., Martin, S. J., Ellerby, L. M., Naiem, S. S., Rabizadeh, S., Salvesen, G. S., Casiano, C. A., Cashman, N. R., Green, D. R., and Bredesen, D. E. (1997). Establishment of a cell-free system of neuronal apoptosis: Comparison of premitochondrial, mitochondrial, and postmitochondrial phases. *J. Neurosci.* **17,** 6165–6178.

Eskes, R., Desagher, S., Antonsson, B., and Martinou, J. C. (2000). Bid induces the oligomerization and insertion of Bax into the outer mitochondrial membrane. *Mol. Cell. Biol.* **20,** 929–935.

Goping, I. S., Gross, A., Lavoie, J. N., Nguyen, M., Jemmerson, R., Roth, K., Korsmeyer, S. J., and Shore, G. C. (1998). Regulated targeting of BAX to mitochondria. *J. Cell. Biol.* **143,** 207–215.

Gross, A., Jockel, J., Wei, M. C., and Korsmeyer, S. J. (1998). Enforced dimerization of BAX results in its translocation, mitochondrial dysfunction and apoptosis. *EMBO. J.* **17,** 3878–3885.

Hockenbery, D., Nunez, G., Milliman, C., Schreiber, R. D., and Korsmeyer, S. J. (1990). Bcl-2 is an inner mitochondrial membrane protein that blocks programmed cell death. *Nature* **348,** 334–336.

Karbowski, M., Norris, K. L., Cleland, M. M., Jeong, S. Y., and Youle, R. J. (2006). Role of Bax and Bak in mitochondrial morphogenesis. *Nature* **443,** 658–662.

Kim, H., Rafiuddin-Shah, M., Tu, H. C., Jeffers, J. R., Zambetti, G. P., Hsieh, J. J., and Cheng, E. H. (2006). Hierarchical regulation of mitochondrion-dependent apoptosis by BCL-2 subfamilies. *Nat. Cell. Biol.* **8,** 1348–1358.

Kuwana, T., Mackey, M. R., Perkins, G., Ellisman, M. H., Latterich, M., Schneiter, R., Green, D. R., and Newmeyer, D. D. (2002). Bid, Bax, and lipids cooperate to form supramolecular openings in the outer mitochondrial membrane. *Cell* **111,** 331–342.

Kuwana, T., Bouchier-Hayes, L., Chipuk, J. E., Bonzon, C., Sullivan, B. A., Green, D. R., and Newmeyer, D. D. (2005). BH3 domains of BH3-only proteins differentially regulate Bax-mediated mitochondrial membrane permeabilization both directly and indirectly. Mol. Cell 17, 525–535.

Letai, A., Bassik, M. C., Walensky, L. D., Sorcinelli, M. D., Weiler, S., and Korsmeyer, S. J. (2002). Distinct BH3 domains either sensitize or activate mitochondrial apoptosis, serving as prototype cancer therapeutics. Cancer Cell 2, 183–192.

Luo, X., Budihardjo, I., Zou, H., Slaughter, C., and Wang, X. (1998). Bid, a Bcl2 interacting protein, mediates cytochrome c release from mitochondria in response to activation of cell surface death receptors. Cell 94, 481–490.

Lutter, M., Fang, M., Luo, X., Nishijima, M., Xie, X., and Wang, X. (2000). Cardiolipin provides specificity for targeting of tBid to mitochondria. Nat. Cell. Biol. 2, 754–761.

McDonnell, T. J., Deane, N., Platt, F. M., Nunez, G., Jaeger, U., McKearn, J. P., and Korsmeyer, S. J. (1989). bcl-2-immunoglobulin transgenic mice demonstrate extended B cell survival and follicular lymphoproliferation. Cell 57, 79–88.

Nunez, G., London, L., Hockenbery, D., Alexander, M., McKearn, J. P., and Korsmeyer, S. J. (1990). Deregulated Bcl-2 gene expression selectively prolongs survival of growth factor-deprived hemopoietic cell lines. J. Immunol. 144, 3602–3610.

Oh, K. J., Barbuto, S., Pitter, K., Morash, J., Walensky, L. D., and Korsmeyer, S. J. (2006). A membrane-targeted BID BCL-2 homology 3 peptide is sufficient for high potency activation of BAX in vitro. J. Biol. Chem. 281, 36999–37008.

Sattler, M., et al. (1997). Structure of Bcl-xL-Bak peptide complex: Recognition between regulators of apoptosis. Science 275, 983–986.

Scorrano, L., Ashiya, M., Buttle, K., Weiler, S., Oakes, S. A., Mannella, C. A., and Korsmeyer, S. J. (2002). A distinct pathway remodels mitochondrial cristae and mobilizes cytochrome c during apoptosis. Dev. Cell 2, 55–67.

Seto, M., Jaeger, U., Hockett, R. D., Graninger, W., Bennett, S., Goldman, P., and Korsmeyer, S. J. (1988). Alternative promoters and exons, somatic mutation and deregulation of the Bcl-2-Ig fusion gene in lymphoma. EMBO J. 7, 123–131.

Suzuki, M., Youle, R. J., and Tjandra, N. (2000). Structure of Bax: Coregulation of dimer formation and intracellular localization. Cell 103, 645–654.

Terrones, O., Antonsson, B., Yamaguchi, H., Wang, H. G., Liu, J., Lee, R. M., Herrmann, A., and Basanez, G. (2004). Lipidic pore formation by the concerted action of proapoptotic BAX and tBID. J. Biol. Chem. 279, 30081–30091.

Terrones, O., Etxebarria, A., Landajuela, A., Landeta, O., Antonsson, B., and Basanez, G. (2008). Bim and tbid are not mechanistically equivalent when assisting Bax to permeabilize bilayer membranes. J. Biol. Chem. 283, 7790–7803.

Tsujimoto, Y., Gorham, J., Cossman, J., Jaffe, E., and Croce, C. M. (1985). The t(14;18) chromosome translocations involved in B-cell neoplasms result from mistakes in VDJ joining. Science 229, 1390–1393.

Uren, R. T., Dewson, G., Chen, L., Coyne, S. C., Huang, D. C., Adams, J. M., and Kluck, R. M. (2007). Mitochondrial permeabilization relies on BH3 ligands engaging multiple prosurvival Bcl-2 relatives, not Bak. J. Cell. Biol. 177, 277–287.

Vaux, D. L., Cory, S., and Adams, J. M. (1988). Bcl-2 gene promotes haemopoietic cell survival and cooperates with c-myc to immortalize pre-B cells. Nature 335, 440–442.

Walensky, L. D., Pitter, K., Morash, J., Oh, K. J., Barbuto, S., Fisher, J., Smith, E., Verdine, G. L., and Korsmeyer, S. J. (2006). A stapled BID BH3 helix directly binds and activates BAX. Mol. Cell 24, 199–210.

Walensky, L. D., Kung, A. L., Escher, I., Malia, T. J., Barbuto, S., Wright, R. D., Wagner, G., Verdine, G. L., and Korsmeyer, S. J. (2004). Activation of apoptosis in vivo by a hydrocarbon-stapled BH3 helix. Science 305, 1466–1470.

Wei, M. C., Lindsten, T., Mootha, V. K., Weiler, S., Gross, A., Ashiya, M., Thompson, C. B., and Korsmeyer, S. J. (2000). tBID, a membrane-targeted death ligand, oligomerizes BAK to release cytochrome c. *Genes. Dev.* **14,** 2060–2071.

Willis, S. N., *et al.* (2007). Apoptosis initiated when BH3 ligands engage multiple Bcl-2 homologs, not Bax or Bak. *Science* **315,** 856–859.

Yethon, J. A., Epand, R. F., Leber, B., Epand, R. M., and Andrews, D. W. (2003). Interaction with a membrane surface triggers a reversible conformational change in Bax normally associated with induction of apoptosis. *J. Biol. Chem.* **278,** 48935–48941.

Youle, R. J., and Strasser, A. (2008). The BCL-2 protein family: Opposing activities that mediate cell death. *Nat. Rev. Mol. Cell Biol.* **9,** 47–59.

Zha, H., Aime-Sempe, C., Sato, T., and Reed, J. C. (1996). Proapoptotic protein Bax heterodimerizes with Bcl-2 and homodimerizes with Bax via a novel domain (BH3) distinct from BH1 and BH2. *J. Biol. Chem.* **271,** 7440–7444.

CHAPTER TWENTY-FOUR

ERM-Mediated Genetic Screens in Mammalian Cells

Dan Liu* *and* Zhou Songyang[†]

Contents

Abstract

Genetic screens have been proven powerful for the identification of components of various signaling pathways. For mammalian cells, methods for genetic screens are limited. We have developed the ERM (enhanced retroviral mutagen) mutagenesis approach that has been shown to be efficient and amenable to genomewide genetic screens in mammalian cells without the need of cDNA library construction. The ERM method offers several advantages, including conditional gene expression and the flexibility to tag endogenous genes with different epitope-tag and marker sequences. This chapter will discuss general design, procedures, and applications of the ERM strategy.

1. Introduction

Cultured mammalian cells offer a highly manipulatable, economical, convenient, and well-established model system for studying mammalian signaling pathways. Both loss-of-function and gain-of-function methods may be

* Cell-based Assay Screening Service Core, Baylor College of Medicine, Houston, Texas
† Verna and Marrs McLean Department of Biochemistry and Molecular Biology, Baylor College of Medicine Houston, Texas

Methods in Enzymology, Volume 446
ISSN 0076-6879, DOI: 10.1016/S0076-6879(08)01624-8

used. Although it is possible to generate loss-of-function mutant somatic cells by chemical mutagens, identification of mutated genes is extremely difficult. Recently developed RNAi genetic screen methods are promising and may serve to provide more insights into signaling networks that are dysregulated in cancer cells (Berns *et al.*, 2004; Paddison *et al.*, 2004; Seyhan *et al.*, 2005).

For gain-of-function studies, the two common approaches are cDNA expression libraries and insertional mutagenesis by retroviruses, both of which isolate candidate genes on the basis of phenotypic changes in the cells (Berns, 1988; D'Andrea *et al.*, 1990; Tsichlis, 1987; Whitehead *et al.*, 1995; Wong *et al.*, 1994). However, spontaneous mutant cells could arise that demand secondary screens for gene identification. In addition, wild-type retrovirus-mediated gene activation is of low frequency, and it is difficult to locate the affected genes. For cDNA libraries, large or cell type–specific cDNAs may be missing, so that current cDNA collections tend to be biased toward shorter cDNAs.

We have developed the ERM (enhanced retroviral mutagen) approach for a genomewide genetic screening strategy that has proven to be efficient and amenable to high-throughput screens in mammalian cells (Fig. 24.1). In particular, the ERM screens offer the following advantages:

1. The screen can be conditional (i.e., gene expression under the control of a inducible promoter).
2. There is no need for cDNA library construction, minimizing difficulties encountered during procedures such as subcloning.

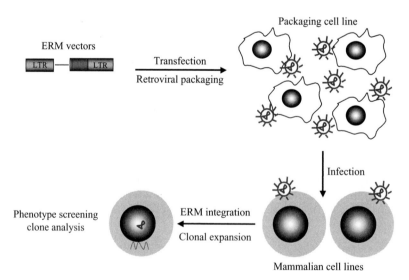

Figure 24.1   Schematic view of ERM-mediated genetic screen in mammalian cells.

3. ERM integration may occur at gene loci with multiple exons and long coding sequences. These genes may be absent from cDNA library collections.

4. The ERM strategy is readily adaptable for a variety of signaling pathways and molecular systems. For example, with different ERM tag sequences, ERM may be used to screen for protein stability, localization, or transcriptional activity.

2. METHODS AND DESIGN

2.1. Key considerations

The past decade or so has seen an explosion in technologies that help to dissect signaling pathways. For example, proteomic approaches such as large-scale immunoprecipitation followed by mass spectrometry sequencing have considerably facilitated our identification of the complex signaling networks that govern a variety of cellular functions (Gingras *et al.*, 2007). Such approaches, however, are often hampered by difficulties in identifying transient or low-affinity interactions. In addition, interactions need to be carefully verified because of possibilities of contamination during complex isolation. In comparison, genetic methods such as the ERM screen offer a powerful and complementary solution for investigating signaling pathways where few players are known.

The ERM-mediated genetic screen is essentially a pooled, nonarrayed screen. Therefore, an enrichment method is needed. For example, in our original screens of transforming genes, we took advantage of the growth properties of transformed cells and were able to isolate such clones to identify the genes mutated by ERM integration (Liu *et al.*, 2000b). Similar considerations will need to be taken when adapting the ERM strategy to study a particular signaling pathway.

2.2. ERM vector design

When we designed the original ERM vectors, we were looking to circumvent several problems associated with convention screens by use of wild-type retroviruses. This was accomplished by engineering the ERM cassette into the *NheI* site of the U3 region within the 3′ LTR of the retroviral pBabe-puro vector (Fig. 24.2A) (Morgenstern and Land, 1990). This design no longer relies on the retroviral LTR promoters that are relatively weak and sometimes suppressed. The ERM cassette contains sequences for the ERM Tag and a consensus splice donor sequence (AAGGTAAGT) under the control of a tetracycline regulatable promoter (*tet-off* or *tet-on*) (Gossen *et al.*, 1993; Paulus *et al.*, 1996). This configuration avoids potential

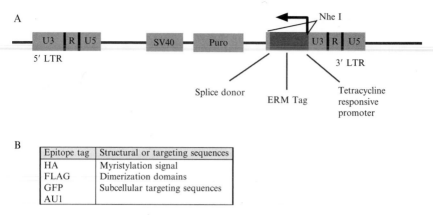

Figure 24.2 (A) ERM vector design. (B) Sequences that have been successfully used in the ERM tag region.

complications from the cryptic splice acceptor sequence present in the 3′ end of the 5′ LTR. ERM Tag allows for expression of a mutagenesis tag. Both epitope tag sequences (such as AU1 or HA), as well as structural or targeting motif sequences (e.g., myristoylation signal), can be included (Fig. 24.2B). Three sets of vectors corresponding to the three reading frames of the ERM Tag are needed to target endogenous genes in all possible reading frames (RF1, RF2, and RF3).

ERM integration and subsequent splicing events may allow the fusion of the 5′ ERM Tag to the endogenous exons in proximity to the integration site. When a fusion transcript is generated, direct cloning of the retrovirally targeted genes can be more easily achieved by RT-PCR and 3′ RACE. The tetracycline-regulated promoter of the ERM vectors would also enable the separation of authentic integration events from spontaneous mutants.

2.3. Cell lines, virus production, and screening procedures

2.3.1. General considerations

To fully use the tetracycline responsive promoter, the *tet* operon binding protein (tTA) will need to be expressed in the cells for the *tet-off* system. For *tet-on*, rtTA will need to be used (Ding *et al.*, 2006; Liu *et al.*, 2000a). The original ERM vector contained the *tet-off* promoter (Liu *et al.*, 2000b). Therefore, we first generated a retroviral vector (MSCV-neo) encoding tTA for tTA retrovirus production. The cells of interest were then infected with tTA viruses and selected in neomycin. Stable tTA-expressing cells were subsequently used for ERM screens.

Retroviral packaging cell lines (e.g., BOSC23 cells [Pear *et al.*, 1993 [for ecotropic viruses and 293-ampho cells [Clontech] for amphotropic viruses) are used for virus production. The ERM vectors are used to transfect the retroviral packaging cell lines either by the calcium phosphate method (Pear et al., 1993) or by use of commercial liposome transfection reagents such as lipofectamine (Invitrogen). The virus supernatant is then collected, and used to infect the cells. The human or mouse genome contains approximately 30,000 genes (Brent, 2005). Therefore, infection of 1 million cells should, in theory, generate enough random insertion events to cover the entire genome.

2.3.2. Flowchart and timetable of ERM screen

Day 1   Exponentially growing packaging cells are plated in 60-mm plates at approximately 50% confluency 1 day before transfection. Depending on the transfection reagents used, the cells are transfected with ~10 to 20 μg of ERM vectors according to manufacturer's protocols (e.g., Lipofectamine 2000 from Invitrogen).

Day 2   Change media for the packaging cells. Plate dividing tTA or rtTA–expressing cells for infection. For adherent cells, 12 h before infection, the cells may be plated at 30 to 40% confluence in a 60-mm dish. For suspension cells, the cells may be plated in 6- or 24-well plates (1×10^6 cells/well) a few hours before infection.

Day 3   The viruses are harvested and filtered through a 0.45-μm syringe filter. One 60-mm plate of packaging cells should yield approximately 4 ml of viruses. The tTA (or rtTA) cells are then spin-infected with the viruses in MOI (multiplicity of infection) of ≤ 1. Generally, the titer obtained ranges between 0.5 and 2×10^{-6} CPU/ml. During infection, Polybrene is added at 4 μg/ml. Spin infection can be carried out at 2500rpm for 1 to 2 h. Media may be changed immediately after infection or left overnight if compatible with the cell growing media. Suspension cells will need to be expanded after infection to minimize crowding.

Day 4   The cells are allowed to recover, and media can be changed if necessary.

Day 5   The cells are replated for genetic screens. For large-scale high-throughput screens, the cells may be plated into 96- or 384-well plates to obtain single colonies. Puromycin (1 to 2 μg/ml) addition to select infected cells is optional at this stage. In fact, it may be preferable not to add selection drugs at this stage because of potential stress on the cells.

After mutant cells grow to colonies, which may take on average 2 weeks, they can be further expanded in puromycin-containing media. And the cell clones are ready for gene identification analysis. At this point, the

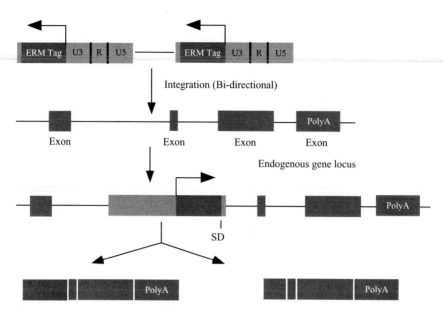

Figure 24.3 Diagram of insertion tagging and activation of endogenous genes by ERM retroviruses. SD, splice donor.

tetracycline or doxycycline sensitivity of the clones may be tested. In general, 1 to 5 μg/ml of doxycycline can be added to the culture.

2.3.3. Identification of ERM-Targeted genes by RT-PCR and 3′ RACE

As shown in Fig. 24.3, ERM retrovirus integration may result in the generation of transcripts that fuse the ERM Tag sequences with endogenous gene sequences. Alternately, transcription through endogenous promoters at gene loci in proximity to ERM integration sites may be upregulated, because of the presence of exogenous ERM enhancer elements. ERM-induced gene activation in either case may be sufficient for phenotypic changes.

For ERM integrations that do not generate fusion transcripts, the integration locus may be ascertained by isolating genomic DNA for inverted PCR. Such approaches have been well established and described in detail elsewhere (Suzuki *et al.*, 2002). When fusion transcripts are generated, these transcripts may contain the 5′ ERM Tag, as well as sequences from the endogenous 3′UTR and poly-A sequences. To identify the gene locus targeted by ERM, individual clones may be expanded and the RNA extracted for RT-PCR and 3′ RACE as described later (Fig. 24.2). The amplified products are then sequenced to identify the gene loci targeted by ERM. On the basis of such information, further molecular and biochemical analyses may be carried out to confirm the functional significance of the gene.

1. Collect enough cells (generally 10^5 to 10^6) for RNA extraction (e.g., with the RNeasy Mini Kit from QIAGEN). The amount of RNA isolated can be estimated on the basis of the number of cells used. Make sure to include control cells for all the steps.

2. Carry out reverse transcription with the isolated RNA, the random primer RT-1   (5′-GCAAATACGACTCACTATAGGGATCCNNNN(GC) ACG-3′, N = AGCT, sequences in bold denote T7 primer sequence), and reverse transcriptase (e.g., Superscript[III] from Invitrogen). Although a random primer with the sequence of 5′-GCAAATACGACTCACTA-TAGGGATCCNNNNNN-3′ may be used, addition of the ACG triplet at the 3′ end of the RT-1 primer should significantly reduce the number of PCR products generated. In our original screen, a myristoylation signal was included in the ERM Tag region. The CGT sequence is not found in the myr sequence; therefore, primer binding should occur outside of the ERM Tag region.

3. Perform PCR reactions by use of the cDNA and primer sets specific for the ERM Tag sequences and sequences contained within the RT-1 primer. PCR conditions may need to optimized, depending on the starting material, the primers, and the polymerase used. In general, one third of the RT products can be used for 30 to 35 cycles of PCR. Sometimes two rounds of nested PCR may need to be performed with 5 to 10 cycles of first-round PCR followed by another 30 cycles.

In our original screens (with the presence of the myristoylation signal and nested T7 sequences within the RT-1 primer), the cDNA was PCR amplified with primers from the myristoylation signal sequence and the T7 primer (Myr1: 5′-ACCATGGGGAGCAGCAAGAGCAAACCAAAA-GACCCCAGCCAACGC-3′). The PCR products were then gel purified and directly sequenced with T7 primer.

3. ERM-MEDIATED GENETIC SCREENS AND RE-PCA

3.1. Re-PCA design

Depending on the tag sequences engineered into the ERM vectors, ERM may be used to dissect a variety signaling pathways (such as cell survival and transformation) (Liu *et al.*, 2000a,b). Recently, we have further modified our ERM approach to incorporate fluorescence complementation that has been developed to study protein–protein interactions in live cells (Ding *et al.*, 2006) (Fig. 24.4). Bimolecular fluorescence complementation (PCA or BiFC) reveals protein–protein interactions when two proteins, fused respectively to the N- and C-terminal fragments of a fluorescent protein (such as YFPn and YFPc), bind together to allow the formation

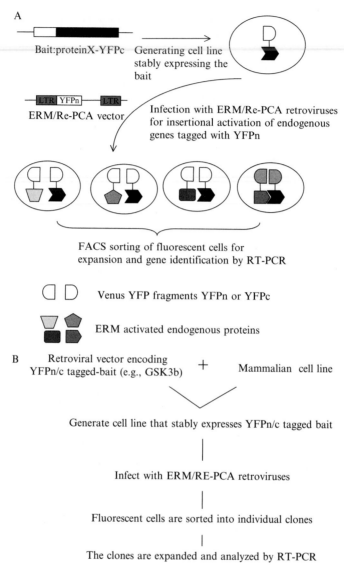

Figure 24.4 (A) Diagram of Re-PCA–mediated genetic screen strategy for protein X. (B) Flowchart of Re-PCA screen.

of a functional fluorescent protein (Hu and Kerppola, 2003; Michnick, 2003). In PCA studies, protein–protein interactions occur in the native cellular environments that may favor physiologically relevant interactions (e.g., posttranslational modification–mediated interactions). In addition, PCA can detect transient interactions in live cells and provide spatial and temporal information about protein–protein interactions.

To dissect the pathways regulated by signaling molecules, we designed genetic screen strategies that use both ERM and PCA (Re-PCA) (Fig. 24.2) (Ding *et al.*, 2006). Here, cell lines are generated that stably express tTA and the protein of interest tagged with a fluorescent protein fragment (bait). The complementary fluorescent protein fragment is engineered into the ERM Tag. ERM screens are then carried out as described previously. Re-PCA is adaptable to a variety of conditions. For example, screens may be carried out in the presence of certain signals (e.g., irradiation or pharmaceutical drugs) to screen for interactions that only occur under these circumstances.

3.2. Methods

1. First, a cell line that stably expresses tTA (or rtTA) and the bait protein are generated. We have empirically determined that the configuration of tagging the bait protein with the C-terminal fragment of YFP (YFPc) works well in most conditions.
2. Next, Re-PCA vectors that contain the Venus YFPn tag are obtained by cloning the N-terminal YFP fragment into ERM Tag region (amino acid sequence 1 to 155 or 1 to 173 of Venus YFP). High-titer Re-PCA retroviruses are then generated and used to infect cycling bait-expressing cells (MOI $\leq$ 1). Again, drug selection is optional at this point.
3. Approximately 2 to 5 days after infection, the cells may be examined by FACS to determine the extent of fluorescence complementation. Cells that exhibit fluorescence are then individually sorted into 96-well plates and allowed to expand into single clones. The individual clones are further analyzed by flow cytometry to confirm stable fluorescence complementation. The positive clones can then be harvested and analyzed by RT-PCR to identify the gene loci targeted by Re-PCA.
4. Certain interactions may be transient, of low affinity, or adversely affect the proliferation and growth of the cells. The *tet-off* (or *tet-on*) promoter may thus be taken advantage of by culturing the cells in the presence (or absence) of tetracycline. Shortly before FACS analysis (~1 day), the cells may be cultured in the absence (or presence) of tetracycline to allow fluorescence complementation to occur. FACS-sorted cells can then be maintained again in the absence (or presence) of tetracycline.

4. CONCLUSIONS

The ERM technology is not limited only to screens that search for genes important for particular signaling pathways. The random nature of retroviral integration makes it an ideal tool for studying global protein changes. ERM-integrated cells essentially represent an *in vivo* library of ERM-tagged proteins whose localization and stability may be investigated.

For instance, fluorescence tags can be engineered into the ERM vectors. The ERM screen may then be coupled with markers for specific cellular compartments. These cells can be subsequently visualized by use of high-throughput fluorescence microscopy to study the changes in protein trafficking and translocation. ERM may, therefore, help to fill the gap in this area that has long been difficult to study genomewide because of technical limitations.

ACKNOWLEDGMENT

This work is supported by NIH grants CA84208 and GM69572. D. L. is supported in part by the American Heart Association. Z. S. is a Leukemia and Lymphoma Society Scholar.

REFERENCES

Berns, A. (1988). Provirus tagging as an instrument to identify oncogenes and to establish synergism between oncogenes [published erratum appears in *Arch. Virol.* 1989;107 (1–2):170]. *Arch. Virol* **102,** 1–18.

Berns, K., Hijmans, E. M., Mullenders, J., Brummelkamp, T. R., Velds, A., Heimerikx, M., Kerkhoven, R. M., Madiredjo, M., Nijkamp, W., Weigelt, B., Agami, R., and Ge, W. (2004). A large-scale RNAi screen in human cells identifies new components of the p53 pathway. *Nature* **428,** 431–437.

Brent, M. R. (2005). Genome annotation past, present, and future: How to define an ORF at each locus. *Genome Res.* **15,** 1777–1786.

D'Andrea, A., Fasman, G., Wong, G., and Lodish, H. (1990). Erythropoietin receptor: Cloning strategy and structural features. *Int. J. Cell Cloning* **8**(Suppl 1), 173–180.

Ding, Z., Liang, J., Lu, Y., Yu, Q., Songyang, Z., Lin, S. Y., and Mills, G. B. (2006). A retrovirus-based protein complementation assay screen reveals functional AKT1-binding partners. *Proc. Natl. Acad. Sci. USA* **103,** 15014–15019.

Gingras, A. C., Gstaiger, M., Raught, B., and Aebersold, R. (2007). Analysis of protein complexes using mass spectrometry. *Nat. Rev. Mol. Cell. Biol.* **8,** 645–654.

Gossen, M., Bonin, A. L., and Bujard, H. (1993). Control of gene activity in higher eukaryotic cells by prokaryotic regulatory elements. *Trends Biochem. Sci.* **18,** 471–475.

Hu, C. D., and Kerppola, T. K. (2003). Simultaneous visualization of multiple protein interactions in living cells using multicolor fluorescence complementation analysis. *Nat. Biotechnol.* **21,** 539–545.

Liu, D., Yang, X., and Songyang, Z. (2000a). Identification of CISK, a new member of the SGK kinase family that promotes IL-3–dependent survival. *Curr. Biol.* **10,** 1233–1236.

Liu, D., Yang, X., Yang, D., and Songyang, Z. (2000b). Genetic screens in mammalian cells by enhanced retroviral mutagens. *Oncogene* **19,** 5964–5972.

Michnick, S. W. (2003). Protein fragment complementation strategies for biochemical network mapping. *Curr. Opin. Biotechnol.* **14,** 610–617.

Morgenstern, J. P., and Land, H. (1990). Advanced mammalian gene transfer: High titre retroviral vectors with multiple drug selection markers and a complementary helper-free packaging cell line. *Nucleic Acids Res.* **18,** 3587–3596.

Paddison, P. J., Silva, J. M., Conklin, D. S., Schlabach, M., Li, M., Aruleba, S., Balija, V., O'Shaughnessy, A., Gnoj, L., Scobie, K., Chang, K., Westbrook, T., *et al.* (2004).

A resource for large-scale RNA-interference-based screens in mammals. *Nature*. **428,** 427–431.

Paulus, W., Baur, I., Boyce, F. M., Breakefield, X. O., and Reeves, S. A. (1996). Self-contained, tetracycline-regulated retroviral vector system for gene delivery to mammalian cells. *J. Virol.* **70,** 62–67.

Pear, W. S., Nolan, G. P., Scott, M. L., and Baltimore, D. (1993). Production of high titer helper-free retroviruses by transient transfection. *Proc. Natl. Acad. Sci. USA* **90,** 8392–8396.

Seyhan, A. A., Vlassov, A. V., Ilves, H., Egry, L., Kaspar, R. L., Kazakov, S. A., and Johnston, B. H. (2005). Complete, gene-specific siRNA libraries: Production and expression in mammalian cells. *Rna* **11,** 837–846.

Suzuki, T., Shen, H., Akagi, K., Morse, H. C., Malley, J. D., Naiman, D. Q., Jenkins, N. A., and Copeland, N. G. (2002). New genes involved in cancer identified by retroviral tagging. *Nat. Genet.* **32,** 166–174.

Tsichlis, P. N. (1987). Oncogenesis by Moloney murine leukemia virus. *Anticancer Res.* **7,** 171–180.

Whitehead, I., Kirk, H., and Kay, R. (1995). Expression cloning of oncogenes by retroviral transfer of cDNA libraries. *Mol. Cell. Biol.* **15,** 704–710.

Wong, B. Y., Chen, H., Chung, S. W., and Wong, P. M. (1994). High-efficiency identification of genes by functional analysis from a retroviral cDNA expression library. *J. Virol.* **68,** 5523–5531.

Author Index

Subject Index

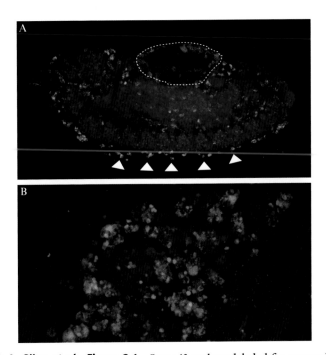

Elizabeth A. Silva *et al.*, Figure 3.1 Stage 13 embryo labeled for macrophages and apoptotic corpses. Macrophages are labeled with the *Crq-eGFP* transgene (green) and the Crq antibody (blue); apoptotic corpses are labeled with 7-AAD (brighter red). (A) This stage is characterized by the completion of germ–band retraction and the initiation of dorsal closure, at which point the amnioserosa is fully exposed on the dorsal side (outlined in white). Macrophages in the CNS along the ventral side of the embryo tend to be smaller and engulf fewer corpses (arrowheads) than in the head and tail regions. Anterior is to the left. (B) Macrophages in the head region engulf between three and seven corpses each (also see Fig. 3.3).

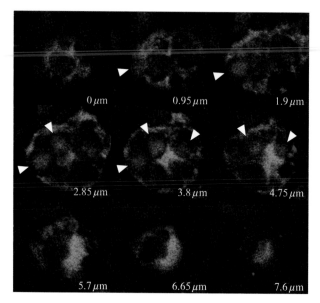

Elizabeth A. Silva *et al.*, Figure 3.3 Confocal stack through a single wild-type macrophage. Z-stack of nine images of a single macrophage taken at intervals of 0.95 μm. The macrophage is marked by the cytoplasmic expression of a UAS-eGFP transgene under the control of the *Crq-Gal4* driver (green) and the Crq antibody (blue); the apoptotic corpses are labeled with 7-AAD (brighter red). Each of three apoptotic corpses engulfed by this macrophage become visible at different levels within the stack (arrowheads). The localization of Crq itself is to the membrane of vesicles surrounding the engulfed corpses.

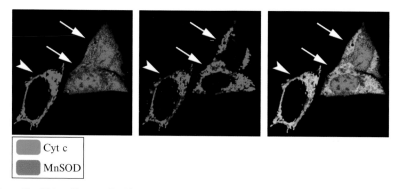

Marcello D'Amelio *et al.*, Figure 15.1 Cytochrome *c* detection in ETNA^{+/+} cells is abolished after STS treatment. Double-labeling immunofluorescence of ETNA^{+/+} cells treated for 12 h with 5 μM staurosporine (STS); images were taken by a confocal microscope. Cytochrome *c* (green), MnSOD (red), and merged patterns are shown. The white arrowhead points to a cell with cytochrome *c* localized into the internal mitochondrial membrane, and the white arrows indicate cells with released cytochrome *c*.

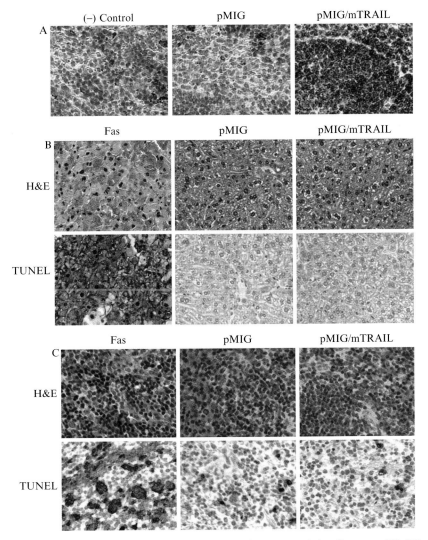

Keli Song et al., Figure 19.2 (A) Immunohistochemistry staining for mouse TRAIL (mTRAIL) protein in spleen tissue from mice 1 month after BMT. (−) Control mice received bone marrow without retroviral infection. pMIG mice received pMIG retroviral-infected bone marrow. pMIG/mTRAIL mice received bone marrow infected with the pMIG/mTRAIL retrovirus. Positive immunohistochemistry staining for mTRAIL in spleen sample of mTRAIL-transduced mouse is shown. (B) Histologic and pathologic examination on liver samples from bone marrow recipient mice. Samples from (top labels left to right) Fas: mice that received 50 μg Fas agonist antibody by I.V. injection. The samples were harvested and processed for pathologic examination study 5 h after the injection; other sample labels are indicated in Fig. 19.2A. Top panel, H&E staining. Bottom panel, TUNEL staining. Liver hemorrhage can be seen in H&E staining of the Fas sample. (C) Spleen samples labeled in the same order as in Fig. 19.2B.

C   0 min   5 min

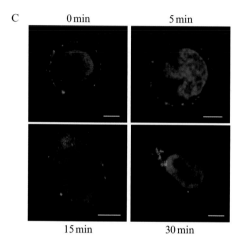

15 min   30 min

Nicholas Harper and Marion MacFarlane, Figure 18.3C   Internalization of TRAIL assessed by Confocal Microscopy using Biotinylated-TRAIL and Streptavidin-labelled Alexa-568. Receptor-mediated internalization of TRAIL was assessed using Biotin-TRAIL as described in Protocol 8. BJAB cells (1×10^6 per sample) were chilled to 4 °C for 1 h followed by treatment with biotinylated-TRAIL (500 ng/ml) for 45 min at 4 °C. Cells were then washed three times with ice-cold PBS (to remove excess TRAIL) and treated with Streptavidin labelled Alexa-568 (*red*) for 1 h at 4 °C. Cells were washed and either fixed after 4 °C treatment (0 min) or released up to 37 °C for 5, 15 or 30 min (to allow TRAIL to internalize) and then fixed in 4% paraformaldehyde for 10 min at room temperature. After fixation, cells were counterstained with the DNA dye Hoechst-33342 (*blue*) for up 5 min to stain nuclei and then visualised using a Zeiss LSM510 with Axiovert 200 confocal microscope. Results shown are of one representative cell from each time point. The *white* bar represents 5 μm. [b]modified with full copyright permission (Kohlhaas *et al.*, 2007).

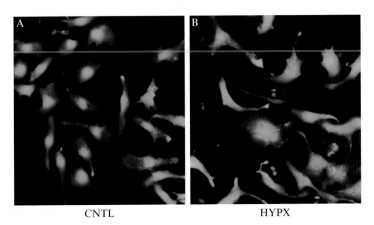

CNTL   HYPX

Danielle Weidman *et al.*, Figure 16.1   Viability staining of postnatal ventricular myocytes by epifluorescence microscopy under normal (CNTL, panel A) and hypoxic
(HYPX, panel B) conditions. Original magnification 100×.

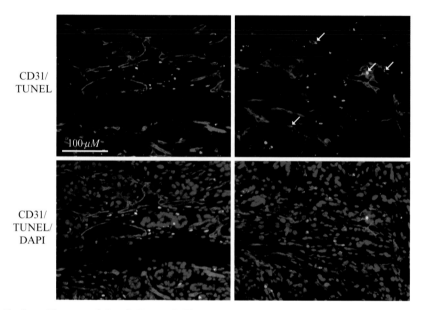

Xuefeng Zhang and Sareh Parangi, Figure 17.1   Sample pictures of CD31/TUNEL/ DAPI staining. Staining was performed on xenograft tumors grown in mouse. Left panel shows tumor from control mouse, and right panel shows tumor from mouse that received treatment that induced endothelial cell apoptosis. Apoptotic endothelial cells can be determined by colocalization of TUNEL signal within the nuclei of CD31-positive endothelial cells, and total endothelial cell number can be determined by counting nuclei localized in endothelial cells, which are labeled by CD31 staining. Please note that to be true positive, TUNEL-positive signals must be localized within nuclei. Figure is adapted from Zhang, X., *et al.* (2005). *Clin. Cancer Res.* **11**, 2337–2344.